PLENUM PRESS DATA DIVISION

NEW YORK

1966

HA
GEN
GAS D

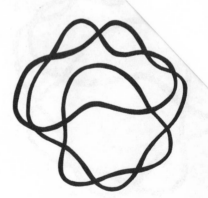

HANDBOOK OF GENERALIZED GAS DYNAMICS

Robert P. Benedict and William G. Steltz

Westinghouse Electric Corporation, Steam Divisions, and
Drexel Institute of Technology, Philadelphia, Pennsylvania

Library of Congress Catalog Card Number 65-25128

© 1966 Plenum Press Data Division
A Division of Consultants Bureau Enterprises, Inc.
227 W. 17th St., New York, N.Y. 10011

Printed in the United States of America

Preface

THE FACT that most books on gas dynamics include separate tables for each simplified flow process casts a shadow of inadequacy over the conventional approach. Why is each process treated as though it were entirely unrelated to the others? Why isn't there, we asked, a generalized approach based on fundamental equations which act as progenitors for the specific equations of all the simplified flow processes, and which provide insight to more general flow processes?

As our solution to the above dilemma, we present a complete treatment of one-dimensional gas dynamics, stressing a fundamental approach. A unified description of this subject is accomplished by means of a single numerical table applicable to the particular gas under study. Separate treatments for the various flow processes are thus combined into one all-encompassing analysis.

These tables are intended for the large group of practicing engineers, of which we are members, who daily must solve routine problems in gas dynamics. Aerodynamic, chemical, and mechanical engineers, as well as students of thermodynamics and gas dynamics, should find these tables useful.

The book is divided into five parts. In Chapter 1, we present a generalized compressible flow function Γ, which is shown to have direct application in the treatment of many simplified one-dimensional flow processes. The particular processes treated are: (a) the familiar adiabatic flows with or without friction, with area variations always allowed; (b) the little-discussed diabatic flows, with or without friction, with area variations allowed under certain conditions; and (c) the discontinuous normal shock process. Moreover, the Γ function is shown to have significance in a generalized flow process having an arbitrary combination of heat transfer, friction, and area variation. A plot of pressure ratio p/p_t vs. Γ is given, and schematic isentropic, Fanno, Rayleigh, isothermal, and normal shock maps are presented in terms of the conventional enthalpy–entropy diagram, and again in terms of the pressure ratio–Γ diagram. Generalized compressible flow tables are developed in Chapter 2 for the various flow processes. These numerical tables are presented in the Appendix in terms of p/p_t and Mach number. Numerical examples are included in Chapter 3, to illustrate the solution of typical problems through the use of the generalized compressible flow tables. The development of a generalized constant-density flow function Γ' appears in Chapter 4 and is shown to have direct application in the treatment of many simplified, workless, one-dimensional flow processes. The particular flow processes treated are: (a) the familiar adiabatic flows, with or without friction, with area variations always allowed; and (b) diabatic flows with or without friction, with area variations always allowed. Moreover, the Γ' function is shown to have significance in a generalized flow process having an arbitrary combination of heat transfer, friction, and area and elevation variation.

Development of the Γ' function is given in some detail. Schematic isentropic, Fanno, Rayleigh, and isothermal flow maps are presented in terms of the conventional enthalpy–entropy diagram, and again in terms of the pressure ratio–Γ' diagram. Numerical examples are included to illustrate the solution of typical problems through use of the generalized constant-density flow function.

The authors would be remiss if they did not mention the cooperation and generosity of the Westinghouse Electric Corporation's Steam Divisions. Computer facilities and time were made available to us toward the completion of this work, and permission to publish these tables was also granted.

The Authors

Lester, Pennsylvania
November, 1965

NOTE: The text in this book is based primarily on the authors' papers:
A Generalized Approach to One-Dimensional Gas Dynamics
(*Trans. ASME, J. Engrg. for Power*, June 1962, p. 49)
Some Generalizations in One-Dimensional Constant Density Fluid Mechanics
(*Trans. ASME, J. Engrg. for Power*, June 1962, p. 44)
Thermodynamics of Compressible Fluids
(*Electro-Technology*, Feb. 1963, p. 85)
Thermodynamics of Constant Density Fluids
(*Electro-Technology*, April 1963, p. 70)

Contents

Appendix

Notation

A = Area, ft^2

c = Specific heat capacity, Btu/lb-deg R

d = Total derivative

e = Base of natural logarithms

F = Specific quantity of heat generated within system boundaries, Btu/lb

g = Gravitational acceleration, ft/sec^2

h = Specific enthalpy, Btu/lb

J = Mechanical equivalent of heat, ft-lb/Btu

K = A constant

M = Mach number

p = Pressure, lb/ft^2

P = Dimensionless pressure function

Q = Specific quantity of heat transferred across system boundaries, Btu/lb

R = Gas constant, ft/deg R

s = Specific entropy, Btu/lb-deg R

T = Absolute temperature, deg R

u = Specific internal energy, Btu/lb

v = Specific volume, ft^3/lb

V = Flow velocity, ft/sec

w = Specific weight, lb/ft^3

$W/\Delta t$ = Flow rate, lb/sec

Z = Elevation, ft

γ = Isentropic exponent or ratio of specific heat capacities of a perfect gas

Γ = Generalized compressible flow function

Γ' = Generalized constant-density flow function

δ = Inexact differential

ρ = Mass density, slugs/ft^3

Subscripts

$1, 2$ = State points

p = Constant pressure

t = Total

Superscript

$*$ = Signifies critical state

Chapter 1

A Generalized Approach to One-Dimensional Gas Dynamics

INTRODUCTION

At the present time, those thermodynamic flow processes amenable to simplified treatment are considered separately in the various texts; one finds chapters on isentropic, adiabatic, and diabatic flows, as though they were entirely unrelated. The normal shock process is told as a story in itself. Quite reasonably then, we are led to expect and are not surprised when we find that separate functional tables exist for each of these processes. Also, the need for a satisfactory approach to variable-area flow processes, for other than the isentropic case, has long been noted; and the usual absence of any treatment of isothermal flow is disturbing. These and other shortcomings, we believe, result from a failure to recognize common bonds which exist among the various simplified flow processes.

First, the concepts of continuity and the critical state will be combined to yield equations for a *generalized flow process*; then, the various flows mentioned above will be analyzed and shown to be special cases of the generalized process. A *generalized compressible flow table* will then be developed, and a number of realistic compressible flow problems will be solved using this table.

Let us note that we are considering here the one-dimensional, steady, workless flow processes of a real gas (i.e., one having viscosity and thermal conductivity) whose equation of state is well represented by the ideal gas relation

$$\frac{p}{\rho} = gRT \qquad (1)$$

and whose specific heat capacity at constant pressure c_p is taken as a constant.

CONTINUITY

The concept of *conservation of mass*, which relates one thermodynamic state to another for the whole gamut of possible one-dimensional flow processes, is expressed as

$$\rho V A = K \qquad (2)$$

where K is a constant. Since a dimensionless form of equation (2) will be more useful, the density and velocity are next evaluated in terms of the total parameters ρ_t, p_t, and T_t, which are based on isentropic stagnation[1] processes from given static states. Thus, the density term may be written as

$$\rho = \rho_t \left(\frac{p}{p_t}\right)^{1/\gamma} \qquad (3)$$

upon recalling that p/ρ^γ is a constant for any isentropic change of state of a gas defined by equation (1). The velocity term also can be written in terms of total and static states by combining the general energy equation (on the basis of a pound of flowing fluid)

$$\delta Q = dh + \frac{V dV}{g} + dZ \qquad (4)$$

with the thermodynamic identity

$$T ds = dh - \frac{dp}{\rho g} \qquad (5)$$

(Note that $\delta Q = 0$ and $ds = 0$ in the isentropic stagnation process, while dZ is negligible in any gaseous flow process.) There results

$$V dV = -\frac{dp}{\rho} \qquad (6)$$

On integrating equation (6) between static and total states, we obtain

$$V = \left[\frac{2\gamma}{\gamma - 1}\left(\frac{p_t}{\rho_t} - \frac{p}{\rho}\right)\right]^{\frac{1}{2}} \qquad (7)$$

---

[1] By the term *isentropic* we imply the special constant-entropy process which is both adiabatic and reversible; by *stagnation* we mean a deceleration (actual or postulated) of the fluid to zero velocity.

When equations (3) and (7) are combined according to equation (2), there results

$$\left(\frac{T_{t2}}{T_{t1}}\right)^{\frac{1}{2}}\left(\frac{p_{t1}}{p_{t2}}\right)\left(\frac{A_1}{A_2}\right)\left\{\left[\frac{p_1}{p_{t1}}\right]^{1/\gamma}\left[1-\left(\frac{p}{p_{t1}}\right)^{(\gamma-1)/\gamma}\right]^{\frac{1}{2}}\right\}$$
$$=\left\{\left[\frac{p_2}{p_{t2}}\right]^{1/\gamma}\left[1-\left(\frac{p_2}{p_{t2}}\right)^{(\gamma-1)/\gamma}\right]^{\frac{1}{2}}\right\} \tag{8}$$

which is an entirely general dimensionless continuity expression.[2]

THE CRITICAL STATE

Now, directing our attention elsewhere, we note that for every thermodynamic process there exists a critical state at which, for a differential change along the process line, $\delta F = 0$ and $dA = 0$. By this we mean that the entropy change, generally given by the Second Law of Thermodynamics as

$$ds = \frac{\delta Q}{T} + \frac{\delta F}{T} \tag{9}$$

always reduces to

$$ds* = \frac{\delta Q}{T*} \tag{10}$$

at the critical state. (The asterisk, here and henceforth, denotes variables pertaining to the critical state.)

Expressions for velocity and pressure ratio at this critical state will prove useful, and are developed next. A general differential form for the critical velocity is derived by combining the general energy equation (4) with the First Law of Thermodynamics, generally given by

$$\delta Q + \delta F = du + p\,dv \tag{11}$$

on a per-pound basis. Recalling that $\delta F = 0$ at the critical state, we have

$$V*\,dV = -\frac{dp}{\rho*} \tag{12}$$

While equation (12) is similar in form to equation (6), its implications are quite different. Equation (12) holds for *any* process at the critical state, while equation (6) is restricted to a *frictionless* process only. Recalling that $dA = 0$ at the critical state, equation (2) can be written as

$$\rho*\,dV + V*\,d\rho = 0 \tag{13}$$

While equation (13) is similar in form to a constant-area differential form of continuity, the implications are quite different; equation (13) holds for any process

[2]The limiting case of $\gamma = 1$ requires special attention and is developed on page 11.

at the critical state. Combining equations (12) and (13), we obtain

$$V* = \left(\frac{dp}{d\rho}\right)^{\frac{1}{2}} \tag{14}$$

which, by employing the equation of state (1) and the thermodynamic identity (5), can also be written as

$$V* = \left\{\frac{p}{\rho}\left[\frac{1}{1/\gamma - (ds/dp)(p/c_p)}\right]\right\}^{\frac{1}{2}} \tag{15}$$

Equation (15) defines the critical velocity for any flow process. By equating the expressions for general velocity and for critical velocity, equations (7) and (15), we obtain

$$\frac{p*}{p_t*} = \left(\frac{1/\gamma - (ds/dp)(p/c_p)}{1/\gamma - (ds/dp)(p/c_p) + (\gamma-1)/2\gamma}\right)^{\gamma/(\gamma-1)} \tag{16}$$

Equation (16) defines the critical pressure ratio for any flow process.

THE GAMMA FUNCTION

We now have considered in some detail two very general concepts: continuity and the critical state. Further generality is achieved by referring continuity [equation (8)] to certain conditions existing at the critical state.

The factors $(T_{t2}/T_{t1})^{\frac{1}{2}}$, p_{t1}/p_{t2}, and A_1/A_2 of equation (8) can be considered to be arbitrary process multipliers of the inlet pressure-ratio function

$$P_1 = \left\{\left(\frac{p_1}{p_{t1}}\right)^{1/\gamma}\left[1-\left(\frac{p_1}{p_{t1}}\right)^{(\gamma-1)/\gamma}\right]^{\frac{1}{2}}\right\} \tag{17}$$

By referring the P_1 of equation (17) to a similar function

$$P* = \left\{\left(\frac{p*}{p_t*}\right)^{1/\gamma}\left[1-\left(\frac{p*}{p_t*}\right)^{(\gamma-1)/\gamma}\right]^{\frac{1}{2}}\right\} \tag{18}$$

pertaining to the critical state, we can rewrite equation (8) as

$$\left(\frac{T_{t2}}{T_{t1}}\right)^{\frac{1}{2}}\left(\frac{p_{t1}}{p_{t2}}\right)\left(\frac{A_1}{A_2}\right)\Gamma_1 = \Gamma_2 \tag{19}$$

The symbol Γ, which equals $P/P*$, represents a *generalized compressible flow function* since it embodies the concepts of continuity and the critical state. By maximizing the function P of equation (17), we obtain the familiar relationship

$$\left(\frac{p}{p_t}\right)_{P_{max}} = \left(\frac{2}{\gamma+1}\right)^{\gamma/(\gamma-1)} \tag{20}$$

When (20) is equated to (16), we find that such a pressure ratio can be realized at the critical state only in those processes where the differential entropy change at the critical state is zero.[3]

When equation (18) is evaluated in terms of equation (20), the Γ function can be uniquely defined as

$$\Gamma = \frac{[p/p_t]^{1/\gamma}[1 - (p/p_t)^{(\gamma-1)/\gamma}]^{\frac{1}{2}}}{[2/(\gamma+1)]^{1/(\gamma-1)}[(\gamma-1)/(\gamma+1)]^{\frac{1}{2}}} \quad (21)$$

where Γ varies between 0 and 1 only, for any and all flow processes. By plotting pressure ratio p/p_t vs. Γ

for particular values of γ, we obtain the *generalized compressible flow chart* of Fig. 1. When the function Γ_1 (completely defined by the inlet pressure ratio p_1/p_{t1} for a specific value of γ) is multiplied by the total temperature, total pressure, and area ratios pertaining to a particular flow process, the function Γ_2 completely defining the exit conditions is obtained as indicated by equation (19), the referred continuity expression.

Since the condition chosen to particularize the Γ function [equation (20)] need not be the critical condition of an arbitrary process, it follows that Γ^* need not always equal 1. Nevertheless, all arbitrary flow processes are precisely represented by the plot of Fig. 1.

---

[3]Under such conditions, the velocity at the critical state [equation (15)] reduces to $V^* = (\gamma p/\rho)^{\frac{1}{2}}$, which is recognized as the acoustic velocity, i.e., the familiar velocity of propagation of a small pressure disturbance.

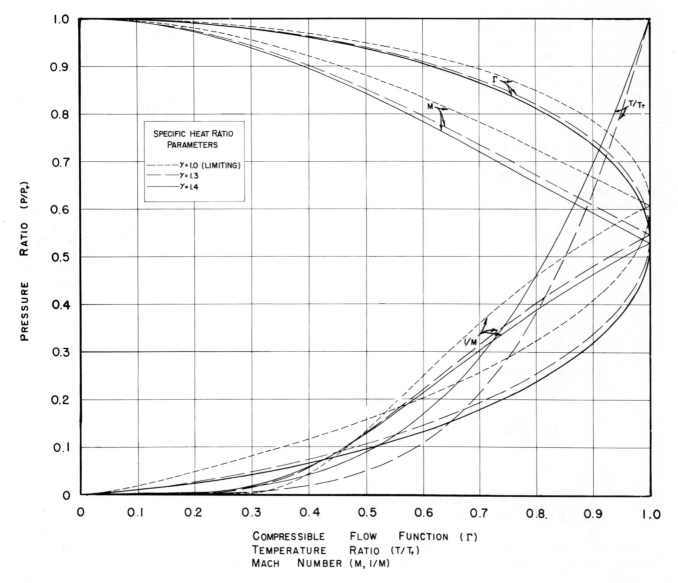

Figure 1. Generalized Compressible Flow Chart.

Other parameters which can be referred to critical-state conditions advantageously are the total pressure, the total temperature, and the area. Thus,

$$\frac{p_t}{p_t^*} = \frac{[p^*/p_t^*]^{1/\gamma}[1 - (p^*/p_t^*)^{(\gamma-1)/\gamma}]^{\frac{1}{2}}[T_t/T_t^*]^{\frac{1}{2}}}{[p/p_t]^{1/\gamma}[1 - (p/p_t)^{(\gamma-1)/\gamma}]^{\frac{1}{2}}[A/A^*]} \qquad (22)$$

$$\frac{T_t}{T_t^*} = \frac{[p/p_t]^{2/\gamma}[1 - (p/p_t)^{(\gamma-1)/\gamma}]}{[p^*/p_t^*]^{2/\gamma}[1 - (p^*/p_t^*)^{(\gamma-1)/\gamma}]}\left[\frac{A}{A^*}\right]^2\left[\frac{p_t}{p_t^*}\right]^2 \qquad (23)$$

and

$$\frac{A}{A^*} = \frac{[p^*/p_t^*]^{1/\gamma}[1 - (p^*/p_t^*)^{(\gamma-1)/\gamma}]^{\frac{1}{2}}[T_t/T_t^*]^{\frac{1}{2}}}{[p/p_t]^{1/\gamma}[1 - (p/p_t)^{(\gamma-1)/\gamma}]^{\frac{1}{2}}[p_t/p_t^*]} \qquad (24)$$

are all consequences of equation (8), the generalized continuity expression.

Finally, we might also express the entropy change to the critical state as follows: The product $(T_{t2}/T_{t1})^{\frac{1}{2}}(p_{t1}/p_{t2})$ in equation (8) is expressed in terms of entropy change by applying the thermodynamic identity (5), thus obtaining

$$\left(\frac{T_{t2}}{T_{t1}}\right)^{\frac{1}{2}}\left(\frac{p_{t1}}{p_{t2}}\right) = e^{(s_2-s_1)/2c_p}\left(\frac{p_{t1}}{p_{t2}}\right)^{(\gamma+1)/2\gamma} \qquad (25)$$

where the subscripts 1 and 2 represent any two thermodynamic states. When this substitution is introduced in the continuity equation (8), there results equation (26), which applies between any arbitrary state and the critical state.

$$s^* - s =$$
$$2c_p\ln\left\{\frac{(p^*/p_t^*)^{1/\gamma}[1 - (p^*/p_t^*)^{(\gamma-1)/\gamma}]^{\frac{1}{2}}}{(p/p_t)^{1/\gamma}[1 - (p/p_t)^{(\gamma-1)/\gamma}]^{\frac{1}{2}}(A/A^*)(p_t/p_t^*)^{(\gamma+1)/2\gamma}}\right\} \qquad (26)$$

We have seen thus far that:
1. A dimensionless form of the continuity equation (8) can be applied to all flow processes.
2. A thermodynamic critical state exists for all flow processes.
3. A generalized compressible flow function Γ is obtained by referring continuity to certain conditions at the isentropic critical state [equation (20)].
4. The Γ function so defined varies only between 0 and 1.

APPLICATION OF GAMMA IN SPECIFIC FLOW PROCESSES

Many practical situations may be approximated by certain processes having restrictions which make possible simplified solutions. Several such flow processes of most general interest will be treated here.

Adiabatic Flow Without Friction

Here we are referring to the one-dimensional, steady flow of a fluid whose thermodynamic properties change solely because of area variations. For note: it follows from the general energy equation, equation (4), since the change in total enthalpy $[dh_t = dh + VdV/g = \delta Q]$ is zero, that the total temperature $[dT_t = dh_t/c_p]$ remains constant throughout the process; further, from the Second Law, equation (9), since the change in entropy $[ds = (\delta Q + \delta F)/T_t]$ is zero, it follows by the thermodynamic identity, equation (5), that the total pressure $[dp_t = (dh_t - T_t\,ds)/v_t]$ remains constant throughout the process. Thus, the referred continuity equation (19) reduces to

$$\left(\frac{A_1}{A_2}\right)\Gamma_1 = \Gamma_2 \qquad (27)$$

Since it is also true that

$$\left(\frac{A_1}{A_2}\right)\left(\frac{A^*}{A_1}\right) = \left(\frac{A^*}{A_2}\right) \qquad (28)$$

it follows that

$$\Gamma_{\text{Isentropic}} = \frac{A^*}{A} \qquad (29)$$

Schematic $h - s$ and $p/p_t - \Gamma$ maps are given in Fig. 2, where subsonic and supersonic processes are indicated.

Observing that the differential change in entropy at the critical state is zero, as it is throughout any isentropic process, we are in a position to present the pertinent equations for isentropic flow, based on the general equations derived above.

$$V^* = \left(\frac{\gamma p}{\rho}\right)^{\frac{1}{2}} \qquad (30)$$

$$\frac{p^*}{p_t^*} = \left(\frac{2}{\gamma+1}\right)^{\gamma/(\gamma-1)} \qquad (31)$$

$$\frac{T_t}{T_t^*} = 1 \qquad (32)$$

$$\frac{p_t}{p_t^*} = 1 \qquad (33)$$

$$s^* - s = 0 \qquad (34)$$

$$\frac{A}{A^*} = \frac{[2/(\gamma+1)]^{1/(\gamma-1)}[(\gamma-1)/(\gamma+1)]^{\frac{1}{2}}}{(p/p_t)^{1/\gamma}[1 - (p/p_t)^{(\gamma-1)/\gamma}]^{\frac{1}{2}}} \qquad (35)$$

where A^* signifies the absolute minimum area beyond which inlet conditions cannot be maintained.

Thus, the Γ function serves well in isentropic flow processes, and the loci of all possible state points in

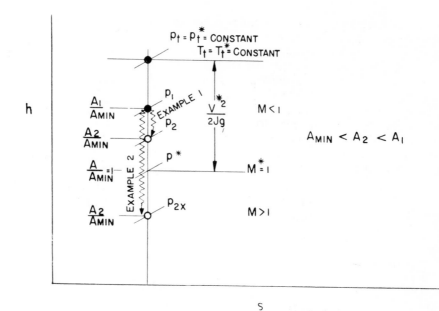

Figure 2. Isentropic Process Maps.

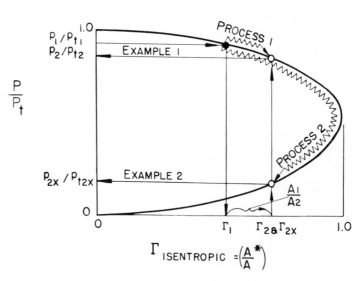

any isentropic process are precisely represented by the single $p/p_t - \Gamma$ curve. Of course, isentropic flows are quite simple to treat (note that the differential change in entropy at the critical state is zero, as it is throughout any isentropic process), and solutions of this type have been known for years. Nevertheless, even in more complex situations, this same procedure will be seen to be valid, and herein lies the generality of the Γ function.

Adiabatic Flow With Friction

Here we are referring to the one-dimensional, steady flow of fluid whose thermodynamic properties change solely because of viscous effects and area variations. In the constant-area case, this type of flow is

called *Fanno flow*, after an early worker in this field. For note: it follows from the general energy equation, equation (4), as before, that the total enthalpy and hence the total temperature remain constant throughout the process; further, since the viscous effects are reflected in entropy production within the system boundaries, by the thermodynamic identity equation, equation (5), the total pressure must always decrease in this type of flow.

The referred continuity equation (19) for Fanno flow reduces to

$$\left(\frac{p_{t1}}{p_{t2}}\right)\Gamma_1 = \Gamma_2 \qquad (36)$$

However, the more general variable-area Fanno-type

flow may always be treated. For note: in any adiabatic flow with friction, the entropy production is only a function of the total-pressure ratio $[s_2 - s_1 = R \ln (p_{t1}/p_{t2})]$, which represents the estimated or empirically determined frictional loss for a given problem, and may always be specified. The effect of an area variation, however, represents an isentropic change of state only, which may be handled separately as described in the section Adiabatic Flow Without Friction.

In this generalized Fanno-type flow, the referred continuity equation (19) becomes

$$\left(\frac{p_{t1}}{p_{t2}}\right)\left(\frac{A_1}{A_2}\right)\Gamma_1 = \Gamma_2 \qquad (37)$$

Since it is also true that

$$\left(\frac{p_{t1}}{p_{t2}}\right)\left(\frac{A_1}{A_2}\right)\left[\left(\frac{p_t^*}{p_{t1}}\right)\left(\frac{A^*}{A_1}\right)\right] = \left[\left(\frac{p_t^*}{p_{t2}}\right)\left(\frac{A^*}{A_2}\right)\right] \qquad (38)$$

it follows that

$$\Gamma_{\text{Fanno-type}} = \left(\frac{p_t^*}{p_t}\right)\left(\frac{A^*}{A}\right) \qquad (39)$$

Schematic $h - s$ and $p/p_t - \Gamma$ maps are given in Fig. 3, where subsonic and supersonic processes are indicated. The numerical evaluation of $\Gamma_{\text{Fanno-type}}$ and $\Gamma_{\text{Isentropic}}$ as functions of p/p_t will be discussed, and practical examples closely following these illustrative diagrams will be given later.

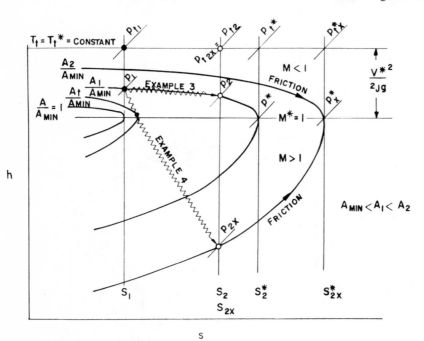

Figure 3. Fanno-Type Process Maps.

Note that the constant-area Fanno plots on the $h-s$ diagram can be interpreted in terms of the more general variable-area Fanno-type flow simply by considering the area effect as an isentropic change of state to a new area ratio line, and then proceeding along the new constant-area Fanno line passing through this revised initial state to the final state.

Compressible-flow Fanno lines are not straight lines on an $h-s$ plot. Thus we note that if a frictional process proceeds far enough, changes of state will occur more rapidly than changes of entropy until a point is reached along each Fanno line where there is no further increase in entropy (i.e., $ds = 0$ at this critical state). With this observation in mind, we are in a position to present the pertinent equations for generalized Fanno-type flow, based on the general equations derived above.

$$V^* = \left(\frac{\gamma p}{\rho}\right)^{\frac{1}{2}} \tag{30}$$

$$\frac{p^*}{p_t^*} = \left(\frac{2}{\gamma + 1}\right)^{\gamma/(\gamma - 1)} \tag{31}$$

$$\frac{T_t}{T_t^*} = 1 \tag{32}$$

$$\frac{p_t}{p_t^*} = \left\{ \frac{[2/(\gamma + 1)]^{1/(\gamma - 1)}[(\gamma - 1)/(\gamma + 1)]^{\frac{1}{2}}}{(p/p_t)^{1/\gamma}[1 - (p/p_t)^{(\gamma - 1)/\gamma}]^{\frac{1}{2}}(A/A^*)} \right\} \tag{40}$$

$$s^* - s =$$
$$2c_p \ln \left\{ \frac{[2/(\gamma + 1)]^{1/2\gamma}[(\gamma - 1)/(\gamma + 1)]^{(\gamma - 1)/4\gamma}}{(A/A^*)^{(\gamma - 1)/2\gamma}(p/p_t)^{(\gamma - 1)/2\gamma^2}[1 - (p/p_t)^{(\gamma - 1)/\gamma}]^{(\gamma - 1)/4\gamma}} \right\}$$
$$\tag{41}$$

The ratio A/A^* is simply written A/A_2, since A^* always signifies the area at which the critical state is attained, and in Fanno-type flow the critical state occurs at the final area. While equation (35) determines the absolute minimum area acceptable for a given inlet pressure ratio, the actual minimum area depends on the state point locus of the particular adiabatic process, and thus cannot be specified in general.

Diabatic Flow Without Friction

Here we are referring to the one-dimensional steady flow of a fluid whose thermodynamic properties change solely because of heat-transfer effects and area variations. In the constant-area case, this type of flow is called *Rayleigh flow*. Here, the total temperature and

therefore the total pressure change because of the allowed heat transfer. However, by the general energy equation (4) and the First Law [equation (11)], we have as a conserved quantity $p + \rho V^2$, conventionally called the *thrust* or *impulse function*. The referred continuity equation (19) for Rayleigh flow must be written as

$$\left(\frac{T_{t2}}{T_{t1}}\right)^{\frac{1}{2}}\left(\frac{p_{t1}}{p_{t2}}\right)\Gamma_1 = \Gamma_2 \tag{42}$$

This constant-area Rayleigh flow may always be treated once the total temperature ratio is specified, for the total pressure ratio may be shown to be a function (implicit, to be sure)[4] of (T_{t1}/T_{t2}) only. Since it is also true that

$$\left(\frac{T_{t2}}{T_{t1}}\right)^{\frac{1}{2}}\left(\frac{p_{t1}}{p_{t2}}\right)\left[\left(\frac{T_{t1}}{T_t^*}\right)^{\frac{1}{2}}\left(\frac{p_t^*}{p_{t1}}\right)\right] = \left[\left(\frac{T_{t2}}{T_t^*}\right)^{\frac{1}{2}}\left(\frac{p_t^*}{p_{t2}}\right)\right] \tag{43}$$

it follows that

$$\Gamma_{\text{Rayleigh}} = \left(\frac{T_t}{T_t^*}\right)^{\frac{1}{2}}\left(\frac{p_t^*}{p_t}\right) \tag{44}$$

The more general variable-area Rayleigh-type flow may sometimes be treated, but no completely general solution is possible. In any diabatic flow, the entropy production is a function of the total-pressure ratio as well as the total-temperature ratio

$$[s_2 - s_1 = c_p \ln(T_{t2}/T_{t1}) + R \ln(p_{t1}/p_{t2})]$$

Now, in Rayleigh flow, we can always specify the total-temperature ratio but we do not necessarily know the total-pressure ratio. In the variable-area case, p_{t1}/p_{t2} depends on the definite physical conditions under which the heat transfer occurs. Only when the heat transfer takes place at specified areas can a variable-area Rayleigh-type flow be treated. With this restriction, the problem reduces to a combination of separate variable-area isentropic processes and constant-area Rayleigh processes. Referred continuity for this semi-generalized Rayleigh-type flow is, of course,

$$\left(\frac{T_{t2}}{T_{t1}}\right)^{\frac{1}{2}}\left(\frac{p_{t1}}{p_{t2}}\right)\left(\frac{A_1}{A_2}\right)\Gamma_1 = \Gamma_2 \tag{19}$$

Schematic $h-s$ and $p/p_t - \Gamma$ maps are given in Fig. 4, where subsonic and supersonic processes are indicated.

Note that compressible-flow Rayleigh lines are not straight lines on an $h-s$ plot. Thus, if a heating process proceeds far enough, a point of maximum enthalpy is reached.[5] If the heating process were to proceed

[4] $$\left(\frac{p_{t1}}{p_{t2}}\right)_{\text{Rayleigh}} = \left(\frac{T_{t1}}{T_{t2}}\right)\left\{\left[\frac{p_2/p_{t2}}{p_1/p_{t1}}\right]^{1/\gamma}\left[\frac{1 - (p_2/p_{t2})^{(\gamma - 1)/\gamma}}{1 - (p_1/p_{t1})^{(\gamma - 1)/\gamma}}\right]\left[\frac{(p_1/p_{t1})^{(\gamma - 1)/\gamma} + [2\gamma/(\gamma - 1)][1 - (p_1/p_{t1})^{(\gamma - 1)/\gamma}]}{(p_2/p_{t2})^{(\gamma - 1)/\gamma} + [2\gamma/(\gamma - 1)][1 - (p_2/p_{t2})^{(\gamma - 1)/\gamma}]}\right]\right\}$$

[5] This may easily be shown to be at the state where the Mach number M reaches $1/\gamma^{\frac{1}{2}}$. Since h is a maximum where T is a maximum, we need only differentiate any T expression with respect to any independent variable, such as M, to obtain the conditions under which h maximizes.

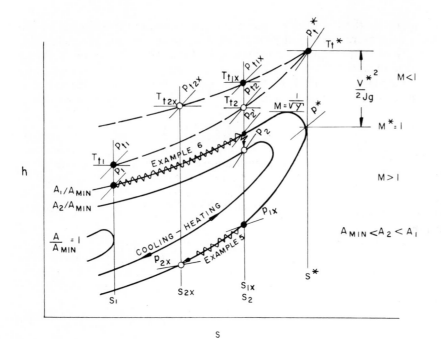

Figure 4. Rayleigh-Type Process Maps.

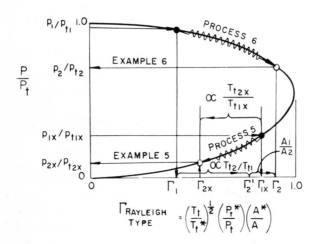

$$\Gamma_{\substack{RAYLEIGH \\ TYPE}} = \left(\frac{T_t}{T_t^*}\right)^{\frac{1}{2}} \left(\frac{P_t^*}{P_t}\right) \left(\frac{A^*}{A}\right)$$

further, the change in enthalpy would actually become negative. As in the Fanno case, changes of state would then occur more rapidly than changes of entropy until a point was reached along each Rayleigh line where there was no further increase in entropy (i.e., $ds = 0$ at this critical state).

With this observation in mind, we are in a position to present the pertinent equations for Rayleigh flow, based on the general equations derived above.

$$V^* = \left(\frac{\gamma p}{\rho}\right)^{\frac{1}{2}} \tag{30}$$

$$\frac{p^*}{p_t^*} = \left(\frac{2}{\gamma + 1}\right)^{\gamma/(\gamma - 1)} \tag{31}$$

$$\frac{T_t}{T_t^*} = \frac{4[(\gamma + 1)/(\gamma - 1)][1 - (p/p_t)^{(\gamma - 1)/\gamma}]}{\{(p/p_t)^{(\gamma - 1)/\gamma} + [2\gamma/(\gamma - 1)][1 - (p/p_t)^{(\gamma - 1)/\gamma}]\}^2} \tag{45}$$

$$\frac{p_t}{p_t^*} = \frac{2[2/(\gamma + 1)]^{1/(\gamma - 1)}}{(p/p_t)^{1/\gamma}\{(p/p_t)^{(\gamma - 1)/\gamma} + [2\gamma/(\gamma - 1)][1 - (p/p_t)^{(\gamma - 1)/\gamma}]\}} \tag{46}$$

$$s^* - s =$$

$$2c_p \ln\left\{\frac{\{(p/p_t)^{(\gamma - 1)/\gamma} + [2\gamma/(\gamma - 1)][1 - (p/p_t)^{(\gamma - 1)/\gamma}]\}^{(\gamma + 1)/2\gamma}}{(p/p_t)^{(\gamma - 1)/2\gamma^2}[1 - (p/p_t)^{(\gamma - 1)/\gamma}]^{\frac{1}{2}}[2/(\gamma - 1)]^{\frac{1}{2}}(\gamma + 1)^{(\gamma + 1)/2\gamma}}\right\}$$

$$\tag{47}$$

The ratio A/A^* is simply written 1, since only the constant-area Rayleigh process is amenable to simplified treatment.

Diabatic Flow With Friction (Isothermal)

Here we are referring to the one-dimensional, steady flow of a fluid whose thermodynamic properties change solely because of heat-transfer effects, viscous effects, and area variations, in such a manner that the static temperature remains constant. Since neither the total temperature nor the total pressure remains constant during such a process, referred continuity for the constant-area isothermal case is the same as for Rayleigh flow, i.e., equation (42). This constant-area isothermal flow may always be treated, once the total temperature ratio is specified (just as in the Rayleigh flow case), for the total pressure ratio may be shown to be an implicit function[6] of (T_{t1}/T_{t2}) only. Since it is also true that

$$\left(\frac{T_{t2}}{T_{t1}}\right)^{\frac{1}{2}}\left(\frac{p_{t1}}{p_{t2}}\right)\left[\left(\frac{T_{t1}}{T_t^*}\right)^{\frac{1}{2}}\left(\frac{p_t^*}{p_{t1}}\right)\right] = \left[\left(\frac{T_{t2}}{T_t^*}\right)^{\frac{1}{2}}\left(\frac{p_t^*}{p_{t2}}\right)\right] \quad (43)$$

it follows that

$$\Gamma_{\text{Isothermal}} = \left(\frac{T_t}{T_t^*}\right)^{\frac{1}{2}}\left(\frac{p_t^*}{p_t}\right)\left[(\gamma)^{1/(\gamma-1)}[(\gamma+1)/(3\gamma-1)]^{(\gamma+1)/2(\gamma-1)}\right] \quad (48)$$

No completely general solution is possible for variable-area isothermal flow. As with Rayleigh-type flow, the total-pressure ratio is dependent on the definite physical conditions under which the heat transfer and viscous effects occur. Only in the very limited cases when both heat transfer and viscous effects are confined to specific areas can a variable-area isothermal flow be treated. With this restriction in mind, the problem reduces to a combination of separate variable-area isentropic processes and constant-area isothermal processes. The referred continuity equation for this semi-generalized isothermal flow is, of course, equation (19).

Schematic $h - s$ and $p/p_t - \Gamma$ maps are given in Fig. 5, where subsonic and supersonic processes are indicated. The numerical evaluation of Γ_{Rayleigh} and $\Gamma_{\text{Isothermal}}$ as functions of p/p_t will be discussed, and practical examples closely following these illustrative diagrams will be given later.

Note that compressible flow isothermal lines are straight lines on an $h - s$ plot; as the combined frictional-heat-transfer process proceeds, changes of state

occur in step with changes of entropy. Hence, we should not expect the differential entropy change to be zero at the thermodynamic critical state; indeed, applying the Second Law, equation (9), and the equation of state, equation (1), we have

$$ds^*_{\text{Isothermal}} = \frac{\delta Q}{T^*} = -\frac{R\,dp}{p^*} \quad (49)$$

Making use of equation (49), we are in a position to present the pertinent equations for isothermal flow, based on the general equations previously derived.

$$V^* = \left(\frac{p}{\rho}\right)^{\frac{1}{2}} \quad (50)$$

$$\frac{p^*}{p_t^*} = \left(\frac{2\gamma}{3\gamma-1}\right)^{\gamma/(\gamma-1)} \quad (51)$$

$$\frac{T_t}{T_t^*} = \left(\frac{2\gamma}{3\gamma-1}\right)\left(\frac{p}{p_t}\right)^{(1-\gamma)/\gamma} \quad (52)$$

$$\frac{p_t}{p_t^*} = \frac{[2\gamma/(3\gamma-1)]^{\gamma/(\gamma-1)}}{\{(p/p_t)^{(\gamma+1)/\gamma}[2\gamma/(\gamma-1)][1-(p/p_t)^{(\gamma-1)/\gamma}]\}^{\frac{1}{2}}} \quad (53)$$

$$s^* - s =$$
$$2c_p \ln\left\{\frac{\{(p/p_t)^{(\gamma+1)/\gamma}[2\gamma/(\gamma-1)][1-(p/p_t)^{(\gamma-1)/\gamma}]\}^{(\gamma+1)/4\gamma}}{(p/p_t)^{1/\gamma}[1-(p/p_t)^{(\gamma-1)/\gamma}]^{\frac{1}{2}}[2\gamma/(\gamma-1)]^{\frac{1}{2}}}\right\} \quad (54)$$

The ratio A/A^* is simply written 1, since only the constant-area isothermal process is amenable to simplified treatment. In the isothermal case, equation (35) no longer describes the limiting area acceptable for a given inlet pressure ratio. In all previous cases, total conditions were assumed constant, whereas in isothermal flow,[7] static conditions are assumed invariant.

Normal Shock

Here we are referring to the one-dimensional, steady, supersonic flow of a fluid whose thermodynamic properties change suddenly to those corresponding to subsonic flow, with no change in total temperature, area, or thrust function. The first two conditions ($T_t = $ a constant and $A = $ a constant) ensure that the final state will be on the Fanno locus passing through the initial state. The last two conditions ($A = $ constant and $p + \rho V^2 = $ a constant) ensure that the final state will be on the Rayleigh locus passing through the initial state. Evidently, the final state of the normal shock must lie at the intersection of these Fanno and Rayleigh lines. But the intersection of these two flow processes always occurs at an entropy greater than the initial

---

[6] $$\left(\frac{p_{t1}}{p_{t2}}\right)_{\text{Isothermal}} = \left(\frac{T_{t1}}{T_{t2}}\right)\left\{\left[\frac{p_2/p_{t2}}{p_1/p_{t1}}\right]^{(3-\gamma)/2\gamma}\left[\frac{1-(p_2/p_{t2})^{(\gamma-1)/\gamma}}{1-(p_1/p_{t1})^{(\gamma-1)/\gamma}}\right]^{\frac{1}{2}}\right\}$$

[7] In isothermal flow, if the inlet state is subsonic, the limiting area represents a minimum area (as in the other flow processes); however, if the inlet state is supersonic, the limiting area represents a maximum area.

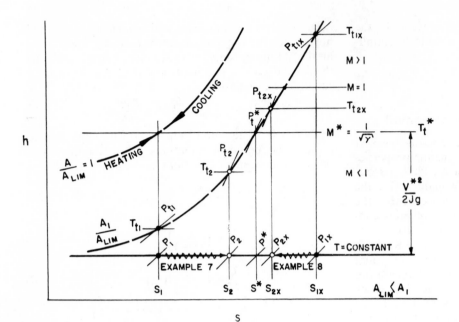

Figure 5. Isothermal Process Maps.

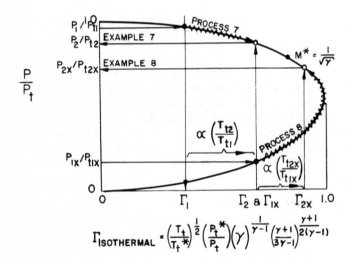

$$\Gamma_{\text{ISOTHERMAL}} = \left(\frac{T_t}{T_t^{\;*}}\right)^{\frac{1}{2}} \left(\frac{P_t^{\;*}}{P_t}\right)(\gamma)^{\frac{1}{\gamma-1}} \left(\frac{\gamma+1}{3\gamma-1}\right)^{\frac{\gamma+1}{2(\gamma-1)}}$$

value.[8] Hence, the total pressure must always decrease across the normal shock. Referred continuity for the normal shock process reduces to

$$\left(\frac{p_{t1}}{p_{t2}}\right)\Gamma_1 = \Gamma_2 \qquad (36)$$

which is of the same form as that for Fanno flow. However, in the case of normal shock, the multiplier

of Γ_1 can be defined explicitly in terms of the (supersonic) inlet pressure ratio only, i.e.,

$$\left(\frac{p_{t1}}{p_{t2}}\right)_{\text{Normal shock}} = \left\{ \left[\frac{4\gamma[1 - (p_1/p_{t1})^{(\gamma-1)/\gamma}]}{(\gamma+1)(\gamma-1)(p_1/p_{t1})^{(\gamma-1)/\gamma}} \right. \right.$$

$$\left. - \left[\frac{\gamma-1}{\gamma+1}\right] \right\}^{1/(\gamma-1)} \left\{ \frac{\gamma-1}{(\gamma+1)[1 - (p_1/p_{t1})^{(\gamma-1)/\gamma}]} \right\}^{\gamma/(\gamma-1)} \qquad (55)$$

---

[8] $(s_2 - s_1)_{\text{Normal shock}} = c_p \ln \left\{ \left[\frac{4\gamma[1 - (p_1/p_{t1})^{(\gamma-1)/\gamma}]}{(\gamma+1)(\gamma-1)(p_1/p_{t1})^{(\gamma-1)/\gamma}} - \left(\frac{\gamma-1}{\gamma+1}\right) \right]^{1/\gamma} \frac{\gamma-1}{(\gamma+1)[1 - (p_1/p_{t1})^{(\gamma-1)/\gamma}]} \right\}$ which is seen to be always positive for

$p_1/p_{t1} < [2/(\gamma+1)]^{\gamma/(\gamma-1)}$, i.e., for a supersonic initial state.

may be obtained by combining the continuity equation (2), the general energy equation (4), and the equation of state (1).

Schematic $h - s$ and $p/p_t - \Gamma$ maps are given in Fig. 6, where a discontinuous normal shock process is indicated. A numerical example closely following these illustrative diagrams is given later.

In Fig. 7, we compare all of these simplified flow processes, indicating accessible states along both static and total loci, as well as those states which are unattainable from the given (supersonic) inlet state.

The general equations, as well as equations for all the processes dealt with in this article, are summarized in terms of pressure ratio in Chart I, and in terms of Mach number in Chart II.

THE LIMITING CASE OF $\gamma = 1$

Equation (3) in this case becomes[9]

$$\rho = \rho_t \left(\frac{p}{p_t}\right) \qquad (3')$$

---

[9]The primed equations denote correspondence with the similarly numbered equations previously given.

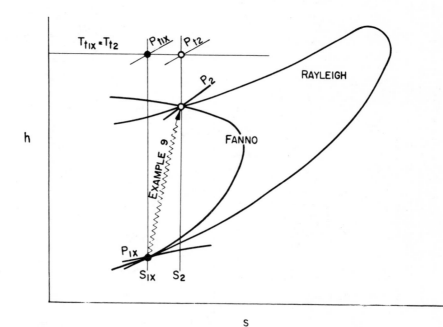

Figure 6. Normal Shock Process Maps.

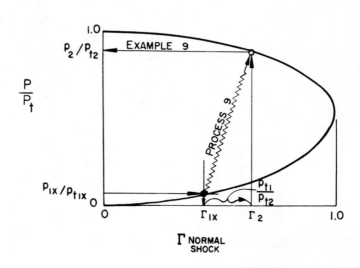

PARAMETER / PROCESS	GENERAL	ADIABATIC FLOW WITHOUT FRICTION (ISENTROPIC FLOW WITH VARIABLE AREA)	ADIABATIC FLOW WITH FRICTION (FANNO TYPE FLOW WITH VARIABLE AREA)	DIABATIC FLOW WITHOUT FRICTION (RAYLEIGH FLOW — CONSTANT AREA)	DIABATIC FLOW WITH FRICTION (ISOTHERMAL FLOW — CONSTANT AREA)
RESTRICTIONS		$T_t = \text{CONSTANT};\ p_t = \text{CONSTANT}$	$T_t = \text{CONSTANT};\ A^* = A_2$	$p + \rho v^2 = \text{CONSTANT};\ A^* = A_1$	$T = \text{CONSTANT};\ A^* = A_1$
CONTINUITY					
REFERRED CONTINUITY					
REFERRED CONTINUITY IN TERMS OF THE GENERALIZED COMPRESSIBLE FLOW FUNCTION Γ					
CRITICAL VELOCITY v^*		$\left(\frac{\gamma p}{\rho}\right)^{\frac{1}{2}}$	$\left(\frac{\gamma p}{\rho}\right)^{\frac{1}{2}}$	$\left(\frac{\gamma p}{\rho}\right)^{\frac{1}{2}}$	$\left(\frac{p}{\rho}\right)^{\frac{1}{2}}$
CRITICAL PRESSURE RATIO p^*/p_t^* (AT CRITICAL VELOCITY)		$\left(\frac{2}{\gamma+1}\right)^{\frac{\gamma}{\gamma-1}}$	$\left(\frac{2}{\gamma+1}\right)^{\frac{\gamma}{\gamma-1}}$	$\left(\frac{2}{\gamma+1}\right)^{\frac{\gamma}{\gamma-1}}$	$\left(\frac{2\gamma}{3\gamma-1}\right)^{\frac{\gamma}{\gamma-1}}$
LIMITING ENTROPY CHANGE $S^* - S$ (TO CRITICAL VELOCITY)		0			
REFERRED TOTAL TEMPERATURE T_t/T_t^* (AT CRITICAL VELOCITY)		1	1		
REFERRED TOTAL PRESSURE p_t/p_t^* (AT CRITICAL VELOCITY)		1			
REFERRED AREA A/A^* (TO CRITICAL VELOCITY)				1	1

Chart I. Summary of General and Specific Equations in Terms of Pressure Ratio.

PARAMETER \ PROCESS	GENERAL	ADIABATIC FLOW WITHOUT FRICTION (ISENTROPIC FLOW WITH VARIABLE AREA)	ADIABATIC FLOW WITH FRICTION (FANNO TYPE FLOW WITH VARIABLE AREA)	DIABATIC FLOW WITHOUT FRICTION (RAYLEIGH TYPE FLOW — CONSTANT AREA)	DIABATIC FLOW WITH FRICTION (ISOTHERMAL FLOW — CONSTANT AREA)
RESTRICTIONS		$T_t = \text{CONSTANT}\,;\; P_t = \text{CONSTANT}$	$T_t = \text{CONSTANT}\,;\; A^* = A_2$	$P + \rho v^2 = \text{CONSTANT}\,; A^* = \text{CONSTANT}$	$T = \text{CONSTANT}\,;\; A^* = A_1$
CONTINUITY	$\left(\dfrac{T_{t2}}{T_{t1}}\right)^{\frac12}\dfrac{P_{t1}}{P_{t2}}\left(\dfrac{A_1}{A_2}\right)\left[\dfrac{M_1\left(\frac{\gamma-1}{2}\right)^{\frac12}}{\left(1+\frac{\gamma-1}{2}M_1^2\right)^{\frac{\gamma+1}{2(\gamma-1)}}}\right]$ $= \dfrac{M_2\left(\frac{\gamma-1}{2}\right)^{\frac12}}{\left(1+\frac{\gamma-1}{2}M_2^2\right)^{\frac{\gamma+1}{2(\gamma-1)}}}$	$\left(\dfrac{A_1}{A_2}\right)\left[\dfrac{M_1\left(\frac{\gamma-1}{2}\right)^{\frac12}}{\left(1+\frac{\gamma-1}{2}M_1^2\right)^{\frac{\gamma+1}{2(\gamma-1)}}}\right]$ $= \dfrac{M_2\left(\frac{\gamma-1}{2}\right)^{\frac12}}{\left(1+\frac{\gamma-1}{2}M_2^2\right)^{\frac{\gamma+1}{2(\gamma-1)}}}$	$\left(\dfrac{P_{t1}}{P_{t2}}\right)\left(\dfrac{A_1}{A_2}\right)\left[\dfrac{M_1\left(\frac{\gamma-1}{2}\right)^{\frac12}}{\left(1+\frac{\gamma-1}{2}M_1^2\right)^{\frac{\gamma+1}{2(\gamma-1)}}}\right]$ $= \dfrac{M_2\left(\frac{\gamma-1}{2}\right)^{\frac12}}{\left(1+\frac{\gamma-1}{2}M_2^2\right)^{\frac{\gamma+1}{2(\gamma-1)}}}$	$\left(\dfrac{T_{t2}}{T_{t1}}\right)^{\frac12}\left(\dfrac{P_{t1}}{P_{t2}}\right)\left[\dfrac{M_1\left(\frac{\gamma-1}{2}\right)^{\frac12}}{\left(1+\frac{\gamma-1}{2}M_1^2\right)^{\frac{\gamma+1}{2(\gamma-1)}}}\right]$ $= \dfrac{M_2\left(\frac{\gamma-1}{2}\right)^{\frac12}}{\left(1+\frac{\gamma-1}{2}M_2^2\right)^{\frac{\gamma+1}{2(\gamma-1)}}}$	$\left(\dfrac{T_{t2}}{T_{t1}}\right)^{\frac12}\left(\dfrac{P_{t1}}{P_{t2}}\right)\left[\dfrac{M_1\left(\frac{\gamma-1}{2}\right)^{\frac12}}{\left(1+\frac{\gamma-1}{2}M_1^2\right)^{\frac{\gamma+1}{2(\gamma-1)}}}\right]$ $= \dfrac{M_2\left(\frac{\gamma-1}{2}\right)^{\frac12}}{\left(1+\frac{\gamma-1}{2}M_2^2\right)^{\frac{\gamma+1}{2(\gamma-1)}}}$
REFERRED CONTINUITY	$\left(\dfrac{T_{t2}}{T_{t1}}\right)^{\frac12}\dfrac{P_{t1}}{P_{t2}}\left(\dfrac{A_1}{A_2}\right)\dfrac{M_1}{\left[\frac{2}{\gamma+1}\left(1+\frac{\gamma-1}{2}M_1^2\right)\right]^{\frac{\gamma+1}{2(\gamma-1)}}}$ $= \dfrac{M_2}{\left[\frac{2}{\gamma+1}\left(1+\frac{\gamma-1}{2}M_2^2\right)\right]^{\frac{\gamma+1}{2(\gamma-1)}}}$	$\left(\dfrac{A_1}{A_2}\right)\dfrac{M_1}{\left[\frac{2}{\gamma+1}\left(1+\frac{\gamma-1}{2}M_1^2\right)\right]^{\frac{\gamma+1}{2(\gamma-1)}}}$ $= \dfrac{M_2}{\left[\frac{2}{\gamma+1}\left(1+\frac{\gamma-1}{2}M_2^2\right)\right]^{\frac{\gamma+1}{2(\gamma-1)}}}$	$\left(\dfrac{P_{t1}}{P_{t2}}\right)\left(\dfrac{A_1}{A_2}\right)\dfrac{M_1}{\left[\frac{2}{\gamma+1}\left(1+\frac{\gamma-1}{2}M_1^2\right)\right]^{\frac{\gamma+1}{2(\gamma-1)}}}$ $= \dfrac{M_2}{\left[\frac{2}{\gamma+1}\left(1+\frac{\gamma-1}{2}M_2^2\right)\right]^{\frac{\gamma+1}{2(\gamma-1)}}}$	$\left(\dfrac{T_{t2}}{T_{t1}}\right)^{\frac12}\left(\dfrac{P_{t1}}{P_{t2}}\right)\dfrac{M_1}{\left[\frac{2}{\gamma+1}\left(1+\frac{\gamma-1}{2}M_1^2\right)\right]^{\frac{\gamma+1}{2(\gamma-1)}}}$ $= \dfrac{M_2}{\left[\frac{2}{\gamma+1}\left(1+\frac{\gamma-1}{2}M_2^2\right)\right]^{\frac{\gamma+1}{2(\gamma-1)}}}$	$\left(\dfrac{T_{t2}}{T_{t1}}\right)^{\frac12}\left(\dfrac{P_{t1}}{P_{t2}}\right)\dfrac{M_1}{\left[\frac{2}{\gamma+1}\left(1+\frac{\gamma-1}{2}M_1^2\right)\right]^{\frac{\gamma+1}{2(\gamma-1)}}}$ $= \dfrac{M_2}{\left[\frac{2}{\gamma+1}\left(1+\frac{\gamma-1}{2}M_2^2\right)\right]^{\frac{\gamma+1}{2(\gamma-1)}}}$
REFERRED CONTINUITY IN TERMS OF THE GENERALIZED COMPRESSIBLE FLOW FUNCTION Γ	$\left(\dfrac{T_{t2}}{T_{t1}}\right)^{\frac12}\dfrac{P_{t1}}{P_{t2}}\left(\dfrac{A_1}{A_2}\right)\;\Gamma_1 = \Gamma_2$	$\left(\dfrac{A_1}{A_2}\right)\;\Gamma_1 = \Gamma_2$	$\left(\dfrac{P_{t1}}{P_{t2}}\right)\left(\dfrac{A_1}{A_2}\right)\;\Gamma_1 = \Gamma_2$	$\left(\dfrac{T_{t2}}{T_{t1}}\right)^{\frac12}\dfrac{P_{t1}}{P_{t2}}\;\Gamma_1 = \Gamma_2$	$\left(\dfrac{T_{t2}}{T_{t1}}\right)^{\frac12}\dfrac{P_{t1}}{P_{t2}}\;\Gamma_1 = \Gamma_2$
CRITICAL MACH NUMBER M^*	$\left[\dfrac{1}{1 - \frac{ds}{dp}\,\frac{\gamma p}{c_p}}\right]^{\frac12}$				$\left(\dfrac{1}{\gamma}\right)^{\frac12}$
CRITICAL PRESSURE RATIO p^*/p_t^* (AT CRITICAL MACH NUMBER)	$\left[\dfrac{\frac{-1}{\gamma}-\frac{ds}{dp}\,\frac{p}{c_p}}{\frac{1}{\gamma}-\frac{ds}{dp}\,\frac{p}{c_p}+\frac{\gamma-1}{2\gamma}}\right]^{\frac{\gamma}{\gamma-1}}$	$\left(\dfrac{2}{\gamma+1}\right)^{\frac{\gamma}{\gamma-1}}$	$\left(\dfrac{2}{\gamma+1}\right)^{\frac{\gamma}{\gamma-1}}$	$\left(\dfrac{2}{\gamma+1}\right)^{\frac{\gamma}{\gamma-1}}$	$\left(\dfrac{2}{3\gamma-1}\right)^{\frac{\gamma}{\gamma-1}}$
LIMITING ENTROPY CHANGE $s^{**}-s$ (TO CRITICAL MACH NUMBER)	$2c_p\ln\left[\dfrac{M^*\left(1+\frac{\gamma-1}{2}M^2\right)^{\frac{\gamma+1}{2(\gamma-1)}}}{M\left(1+\frac{\gamma-1}{2}M^{*2}\right)^{\frac{\gamma+1}{2(\gamma-1)}}}\left(\frac{P_t}{P_t^*}\right)\left(\frac{A}{A^*}\right)\right]$	0	$2c_p\ln\left[\dfrac{\left[\frac{2}{\gamma+1}\left(1+\frac{\gamma-1}{2}M^2\right)\right]^{\frac{\gamma+1}{4\gamma}}\frac{\gamma-1}{2\gamma}}{\left(\frac{A}{A_2}\right)}\right]$	$2c_p\ln\left[\dfrac{\left[\frac{2}{\gamma+1}\left(1+\frac{\gamma-1}{2}M^2\right)\right]^{\frac{\gamma+1}{2\gamma}}}{M}\right]$	$2c_p\ln\left[\gamma M^2\right]^{\frac{1-\gamma}{4\gamma}}$
REFERRED TOTAL TEMPERATURE T_t/T_t^* (AT CRITICAL MACH NO.)	$\dfrac{M^2}{M^{*2}}\left[\dfrac{1+\frac{\gamma-1}{2}M^{*2}}{1+\frac{\gamma-1}{2}M^2}\right]^{\frac{\gamma+1}{2(\gamma-1)}}\left(\dfrac{P_t}{P_t^*}\right)^2\left(\dfrac{A}{A^*}\right)^2$			$\dfrac{2(\gamma+1)M^2\left(1+\frac{\gamma-1}{2}M^2\right)}{(1+\gamma M^2)^2}$	$\left(\dfrac{2\gamma}{3\gamma-1}\right)\left(1+\frac{\gamma-1}{2}M^2\right)$
REFERRED TOTAL PRESSURE P_t/P_t^* (AT CRITICAL MACH NO.)	$\dfrac{M^*}{M}\left[\dfrac{1+\frac{\gamma-1}{2}M^2}{1+\frac{\gamma-1}{2}M^{*2}}\right]^{\frac{\gamma+1}{2(\gamma-1)}}\left(\dfrac{A}{A^*}\right)$		$\dfrac{\left[\frac{2}{\gamma+1}\left(1+\frac{\gamma-1}{2}M^2\right)\right]^{\frac{\gamma+1}{2(\gamma-1)}}}{M\left(\frac{A}{A_2}\right)}$	$\dfrac{1}{(\gamma+1)}\left[\frac{2}{\gamma+1}\left(1+\frac{\gamma-1}{2}M^2\right)\right]^{\frac{\gamma}{\gamma-1}}(1+\gamma M^2)$	$\dfrac{\left[\frac{2\gamma}{3\gamma-1}\left(1+\frac{\gamma-1}{2}M^2\right)\right]^{\frac{\gamma}{\gamma-1}}}{M\gamma^{\frac{\gamma}{\gamma-1}}}$
REFERRED AREA A/A^* (TO CRITICAL MACH NUMBER)	$\dfrac{M^*}{M}\left[\dfrac{1+\frac{\gamma-1}{2}M^2}{1+\frac{\gamma-1}{2}M^{*2}}\right]^{-\frac12}\left(\dfrac{P_t}{P_t^*}\right)$	$\dfrac{1}{M}\left[\frac{2}{\gamma+1}\left(1+\frac{\gamma-1}{2}M^2\right)\right]^{\frac{\gamma+1}{2(\gamma-1)}}$	$\dfrac{A}{A_2}$		

Chart II. Summary of General and Specific Equations in Terms of Mach Number.

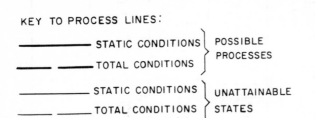

COMPARISON OF VARIOUS CONSTANT-AREA
PROCESSES FOR A SUPERSONIC INLET STATE

KEY TO PROCESS LINES:

———————— STATIC CONDITIONS ⎫ POSSIBLE
————— · ————— TOTAL CONDITIONS ⎭ PROCESSES

—————————— STATIC CONDITIONS ⎫ UNATTAINABLE
————— ————— TOTAL CONDITIONS ⎭ STATES

○ INLET STATE (ALSO EXIT STATE FOR NORMAL SHOCK)
● CRITICAL STATE FOR VARIOUS PROCESSES.

Figure 7. Comparison of Various Constant-Area Processes for a Supersonic Inlet State.

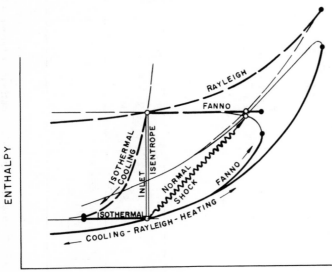

which together with equation (6) yields by normal integration

$$V = \left[2\frac{p_t}{\rho_t} \ln \frac{p_t}{p} \right]^{\frac{1}{2}} \qquad (7')$$

Equation (7') also can be obtained by the more general and powerful method of limits as follows: by combining equations (3) and (7), we obtain

$$\frac{V^2}{2\gamma} = \frac{(p_t/\rho_t)[1 - (p/p_t)^{(\gamma-1)/\gamma}]}{\gamma - 1} \qquad (56)$$

which is seen to be indeterminate at $\gamma = 1$ (i.e., $V^2/2\gamma = 0/0$). When L'Hopital's rule (i.e., derivative of numerator over derivative of denominator) is applied, there results

$$\lim_{\gamma \to 1} \frac{V^2}{2\gamma} = \frac{\lim\limits_{\gamma \to 1} (d/d\gamma)\{(p_t/\rho_t)[1 - (p/p_t)^{(\gamma-1)/\gamma}]\}}{\lim\limits_{\gamma \to 1} (d/d\gamma)(\gamma - 1)} \qquad (57)$$

or

$$\lim_{\gamma \to 1} \frac{V^2}{2\gamma} = \lim_{\gamma \to 1} \left(\frac{p_t}{\rho_t}\right)\left[\frac{(p/p_t)^{(\gamma-1)/\gamma} \ln (p_t/p)}{\gamma^2}\right] \qquad (58)$$

which on applying limits yields equation (7'). Continuity [equation (8)] now can be expressed as

$$\left(\frac{T_{t2}}{T_{t1}}\right)^{\frac{1}{2}}\left(\frac{p_{t1}}{p_{t2}}\right)\left(\frac{A_1}{A_2}\right)\left\{\left(\frac{p_1}{p_{t1}}\right)\left[\ln \frac{p_{t1}}{p_1}\right]^{\frac{1}{2}}\right\} = \left\{\left(\frac{p_2}{p_{t2}}\right)\left[\ln \frac{p_{t2}}{p_2}\right]^{\frac{1}{2}}\right\} \qquad (8')$$

Equations (14) and (7') combine to yield

$$\frac{p^*}{p_t^*} = e^{-\frac{1}{2}} \qquad (16')$$

which defines the critical pressure ratio for any flow process in the limit as $\gamma \to 1$. Equation (16') also can be determined by the method of limits as follows. Equation (31) for isentropic, Fanno, and Rayleigh processes is seen to be indeterminate when expressed in the form

$$\lim_{\gamma \to 1} \ln \left(\frac{p^*}{p_t^*}\right) = \lim_{\gamma \to 1} \left\{ \frac{\gamma \ln [2/(\gamma + 1)]}{\gamma - 1} \right\} \qquad (59)$$

However, applying L'Hopital's rule, we have

$$\lim_{\gamma \to 1} \ln \left(\frac{p^*}{p_t^*}\right) = \lim_{\gamma \to 1} \left[-\left(\frac{\gamma}{\gamma + 1}\right) + \ln \left(\frac{2}{\gamma + 1}\right) \right] \qquad (60)$$

which reduces to

$$\ln \left(\frac{p^*}{p_t^*}\right)_{\substack{\text{Isentropic} \\ \text{Fanno} \\ \text{Rayleigh}}} = -\frac{1}{2} \qquad (61)$$

Similarly, equation (51) for isothermal flow is seen to be indeterminate when expressed in the form

$$\lim_{\gamma \to 1} \ln \left(\frac{p^*}{p_t^*}\right) = \lim_{\gamma \to 1} \left\{ \frac{\gamma \ln [2\gamma/(3\gamma - 1)]}{\gamma - 1} \right\} \qquad (62)$$

However, by the methods of limits we have

$$\lim_{\gamma \to 1} \ln \left(\frac{p^*}{p_t^*}\right) = \lim_{\gamma \to 1} \left[-\left(\frac{3\gamma}{3\gamma - 1}\right) + 1 + \ln \left(\frac{2\gamma}{3\gamma - 1}\right) \right] \qquad (63)$$

which also reduces to

$$\ln \left(\frac{p^*}{p_t^*}\right)_{\text{Isothermal}} = -\frac{1}{2} \qquad (61)$$

Thus, equation (16′) is obtained by the method of limits as well as by thermodynamic reasoning.

If this development is patterned after that in the section entitled The Gamma Function, the reference pressure ratio function for continuity becomes

$$\left(\frac{p^*}{p_t^*}\right) \left(\ln \frac{p_t^*}{p^*}\right)^{\frac{1}{2}} = e^{-\frac{1}{2}} (\ln e^{\frac{1}{2}})^{\frac{1}{2}} = \left(\frac{1}{2e}\right)^{\frac{1}{2}} \qquad (18')$$

which is seen to be a constant in the $\gamma = 1$ case.
Consequently, we obtain

$$\left(\frac{T_{t2}}{T_{t1}}\right)^{\frac{1}{2}} \left(\frac{p_{t1}}{p_{t2}}\right) \left(\frac{A_1}{A_2}\right) \Gamma_1 = \Gamma_2 \qquad (19')$$

where

$$\Gamma = (2e)^{\frac{1}{2}} \left[\left(\frac{p}{p_t}\right) \left[\ln \frac{p_t}{p}\right]^{\frac{1}{2}} \right] \qquad (21')$$

and where Γ varies only from 0 to 1 for any and all flow processes.

It is possible to develop expressions for M, (T/T_t), $(T_t/T_t^*)_{\text{Rayleigh}}$, $(T_t/T_t^*)_{\text{Isothermal}}$, and $(p_{t1}/p_{t2})_{\text{Normal shock}}$ by the method of limits, yielding

$$M = \left[2 \ln \left(\frac{p_t}{p}\right) \right]^{\frac{1}{2}} \qquad (64)$$

$$\frac{T}{T_t} = 1 \qquad (65)$$

$$\left(\frac{T_t}{T_t^*}\right)_{\text{Rayleigh}} = \frac{8 \ln (p_t/p)}{[1 + 2 \ln (p_t/p)]^2} \qquad (45')$$

$$\left(\frac{T_t}{T_t^*}\right)_{\text{Isothermal}} = 1 \qquad (52')$$

$$\left(\frac{p_{t1}}{p_{t2}}\right)_{\text{Normal shock}} = \frac{e^{[(M_1^4 - 1)/2M_1^2]}}{M_1^2} \qquad (55')$$

Chapter 2

Generalized Compressible Flow Tables

THE GAMMA FUNCTION FOR NUMERICAL SOLUTIONS

We have seen that a generalized compressible flow function Γ may be defined and that it has significance in any arbitrary, one-dimensional, workless flow process. The individual processes of most general interest have been discussed in some detail and the role of the Γ function in describing and facilitating the solution of these typical processes has been presented. Of more practical use would be a convenient means of obtaining numerical answers to the problems at hand in a manner that would utilize the generalized compressible flow function Γ.

We find, in the published literature and texts concerned with the thermodynamics of moving fluids, separate and distinct analyses and problem-solution methods. For example, one finds chapters on isentropic, adiabatic, and diabatic flows, as well as the normal-shock process, while the isothermal process is quite often ignored. It would therefore be quite valuable to have a means available for the solution of each of these basic processes in a similar manner, using similar procedures, and (if possible) a common reference scheme (e.g., a table or plot). We shall now proceed to develop a *generalized compressible flow table* which provides the means to satisfy these requirements.

The Γ function has been carried through in its development as a function of pressure ratio, p/p_t. It is then just one step to convert this independent variable to Mach number M,[10] as they are both state functions and are intimately related. Hence, we now have a choice of two independent variables for our generalized compressible flow table. Upon further examination of the Γ function, equation (21), it is seen to be also a function (however weak) of the ratio of specific heats, γ, and this constitutes one other independent variable

that must be considered. As we can usually consider γ to be constant over the range of thermodynamic variables for a particular problem, we may compile our table for various constant values of γ, essentially reserving a particular value for a particular gas.

The question arises as to what to include in our table. Of the physical variables such as pressure, temperature, and density, the first two mentioned are the most easily measurable and are logical candidates for the table. We have seen that p/p_t will be a prime independent variable; thus, we shall include temperature, in the form of T/T_t, as one of our table entries. The density may easily be derived, if desired, from equation (1), once the pressure and temperature are defined. Let us now consider the various processes and determine which parameters will be necessary to complete the generalized compressible flow table.

Adiabatic Flow Without Friction

The flow characteristics of this process change solely because of area variations as the total temperature and total pressure remain constant; hence referred continuity [equation (19)] reduces to

$$\left(\frac{A_1}{A_2}\right)\Gamma_1 = \Gamma_2 \qquad (27)$$

It has been noted that $\Gamma_{\text{Isentropic}} = A^*/A$; it may also be easily shown that $\Gamma = (W/\Delta t)/(W/\Delta t)^*_{\text{Isentropic}}$. As our means of isentropic-flow solution requires only the Γ function, we need only this column in the generalized compressible flow table. One simply modifies Γ_1 by the isentropic multiplier A_1/A_2 to obtain Γ_2 (see Examples 1 and 2 in Chapter 3).

Adiabatic Flow With Friction

In this flow process we consider changes in fluid properties that arise due to variations in area and total pressure; total temperature remains invariant. Referred continuity [equation (19)] for the more

---

[10]M is the ratio of directed velocity to acoustic velocity; analytically,

$$M = \left\{ \left[\left(\frac{p}{p_t}\right)^{(1-\gamma)/\gamma} - 1 \right] \frac{2}{\gamma - 1} \right\}^{\frac{1}{2}}$$

general Fanno-type process reduces to

$$\left(\frac{p_{t1}}{p_{t2}}\right)\left(\frac{A_1}{A_2}\right)\Gamma_1 = \Gamma_2 \qquad (37)$$

Hence, when developing a generalized compressible flow table by which Fanno-type flows may be solved, we again need only the Γ column, since area ratio and total-pressure ratio may always be specified; no other special column is required. One simply modifies Γ_1 by the Fanno multiplier, $(p_{t1}/p_{t2})(A_1/A_2)$, to obtain Γ_2 (see Examples 3 and 4 in Chapter 3).

Diabatic Flow Without Friction

Here we consider changes in thermodynamic properties that are effected solely by heat transfer and (under certain conditions) area variation. For the Rayleigh case (constant area), total temperature and total pressure do not remain constant and the referred continuity relation [equation (19)] is written

$$\left(\frac{T_{t2}}{T_{t1}}\right)^{\frac{1}{2}}\left(\frac{p_{t1}}{p_{t2}}\right)\Gamma_1 = \Gamma_2 \qquad (42)$$

As the total-pressure and total-temperature ratios are related by an implicit function, we need specify only one of the two, the other then being immediately defined. Hence, when developing a generalized compressible flow table by which Rayleigh flows may be solved, we will naturally make use of the Γ column. But, in addition, we must include a special column tabulating either T_t/T_t^* or p_t^*/p_t to define the process completely, since the two ratios making up the Rayleigh multiplier $(T_{t2}/T_{t1})^{\frac{1}{2}}(p_{t1}/p_{t2})$, while not independent variables, are yet not explicit functions one of the other, as previously mentioned. Choosing the referred temperature as the more significant parameter in Rayleigh flow (which is predominantly a heat-transfer problem), one simply modifies $(T_{t1}/T_t^*)_{\text{Rayleigh}}$ by the multiplier T_{t2}/T_{t1} to obtain $(T_{t2}/T_t^*)_{\text{Rayleigh}}$. Since $(T_{t2}/T_t^*)_{\text{Rayleigh}}$ is a function only of the exit-pressure ratio, it must necessarily appear in the same row as the Γ_2 function in the table. Note that Γ_2/Γ_1, when divided by $(T_{t2}/T_{t1})^{\frac{1}{2}}$, yields the unspecified total-pressure ratio p_{t1}/p_{t2}.

The more general variable-area Rayleigh-type flow process may be treated if the heat transfer takes place at certain specified areas. The problem then reduces to a series combination of the isentropic and Rayleigh processes (see Examples 5 and 6 in Chapter 3).

Diabatic Flow With Friction (Isothermal)

Here we consider changes in thermodynamic properties (except for static temperature, which is constant) to be influenced by heat transfer, viscous effects, and (under certain conditions) area variations.

Referred continuity for the constant-area case [equation (19)] is again written

$$\left(\frac{T_{t2}}{T_{t1}}\right)^{\frac{1}{2}}\left(\frac{p_{t1}}{p_{t2}}\right)\Gamma_1 = \Gamma_2 \qquad (42)$$

An analysis nearly identical to that described under the Rayleigh process requires another special column —the parameter $(T_t/T_t^*)_{\text{Isothermal}}$—to facilitate this problem solution. Again, the total-pressure and total-temperature ratios are implicit functions one of the other, and hence remarks similar to those for the Rayleigh case apply here as well (see Examples 7 and 8 in Chapter 3).

Normal Shock

Referred continuity [equation (19)] for this discontinuous-flow phenomenon reduces to

$$\left(\frac{p_{t1}}{p_{t2}}\right)\Gamma_1 = \Gamma_2 \qquad (36)$$

where the total-pressure ratio p_{t1}/p_{t2} is an explicit function of the inlet-pressure ratio only [equation (55)]. Hence, when developing a generalized compressible flow table, to include the solution of normal-shock processes we will naturally make use of the Γ column. In addition, we will include a column tabulating $(p_{t1}/p_{t2})_{\text{Normal shock}}$. One simply modifies Γ_1 by the ratio $(p_{t1}/p_{t2})_{\text{Normal shock}}$, found in the same row as Γ_1, to obtain Γ_2 (see Example 9 in Chapter 3).

THE GENERALIZED COMPRESSIBLE FLOW TABLE

The generalized compressible flow function Γ is tabulated against the independent parameters p/p_t and M, the state point function T/T_t, and also against the specific process functions $(T_t/T_t^*)_{\text{Rayleigh}}$, $(T_t/T_t^*)_{\text{Isothermal}}$, and $(p_{t1}/p_{t2})_{\text{Normal shock}}$ to give the generalized compressible flow tables. Computer-generated numerical flow tables, Tables I to XII, appear in the Appendix in the following order:

Table I.	$\gamma = 1.0$	
Table II.	$\gamma = 1.1$	
Table III.	$\gamma = 1.2$	
Table IV.	$\gamma = 1.3$	$f(p/p_t)$
Table V.	$\gamma = 1.4$	
Table VI.	$\gamma = 1.67$	
Table VII.	$\gamma = 1.0$	
Table VIII.	$\gamma = 1.1$	
Table IX.	$\gamma = 1.2$	
Table X.	$\gamma = 1.3$	$f(M)$
Table XI.	$\gamma = 1.4$	
Table XII.	$\gamma = 1.67$	

Chapter 3

Numerical Examples in Gas Dynamics

ISENTROPIC FLOW

Example 1

Consider the subsonic expansion of air ($\gamma = 1.4$) from an inlet pressure ratio of 0.980 through a duct of area ratio 1.6937. The inlet total temperature is 540°R; the inlet static pressure is 20 psia.

Find the total and static pressures, the total and static temperatures, and the Mach number at exit. (See Fig. 8.)

Solution by Table V. From

$$\frac{p_1}{p_{t1}} = 0.980$$

we obtain

$$\Gamma_1 = \frac{A^*}{A_1} = 0.28894$$

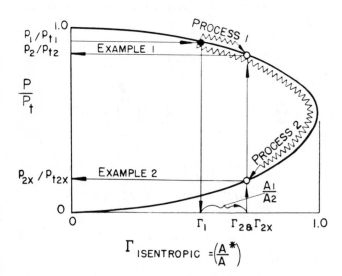

Figure 8. Isentropic Flow—Examples 1 and 2 (Note: Figures 8 through 12 are bottom halves of Figures 2 through 6).

But

$$\Gamma_2 = \Gamma_1\left(\frac{A_1}{A_2}\right) = 0.28894 \times 1.6937 = 0.48938 = \frac{A^*}{A_2}$$

From

$$\Gamma_2 = 0.48938$$

we obtain

$$\frac{p_2}{p_{t2}} = 0.940 \qquad \frac{T_2}{T_{t2}} = 0.98248 \qquad M_2 = 0.29863$$

whence

$$p_{t2} = p_{t1} = \frac{20 \text{ psia}}{0.980} = 20.408 \text{ psia}$$

$$p_2 = 0.940 \times 20.408 \text{ psia} = 19.184 \text{ psia}$$

$$T_{t2} = T_{t1} = 540°R$$

$$T_2 = 0.98248 \times 540°R = 530.54°R$$

$$M_2 = 0.29863$$

Example 2

Consider the supersonic expansion of steam ($\gamma = 1.3$) from an inlet Mach number of 0.15 through a convergent–divergent nozzle of overall area ratio 2.99858. The inlet total temperature is 1000°R; the inlet static pressure is 100 psia.

Find the total and static pressures, the total and static temperatures, and the Mach number at exit. (See Fig. 8.)

Solution by Table X. From

$$M_1 = 0.15$$

we obtain

$$\frac{p_1}{p_{t1}} = 0.98551 \qquad \Gamma_1 = \frac{A^*}{A_1} = 0.25302$$

But

$$\Gamma_{2x} = \Gamma_1\left(\frac{A_1}{A_2}\right) = 0.25302 \times 2.99858 = 0.75870 = \frac{A^*}{A_2}$$

From

$$\Gamma_{2x} = 0.75870$$

we obtain

$$\frac{p_{2x}}{p_{t2x}} = 0.22675 \qquad \frac{T_{2x}}{T_{t2x}} = 0.71004 \qquad M_{2x} = 1.65$$

whence

$$p_{t2x} = p_{t1} = \frac{100 \text{ psia}}{0.98551} = 101.470 \text{ psia}$$

$$p_{2x} = 0.22675 \times 101.470 \text{ psia} = 23.008 \text{ psia}$$

$$T_{t2x} = T_{t1} = 1000°R$$

$$T_{2x} = 0.71004 \times 1000°R = 710.04°R$$

$$M_{2x} = 1.65$$

FANNO AND FANNO-TYPE FLOW

Example 3. Fanno Flow

Consider the adiabatic flow of air ($\gamma = 1.4$) from an inlet pressure ratio of 0.960 through a duct of constant area. The inlet total temperature is 540°R; the inlet static pressure is 50 psia.

Find the total and static pressures, the total and static temperatures, and the Mach number at exit if the total-pressure ratio across the duct is 1.8609. (See Fig. 9.)

Solution by Table V. From

$$\frac{p_1}{p_{t1}} = 0.960$$

we obtain

$$\Gamma_1 = \left(\frac{p_t^*}{p_{t1}}\right)\left(\frac{A^*}{A_1}\right) = 0.40412$$

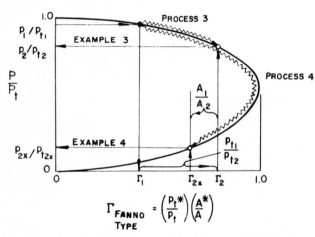

$$\Gamma_{\text{FANNO} \atop \text{TYPE}} = \left(\frac{p_t^*}{p_t}\right)\left(\frac{A^*}{A}\right)$$

Figure 9. Fanno-Type Flow—Examples 3 and 4.

But

$$\Gamma_2 = \Gamma_1\left(\frac{p_{t1}}{p_{t2}}\right)\left(\frac{A_1}{A_2}\right) = 0.40412 \times 1.8609 \times 1$$

$$= 0.75203 = \left(\frac{p_t^*}{p_{t2}}\right)\left(\frac{A^*}{A_2}\right)$$

From

$$\Gamma_2 = 0.75203$$

we obtain

$$\frac{p_2}{p_{t2}} = 0.840 \qquad \frac{T_2}{T_{t2}} = 0.95141 \qquad M_2 = 0.50536$$

whence

$$p_{t1} = \frac{50 \text{ psia}}{0.960} = 52.083 \text{ psia}$$

$$p_{t2} = \frac{52.083 \text{ psia}}{1.8609} = 27.988 \text{ psia}$$

$$p_2 = 0.840 \times 27.988 \text{ psia} = 23.510 \text{ psia}$$

$$T_{t2} = T_{t1} = 540°R$$

$$T_2 = 0.95141 \times 540°R = 513.76°R$$

$$M_2 = 0.50536$$

Example 4. Fanno-Type Flow

Consider the adiabatic flow of steam ($\gamma = 1.3$) from an inlet Mach number of 0.25 through a convergent–divergent nozzle of overall area ratio 0.73527. The inlet total temperature is 1000°R; the inlet static pressure is 100 psia.

Find the total and static pressures, the total and static temperatures, and the Mach number at exit if the total-pressure ratio across the duct is 1.8609. (See Fig. 9.)

Solution by Table X. From

$$M_1 = 0.25$$

we obtain

$$\frac{p_1}{p_{t1}} = 0.96037 \qquad \Gamma_1 = \left(\frac{p_t^*}{p_{t1}}\right)\left(\frac{A^*}{A_1}\right) = 0.41217$$

But

$$\Gamma_{2x} = \Gamma_1\left(\frac{p_{t1}}{p_{t2x}}\right)\left(\frac{A_1}{A_2}\right)$$

$$= 0.41217 \times 1.8609 \times 0.73527 = 0.56396$$

From

$$\Gamma_{2x} = 0.56396$$

we obtain

$$\frac{p_{2x}}{p_{t2x}} = 0.13046 \qquad \frac{T_{2x}}{T_{t2x}} = 0.62500 \qquad M_{2x} = 2.0$$

whence

$$p_{t1} = \frac{100 \text{ psia}}{0.96037} = 104.126 \text{ psia}$$

$$p_{t2x} = \frac{104.126 \text{ psia}}{1.8609} = 55.955 \text{ psia}$$

$$p_{2x} = 0.13046 \times 55.955 \text{ psia} = 7.300 \text{ psia}$$

$$T_{t2x} = T_{t1} = 1000°R$$

$$T_{2x} = 0.6250 \times 1000°R = 625.00°R$$

$$M_{2x} = 2.0$$

RAYLEIGH AND RAYLEIGH-TYPE FLOW

Example 5. Rayleigh Flow

Consider the frictionless flow of air ($\gamma = 1.4$) from an inlet pressure ratio of 0.300 through a duct of constant area. The inlet total temperature is 540°R; the inlet static pressure is 5 psia.

Find the total and static pressures, the total and static temperatures, and the Mach number at exit if the total-temperature ratio across the duct is 0.79856. (See Fig. 10.)

Solution by Table V. From

$$\frac{p_{1x}}{p_{t1x}} = 0.300$$

we obtain

$$\Gamma_{1x} = \left(\frac{T_{t1x}}{T_t^*}\right)^{\frac{1}{2}} \left(\frac{p_t^*}{p_{t1x}}\right) = 0.88214 \qquad \frac{T_{t1x}}{T_t^*} = 0.92614$$

But

$$\frac{T_{t2x}}{T_t^*} = \left(\frac{T_{t1x}}{T_t^*}\right)\left(\frac{T_{t2x}}{T_{t1x}}\right) = 0.92614 \times 0.79856 = 0.73958$$

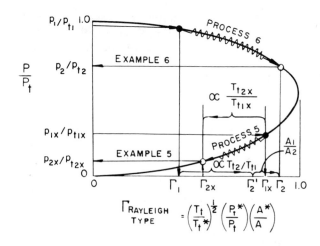

Figure 10. Rayleigh-Type Flow—Examples 5 and 6.

Corresponding to $T_{t2x}/T_t^* = 0.73958$ is

$$\Gamma_{2x} = \Gamma_{1x}\left(\frac{T_{t2x}}{T_{t1x}}\right)^{\frac{1}{2}}\left(\frac{p_{t1x}}{p_{t2x}}\right)$$

$$= 0.45606 = \left(\frac{T_{t2x}}{T_t^*}\right)^{\frac{1}{2}}\left(\frac{p_t^*}{p_{t2x}}\right)$$

From

$$\Gamma_{2x} = 0.45606$$

we obtain

$$\frac{p_{2x}}{p_{t2x}} = 0.080 \qquad \frac{T_{2x}}{T_{t2x}} = 0.48596 \qquad M_{2x} = 2.29978$$

whence

$$\frac{p_{t1x}}{p_{t2x}} = \left(\frac{\Gamma_{2x}}{\Gamma_{1x}}\right)\left(\frac{T_{t1x}}{T_{t2x}}\right)^{\frac{1}{2}} = \frac{0.45606}{(0.88214)(0.79856)^{\frac{1}{2}}} = 0.57854$$

$$p_{t1x} = \frac{5 \text{ psia}}{0.300} = 16.667 \text{ psia}$$

$$p_{t2x} = \frac{16.667 \text{ psia}}{0.57854} = 28.809 \text{ psia}$$

$$p_{2x} = 0.080 \times 28.809 \text{ psia} = 2.305 \text{ psia}$$

$$T_{t2x} = 0.79856 \times 540°R = 431.22°R$$

$$T_{2x} = 0.48596 \times 431.22°R = 209.56°R$$

$$M_{2x} = 2.29978$$

Example 6. Rayleigh-Type Flow

Consider the frictionless flow of air ($\gamma = 1.4$) from an inlet Mach number of 0.25 through a duct of area ratio 1.45301. The inlet total temperature is 460°R; the inlet static pressure is 10 psia.

Find the total and static pressures, the total and static temperatures, and the Mach number at exit if the total-temperature ratio across the duct is 2.05976 and all heat is transferred at the inlet area. (See Fig. 10.)

Solution by Table XI. From

$$M_1 = 0.25$$

we obtain

$$\frac{p_1}{p_{t1}} = 0.95745 \qquad \Gamma_1 = \left(\frac{T_{t1}}{T_t^*}\right)^{\frac{1}{2}}\left(\frac{p_t^*}{p_{t1}}\right) = 0.41620$$

$$\frac{T_{t1}}{T_t^*} = 0.25684$$

But

$$\frac{T_{t2}}{T_t^*} = \left(\frac{T_{t1}}{T_t^*}\right)\left(\frac{T_{t2}}{T_{t1}}\right) = 0.25684 \times 2.05976 = 0.52903$$

Corresponding to $T_{t2}/T_t^* = 0.52903$ is

$$\Gamma_2' = \Gamma_1 \left(\frac{T_{t2}}{T_{t1}}\right)^{\frac{1}{2}} \left(\frac{p_t^*}{p_{t2}}\right) = 0.62888$$

$$= \left(\frac{T_{t2}}{T_t^*}\right)^{\frac{1}{2}} \left(\frac{p_t^*}{p_{t2}}\right)$$

Now

$$\Gamma_2 = \Gamma_2' \left(\frac{A_1}{A_2}\right) = 0.62888 \times 1.45301 = 0.91377$$

From

$$\Gamma_2 = 0.91377$$

we obtain

$$\frac{p_2}{p_{t2}} = 0.72093 \qquad \frac{T_2}{T_{t2}} = 0.91075 \qquad M_2 = 0.70$$

whence

$$\frac{p_{t1}}{p_{t2}} = \frac{\Gamma_2'}{\Gamma_1}\left(\frac{T_{t1}}{T_{t2}}\right)^{\frac{1}{2}} = \frac{0.62888}{(0.41620)(2.05976)^{\frac{1}{2}}} = 1.05282$$

$$p_{t1} = \frac{10\ \text{psia}}{0.95745} = 10.444\ \text{psia}$$

$$p_{t2} = \frac{10.444\ \text{psia}}{1.05282} = 9.920\ \text{psia}$$

$$p_2 = 0.72093 \times 9.920\ \text{psia} = 7.152\ \text{psia}$$

$$T_{t2} = 2.05976 \times 460°R = 947.49°R$$

$$T_2 = 0.91075 \times 947.49°R = 862.93°R$$

$$M_2 = 0.70$$

ISOTHERMAL FLOW

Example 7

Consider the isothermal flow of air ($\gamma = 1.4$) from an inlet pressure ratio of 0.960 through a duct of constant area. The inlet total temperature is 540°R; the inlet static pressure is 15 psia.

Find the total and static pressures, the total and static temperatures, and the Mach number at exit if the total-temperature ratio across the duct is 1.112994. (See Fig. 11.)

Solution by Table V. From

$$\frac{p_1}{p_{t1}} = 0.960$$

we obtain

$$\Gamma_1 = \left(\frac{T_{t1}}{T_t^*}\right)^{\frac{1}{2}} \left(\frac{p_t^*}{p_{t1}}\right)(\gamma)^{1/(\gamma-1)}\left(\frac{\gamma+1}{3\gamma-1}\right)^{(\gamma+1)/2(\gamma-1)} = 0.40412$$

$$\frac{T_{t1}}{T_t^*} = 0.88527$$

But

$$\frac{T_{t2}}{T_t^*} = \left(\frac{T_{t1}}{T_t^*}\right)\left(\frac{T_{t2}}{T_{t1}}\right) = 0.88527 \times 1.112994 = 0.98530$$

Corresponding to $T_{t2}/T_t^* = 0.98530$ is

$$\Gamma_2 = \Gamma_1 \left(\frac{T_{t2}}{T_{t1}}\right)^{\frac{1}{2}} \left(\frac{p_{t1}}{p_{t2}}\right) = 0.96079$$

From

$$\Gamma_2 = 0.96079$$

we obtain

$$\frac{p_2}{p_{t2}} = 0.660 \qquad \frac{T_2}{T_{t2}} = 0.88806 \qquad M_2 = 0.79389$$

whence

$$\frac{p_{t1}}{p_{t2}} = \left(\frac{\Gamma_2}{\Gamma_1}\right)\left(\frac{T_{t1}}{T_{t2}}\right)^{\frac{1}{2}} = \frac{0.96079}{(0.40412)(1.112994)^{\frac{1}{2}}} = 2.25357$$

$$p_{t1} = \frac{15\ \text{psia}}{0.960} = 15.625\ \text{psia}$$

$$p_{t2} = \frac{15.625\ \text{psia}}{2.25357} = 6.933\ \text{psia}$$

$$p_2 = 0.660 \times 6.933\ \text{psia} = 4.576\ \text{psia}$$

$$T_{t2} = 1.112994 \times 540°R = 601.02°R$$

$$T_2 = T_1 = 0.88806 \times 601.02°R = 533.74°R$$

$$M_2 = 0.79389$$

Example 8

Consider the isothermal flow of air ($\gamma = 1.4$) from an inlet Mach number of 1.30 through a duct of constant area. The inlet total temperature is 700°R; the inlet static pressure is 3 psia.

Find the total and static pressures, the total and static temperatures, and the Mach number at exit if the total-temperature ratio across the duct is 0.86846. (See Fig. 11.)

Solution by Table XI. From

$$M_1 = 1.30$$

we obtain

$$\frac{p_{1x}}{p_{t1x}} = 0.36091$$

$$\Gamma_{1x} = \left(\frac{T_{t1x}}{T_t^*}\right)^{\frac{1}{2}} \left(\frac{p_t^*}{p_{t1x}}\right)(\gamma)^{1/(\gamma-1)}\left(\frac{\gamma+1}{3\gamma-1}\right)^{(\gamma+1)/2(\gamma-1)}$$

$$= 0.93782$$

$$\frac{T_{t1x}}{T_t^*} = 1.17075$$

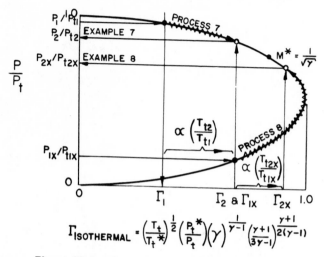

$$\Gamma_{\text{ISOTHERMAL}} = \left(\frac{T_t}{T_t^*}\right)^{\frac{1}{2}}\left(\frac{P_t^*}{P_t}\right)(\gamma)^{\frac{1}{\gamma-1}}\left(\frac{\gamma+1}{3\gamma-1}\right)^{\frac{\gamma+1}{2(\gamma-1)}}$$

Figure 11. Isothermal Flow—Examples 7 and 8.

But

$$\frac{T_{t2x}}{T_t^*} = \left(\frac{T_{t1x}}{T_t^*}\right)\left(\frac{T_{t2x}}{T_{t1x}}\right) = 1.17075 \times 0.86846 = 1.01675$$

Corresponding to $T_{t2x}/T_t^* = 1.01675$ is

$$\Gamma_{2x} = \Gamma_{1x}\left(\frac{T_{t2x}}{T_{t1x}}\right)^{\frac{1}{2}}\left(\frac{p_{t1x}}{p_{t2x}}\right) = 0.99121$$

From

$$\Gamma_{2x} = 0.99121$$

we obtain

$$\frac{p_{2x}}{p_{t2x}} = 0.59126 \qquad \frac{T_{2x}}{T_{t2x}} = 0.86059 \qquad M_{2x} = 0.90$$

whence

$$\frac{p_{t1x}}{p_{t2x}} = \left(\frac{\Gamma_{2x}}{\Gamma_{1x}}\right)\left(\frac{T_{t1x}}{T_{t2x}}\right)^{\frac{1}{2}} = \frac{0.99121}{(0.93782)(0.86846)^{\frac{1}{2}}} = 1.13415$$

$$p_{t1x} = \frac{3\text{ psia}}{0.36091} = 8.312\text{ psia}$$

$$p_{t2x} = \frac{8.312\text{ psia}}{1.13415} = 7.329\text{ psia}$$

$$p_{2x} = 0.59126 \times 7.329\text{ psia} = 4.333\text{ psia}$$

$$T_{t2x} = 0.86846 \times 700°\text{R} = 607.92°\text{R}$$

$$T_{2x} = T_{1x} = 0.86059 \times 607.92°\text{R} = 523.17°\text{R}$$

$$M_{2x} = 0.90$$

NORMAL SHOCK

Example 9

Consider air ($\gamma = 1.4$) at an initial Mach number of 3.0 to undergo a normal-shock process. The inlet total temperature is 900°R; the inlet static pressure is 5 psia.

Find the total and static pressures, the total and static temperatures, and the Mach number immediately after the shock. (See Fig. 12.)

Solution by Table XI. From

$$M_{1x} = 3.0$$

we obtain

$$\frac{p_{1x}}{p_{t1x}} = 0.02722 \qquad \Gamma_{1x} = \left(\frac{p_t^*}{p_{t1x}}\right) = 0.23615 \qquad \frac{p_{t1x}}{p_{t2}} = 3.04559$$

But

$$\Gamma_2 = \Gamma_{1x}\left(\frac{p_{t1x}}{p_{t2}}\right) = 0.23615 \times 3.04559$$

$$= 0.71922 = \left(\frac{p_t^*}{p_{t2}}\right)$$

From

$$\Gamma_2 = 0.71922$$

we obtain, by linear interpolation

$$\frac{p_2}{p_{t2}} = 0.85676 \qquad \frac{T_2}{T_{t2}} = 0.95679 \qquad M_2 = 0.47519$$

whence

$$p_{t1x} = \frac{5\text{ psia}}{0.02722} = 183.688\text{ psia}$$

$$p_{t2} = \frac{183.688\text{ psia}}{3.04559} = 60.313\text{ psia}$$

$$p_2 = 0.85676 \times 60.313\text{ psia} = 51.674\text{ psia}$$

$$T_{t2} = T_{t1} = 900°\text{R}$$

$$T_2 = 0.95679 \times 900°\text{R} = 861.11°\text{R}$$

$$M_2 = 0.47519$$

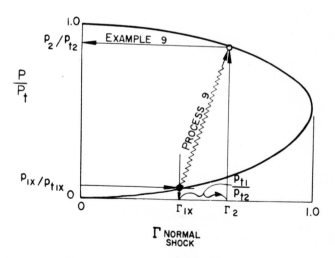

Figure 12. Normal Shock—Example 9.

Chapter 4

Some Generalizations in One-Dimensional Constant-Density Fluid Dynamics

INTRODUCTION

The constant-density flow problem has received little attention in the literature, probably because of the simplified treatments possible. Consequently, the characteristics of constant-density isentropic, Fanno, Rayleigh, and isothermal flows are not as familiar as they might be. We will examine these processes in some detail and show them to be progeny of a *generalized constant-density flow process*. The analyses presented in this chapter very closely follow the procedures and philosophy of Chapter 1 and serve as an application of its methods.

We approach this problem by considering a real fluid (i.e., one having viscosity and thermal conductivity) whose equation of state is well represented by the relation $w =$ constant, and whose specific heat capacity c is taken as a constant. We show how the usual flow processes are considered special cases of a generalized process, which in turn is shown to be represented by a single curve plotted on a dimensionless $p/p_t - \Gamma'$ chart.

CONTINUITY

When searching for a unifying principle relating one thermodynamic state to another for the whole range of workless flow processes (from isentropic[11] through Fanno and Rayleigh), we come naturally to the far-reaching concept of conservation of mass embodied by the continuity equation, which in the constant-density case is simply

$$VA = \text{constant} \qquad (66)$$

A dimensionless form of equation (66) will prove to be more useful; hence velocity is next evaluated in terms of the total pressure p_t based on isentropic stagnation processes from given static states. A general expression for velocity is obtained by combining the general energy equation

$$\delta Q = dh + \frac{V\,dV}{g} + dZ \qquad (67)$$

where δQ and dZ equal 0 when equation (67) is applied between corresponding static and total states, with the thermodynamic identity

$$T\,ds = dh - \frac{dp}{w} \qquad (68)$$

where $ds = 0$ in the isentropic stagnation process.

There results

$$V\,dV = -\frac{g}{w}\,dp \qquad (69)$$

and upon integrating equation (69) between static and total states we obtain

$$V = \left[\frac{2g}{w}(p_t - p)\right]^{\frac{1}{2}} \qquad (70)$$

The concept of total pressure p_t has been introduced and serves well in the velocity expression (70); however, the concept of total temperature has no significance in the constant-density flow process since in this regime temperature is independent of the flow velocity. This is a consequence of the First Law of Thermodynamics

$$\delta Q + \delta F = du + p\,dv \qquad (71)$$

which requires a constant internal energy ($du = c\,dT = 0$) and hence a constant temperature in the isentropic, constant-density case.

Substituting equation (70) in the continuity equation (66), we obtain

$$\left[\frac{2g}{w}(p_t - p)\right]^{\frac{1}{2}} A = K \qquad (72)$$

---

[11]By this term we mean the special constant-entropy process which is both adiabatic and reversible (in addition this process will be shown to be isothermal). It is in this restricted sense that the term isentropic is used throughout this chapter.

or

$$p_t \left(1 - \frac{p}{p_t}\right) A^2 = K' \qquad (73)$$

where K and K' are different constants.

Hence, between any two thermodynamic states joined by any arbitrary flow process, we write the entirely general dimensionless continuity expression

$$\left(\frac{p_{t1}}{p_{t2}}\right)\left(\frac{A_1}{A_2}\right)^2 \left(1 - \frac{p_1}{p_{t1}}\right) = \left(1 - \frac{p_2}{p_{t2}}\right) \qquad (74)$$

and find it satisfactory as to its generality and simplicity.

THE Γ' FUNCTION

Thus far we have obtained a dimensionless form of the continuity relation [equation (74)]. We seek a function representative of all constant-density flow processes which varies from 0 to 1 and which is a function of p/p_t only. If we view the factors p_{t1}/p_{t2} and $(A_1/A_2)^2$ as arbitrary process multipliers, we note that the remaining term, $(1 - p/p_t)$, satisfies these requirements. Thus, we explicitly define the *constant-density flow function* as

$$\Gamma' = 1 - \frac{p}{p_t} \qquad (75)$$

which is, of course, represented by a straight line on a $p/p_t - \Gamma'$ plot as shown in Fig. 13. Then continuity [equation (74)] may be written as

$$\left(\frac{p_{t1}}{p_{t2}}\right)\left(\frac{A_1}{A_2}\right)^2 \Gamma'_1 = \Gamma'_2 \qquad (76)$$

Finally, we might advantageously refer the inlet area to the area corresponding to an isentropic expansion from inlet conditions to zero pressure (this assumes that the substance remains a fluid and does not change phase).

Thus, from the continuity equation (74),

$$\frac{A_1}{A_{\text{Limiting}}} = \left[\frac{1 - (w/p_{t1})(Z_2 - Z_1)}{1 - (p_1/p_{t1})}\right]^{\frac{1}{2}} \qquad (77)$$

where A_{Limiting} signifies the area beyond which inlet conditions could not be maintained. Equation (77) is valid for all processes.

* * *

Thus far we have seen that a dimensionless form of the continuity relation [equation (76)] may be defined and applied to all flow processes and that a generalized constant-density flow function Γ' may be defined and as such varies between 0 and 1 only.

In the remainder of this analysis, we will be concerned with the characteristics of the individual constant-density flow processes and with numerical examples illustrating the use of the Γ' function.

APPLICATION OF Γ' IN SPECIFIC FLOW PROCESSES

Many practical situations may be approximated by certain processes having restrictions making possible simplified solutions. Several such flow processes of most general interest will be treated in this section.

Adiabatic Flow Without Friction

Here we are referring to the one-dimensional, steady flow of a fluid whose thermodynamic properties change solely because of area and elevation variations. By general energy equation (67), since the change in total enthalpy

$$dh_t = dh + \frac{V\,dV}{g} = \delta Q - dZ$$

in general is not zero, the thermodynamic identity, equation (68) shows that the total pressure

$$dp_t = dp + \frac{wV\,dV}{g} = (dh_t - T\,ds)w$$

in general is not constant throughout the process. Thus, continuity is expressed by

$$\left(\frac{p_{t1}}{p_{t2}}\right)\left(\frac{A_1}{A_2}\right)^2 \Gamma'_1 = \Gamma'_2 \qquad (76)$$

where

$$\left(\frac{p_{t1}}{p_{t2}}\right) = \frac{1}{1 - (w/p_{t1})(Z_2 - Z_1)} \qquad (78)$$

Schematic $h - s$ and $p/p_t - \Gamma'$ maps are given in Fig. 13; a numerical example closely following these diagrams follows in a later section.

Adiabatic Flow With Friction

Here we are referring to the one-dimensional, steady flow of a fluid whose thermodynamic properties change solely because of viscous effects and area and elevation variations. In the constant-area, constant-elevation case we have constant-density Fanno flow. For note: by the general energy equation (67), the total enthalpy remains constant throughout the process, while the viscous effects are reflected in entropy production within the system boundaries; and by the thermodynamic identity equation (68), the total pressure must always decrease in this type of flow.

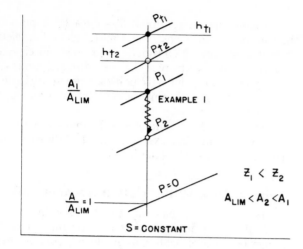

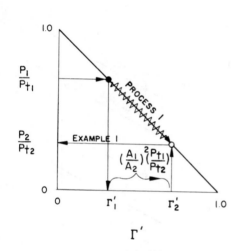

Figure 13. Isentropic Process Maps.

Continuity [equation (76)] for Fanno flow reduces to

$$\left(\frac{p_{t1}}{p_{t2}}\right)\Gamma_1' = \Gamma_2' \qquad (79)$$

where the total pressure ratio p_{t1}/p_{t2} represents the estimated or empirically determined frictional loss for a given problem, and may always be specified.

However, the more general variable-area, variable-elevation, Fanno-type flow may always be treated. Again, as in the case of adiabatic flow without friction, neither the total enthalpy nor the total pressure necessarily remains constant and continuity must be expressed by

$$\left(\frac{p_{t1}}{p_{t2}}\right)\left(\frac{A_1}{A_2}\right)^2\Gamma_1' = \Gamma_2' \qquad (76)$$

The effects of area and elevation variations represent an isentropic change of state only, and may be handled as described in the preceding section. In the generalized Fanno-type flow process the entropy increase may be obtained by combining the First Law [equation (71)] and the Second Law of Thermodynamics, generally given as

$$ds = \frac{\delta Q}{T} + \frac{\delta F}{T} \qquad (80)$$

Upon utilizing the internal energy relation [$du = c\,dT$] and the definition of enthalpy [$dh = du + d(pv)$], there results

$$s_2 - s_1 = c \ln\left[1 + \frac{p_{t1}}{cT_1 w}\left(1 - \frac{p_{t2}}{p_{t1}}\right) - \frac{(Z_2 - Z_1)}{cT_1}\right] \qquad (81)$$

The constant-area, constant-elevation, $h - s$ Fanno plots can be interpreted in terms of the more general Fanno-type flow simply by considering the area and elevation effects as an isentropic change of state, and then proceeding along the new constant-enthalpy Fanno line passing through this revised initial state to the final state.

Schematic $h - s$ and $p/p_t - \Gamma'$ maps are given in Fig. 14; a numerical example closely following these diagrams follows in a later section.

Diabatic Flow Without Friction

Here we are referring to the one-dimensional, steady flow of a fluid whose thermodynamic properties change because of heat-transfer effects and area and elevation variations. In the constant-area, constant-elevation case, we have the constant-density Rayleigh flow. For note: by the general energy equation (67) and the First Law [equation (71)], we have as the conserved quantity

$$p + \frac{w}{2g}V^2 = \text{constant}$$

which is seen to be [from equation (70)] the definition of total pressure. Continuity, then, for Rayleigh flow is simply the identity

$$\Gamma_1' = \Gamma_2' \qquad (82)$$

indicating that no changes occur in either static or total pressure in the constant-density Rayleigh flow process. However, the more general variable-area, variable-elevation Rayleigh-type flow may always be treated, and the total pressure may not remain constant throughout the process.

Thus, continuity must be expressed by

$$\left(\frac{p_{t1}}{p_{t2}}\right)\left(\frac{A_1}{A_2}\right)^2\Gamma_1' = \Gamma_2' \qquad (76)$$

27

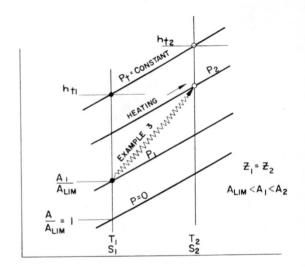

Again, the conventional $h - s$ Rayleigh plots can be interpreted in terms of the more general Rayleigh-type flow simply by considering the area and elevation effects as an isentropic change of state, and then proceeding along the new constant-pressure Rayleigh line passing through this revised initial state to the final state.

Schematic $h - s$ and $p/p_t - \Gamma'$ maps are given in Fig. 15; a numerical example closely following these diagrams follows in a later section.

Diabatic Flow With Friction (Isothermal)

Here we are referring to the one-dimensional, steady flow of a fluid whose thermodynamic properties change because of heat-transfer effects, viscous effects, and area and elevation variations in such a manner

Figure 14. Fanno-Type Process Maps.

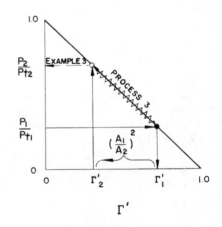

where p_{t1}/p_{t2} is given by equation (78). These relationships are seen to be the same as for the isentropic process, and only the change in entropy distinguishes these two processes. In any constant-density flow process, the entropy production is simply a function of the temperature ratio, as seen by combining the First [equation (71)] and Second [equation (80)] Laws, to obtain

$$s_2 - s_1 = c \ln\left(\frac{T_2}{T_1}\right) \qquad (83)$$

In Rayleigh-type flow we may always specify the temperature ratio of a given problem (being primarily a heat transfer problem), and regardless of the physical areas at which the heat transfer occurs, the final thermodynamic conditions are fully defined.

Figure 15. Rayleigh-Type Process Maps.

that the temperature remains constant. It should be noted that the restriction of constant temperature also requires that entropy remain invariant as well [this can be seen from equation (83)]. Note that as the contribution of viscous effects ($\delta F/T$) to entropy production are within the system boundaries and are always positive, to satisfy the restriction of entropy remaining constant (i.e., the process being isothermal) the heat-transfer effects ($\delta Q/T$) must be negative and hence we must cool (i.e., abstract energy from) the fluid.

A combination of the First [equation (71)] and Second [equation (80)] Laws now shows that the total pressure [$dp_t = (dh_t - T\,ds)w$] must always decrease in this type of flow process.

The continuity equation for the constant-area, constant-elevation case is written as

$$\left(\frac{p_{t1}}{p_{t2}}\right)\Gamma_1' = \Gamma_2' \qquad (79)$$

which is seen to be the same as for the constant-area Fanno process; the change in entropy is again seen to be the only distinguishing feature of these two processes. This constant-area case may be treated once the total-pressure ratio, or enthalpy ratio, is specified. The more general variable-area, variable-elevation isothermal flow may also be treated once the total-pressure or total-enthalpy ratio is specified. Continuity is then expressed as

$$\left(\frac{p_{t1}}{p_{t2}}\right)\left(\frac{A_1}{A_2}\right)^2 \Gamma_1' = \Gamma_2' \qquad (76)$$

Schematic $h-s$ and $p/p_t - \Gamma'$ maps are given in Fig. 16; a numerical example closely following these diagrams follows in a later section.

* * *

The generalized constant-density flow function Γ' is shown to serve in isentropic, Fanno-type, Rayleigh-type, and isothermal processes; and the loci of state points for these as well as for any arbitrary flow process are precisely represented by the single $p/p_t - \Gamma'$ curve. In Fig. 17 we compare all of these simplified flow processes, indicating accessible states along both static and total loci, as well as those states which are unattainable from the given inlet state.

NUMERICAL EXAMPLES

Example 1. Isentropic Flow

Consider the expansion of a liquid ($w = 64.34\ \text{lb/ft}^3$) from an inlet pressure ratio of 0.8 through a duct of area ratio 2 having an elevation increase of 2 ft. The inlet static pressure is 1024 psfa.

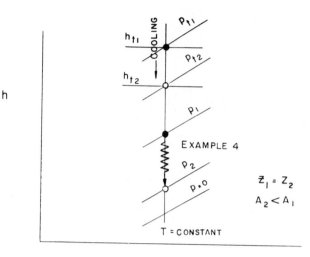

Figure 16. Isothermal Process Maps.

Find the total and static pressure at exit, the velocities at inlet and exit, and the entropy change. (See Fig. 13.)

Solution. From $p_1/p_{t1} = 0.8$ we obtain

$$\Gamma_1' = 1 - p_1/p_{t1} = 0.2$$

Now

$$p_{t1} = \frac{1024\ \text{psfa}}{0.8} = 1280\ \text{psfa}$$

$$p_{t2} = p_{t1}\left[1 - \frac{w}{p_{t1}}(Z_2 - Z_1)\right]$$

$$= 1280\ \text{psfa}\left(1 - \frac{64.34 \times 2}{1280}\right) = 1151.32\ \text{psfa}$$

Then

$$\Gamma_2' = \Gamma_1'\left(\frac{p_{t1}}{p_{t2}}\right)\left(\frac{A_1}{A_2}\right)^2 = 0.2 \times \frac{1280}{1151.32} \times 4.0 = 0.88941$$

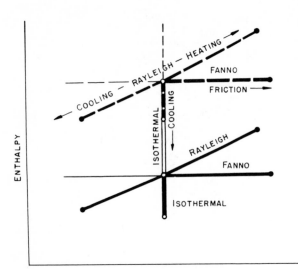

KEY TO PROCESS LINES:

———————— STATIC CONDITIONS	} POSSIBLE PROCESSES
— — — — TOTAL CONDITIONS	
———————— STATIC CONDITIONS	} UNATTAINABLE STATES
———————— TOTAL CONDITIONS	

⁕ INLET STATE

● TYPICAL EXIT STATES

Figure 17. Comparison of Various Constant-Area, Constant-Elevation Processes.

From $\Gamma'_2 = 0.88941$ we obtain

$$\frac{p_2}{p_{t2}} = 0.11059$$

whence

$$p_2 = 0.11059 \times 1151.32 \text{ psfa} = 127.32 \text{ psfa}$$

$$V_1 = \left[\frac{2g}{w}(p_{t1} - p_1)\right]^{\frac{1}{2}} = (1280 - 1024)^{\frac{1}{2}} = 16 \text{ ft/sec}$$

$$V_2 = (A_1/A_2)V_1 = 2 \times 16 \text{ ft/sec} = 32 \text{ ft/sec}$$

$$s_2 - s_1 = 0$$

Example 2. Fanno-Type Flow

Consider the adiabatic flow of a liquid ($w = 64.34 \text{ lb/ft}^3$, $c = 1 \text{ Btu/lb-deg R}$) from an inlet pressure ratio of 0.9 through a duct of area ratio 0.6 having an elevation increase of 50 ft. The inlet static pressure is 230,400 psfa and the inlet temperature is 500°R.

Find the total and static pressure at exit, the velocities at inlet and exit, and the entropy change, if the total-pressure ratio across the duct is 2. (See Fig. 14.)

Solution. From $p_1/p_{t1} = 0.9$ we obtain

$$\Gamma'_1 = 1 - p_1/p_{t1} = 0.1$$

But

$$\Gamma'_2 = \Gamma'_1(p_{t1}/p_{t2})(A_1/A_2)^2 = 0.1 \times 2 \times 0.36 = 0.072$$

From $\Gamma'_2 = 0.072$ we obtain

$$\frac{p_2}{p_{t2}} = 0.928$$

whence

$$p_{t1} = \frac{230,400 \text{ psfa}}{0.9} = 256,000 \text{ psfa}$$

$$p_{t2} = \frac{256,000 \text{ psfa}}{2} = 128,000 \text{ psfa}$$

$$p_2 = 0.928 \times 128,000 \text{ psfa} = 118,784 \text{ psfa}$$

$$V_1 = \left[\frac{2g}{w}(p_{t1} - p_1)\right]^{\frac{1}{2}} = (256,000 - 230,400)^{\frac{1}{2}} = 160 \text{ ft/sec}$$

$$V_2 = (A_1/A_2)V_1 = 0.6 \times 160 \text{ ft/sec} = 96 \text{ ft/sec}$$

$$s_2 - s_1 = c \ln\left[1 + \frac{p_{t1}}{cT_1Jw}\left(1 - \frac{p_{t2}}{p_{t1}}\right) - \frac{(Z_2 - Z_1)}{cT_1}\right]$$

$$= 0.00498 \text{ Btu/lb-deg R}$$

Example 3. Rayleigh-Type Flow

Consider the frictionless flow of a liquid ($w = 64.34 \text{ lb/ft}^3$, $c = 1 \text{ Btu/lb-deg R}$) from an inlet pressure ratio of 0.2 through a constant-elevation duct of area ratio 0.5. The inlet static pressure is 144 psfa.

Find the total and static pressures at exit, the velocities at inlet and exit, and the entropy change, if the temperature ratio across the duct is 1.2. (See Fig. 15.)

Solution. From $p_1/p_{t1} = 0.2$ we obtain

$$\Gamma'_1 = 1 - \frac{p_1}{p_{t1}} = 0.8$$

but

$$\Gamma'_2 = \Gamma'_1\left(\frac{A_1}{A_2}\right)^2 = 0.8 \times 0.25 = 0.2$$

from $\Gamma'_2 = 0.2$ we obtain

$$p_2/p_{t2} = 0.8$$

whence

$$p_{t2} = p_{t1} = \frac{144 \text{ psfa}}{0.2} = 720 \text{ psfa}$$

$$p_2 = 0.8 \times 720 \text{ psfa} = 576 \text{ psfa}$$

$$V_1 = \left[\frac{2g}{w}(p_{t1} - p_1)\right]^{\frac{1}{2}} = (720 - 144)^{\frac{1}{2}} = 24 \text{ ft/sec}$$

$$V_2 = (A_1/A_2)V_1 = 0.5 \times 24 \text{ ft/sec} = 12 \text{ ft/sec}$$

$$s_2 - s_1 = c \ln(T_2/T_1) = \ln(1.2) = 0.18232 \text{ Btu/lb-deg R}.$$

Example 4. Isothermal Flow

Consider the isothermal flow of a liquid ($w = 60.319 \text{ lb/ft}^3$) from an inlet pressure ratio of 0.75

through a constant-elevation duct of area ratio 1.4. The inlet static pressure is 4500 psfa.

Find the total and static pressures at exit, the velocities at inlet and exit, and the entropy change, if the total-pressure ratio across the duct is 1.2. (See Fig. 16.)

Solution. From $p_1/p_{t1} = 0.75$ we obtain

$$\Gamma'_1 = 1 - \frac{p_1}{p_{t1}} = 0.25$$

but

$$\Gamma'_2 = \Gamma'_1 \frac{p_{t1}}{p_{t2}} \left(\frac{A_1}{A_2}\right)^2 = 0.25 \times 1.2 \times 1.96 = 0.588$$

from $\Gamma'_2 = 0.588$ we obtain

$$\frac{p_2}{p_{t2}} = 0.412$$

whence

$$p_{t1} = \frac{4500 \text{ psfa}}{0.75} = 6000 \text{ psfa}$$

$$p_{t2} = \frac{6000 \text{ psfa}}{1.2} = 5000 \text{ psfa}$$

$$p_2 = 0.412 \times 5000 \text{ psfa} = 2060 \text{ psfa}$$

$$V_1 = \left[\frac{2g}{w}(p_{t1} - p_1)\right]^{\frac{1}{2}}$$

$$= \left[\frac{2 \times 32.17}{60.319} \times (6000 - 4500)\right]^{\frac{1}{2}} = 40 \text{ ft/sec}$$

$$V_2 = (A_1/A_2)V_1 = 1.4 \times 40 \text{ ft/sec} = 56 \text{ ft/sec}$$

$$s_2 - s_1 = c \ln(T_2/T_1) = 0$$

Appendix

Appendix

TABLE I.   Isentropic Exponent = 1.00

p/p_t	M	T/T_t	GAMMA	ISOTHERMAL T_t/T_t^*	RAYLEIGH T_t/T_t^*	SHOCK p_{t1}/p_{t2}
1.0000	.00000	1.	.00000	1.	.00000	1.
.9995	.03163	1.	.05212	1.	.00399	1.
.9990	.04473	1.	.07368	1.	.00797	1.
.9985	.05479	1.	.09020	1.	.01194	1.
.9980	.06328	1.	.10412	1.	.01589	1.
.9975	.07075	1.	.11636	1.	.01983	1.
.9970	.07752	1.	.12742	1.	.02375	1.
.9965	.08374	1.	.13758	1.	.02766	1.
.9960	.08953	1.	.14702	1.	.03156	1.
.9955	.09498	1.	.15588	1.	.03544	1.
.9950	.10013	1.	.16425	1.	.03931	1.
.9945	.10503	1.	.17221	1.	.04316	1.
.9940	.10971	1.	.17980	1.	.04701	1.
.9935	.11420	1.	.18707	1.	.05084	1.
.9930	.11853	1.	.19405	1.	.05465	1.
.9925	.12271	1.	.20079	1.	.05845	1.
.9920	.12675	1.	.20730	1.	.06224	1.
.9915	.13066	1.	.21359	1.	.06602	1.
.9910	.13447	1.	.21970	1.	.06978	1.
.9905	.13817	1.	.22564	1.	.07353	1.
.9900	.14178	1.	.23141	1.	.07727	1.
.9895	.14530	1.	.23704	1.	.08099	1.
.9890	.14873	1.	.24252	1.	.08470	1.
.9885	.15210	1.	.24788	1.	.08840	1.
.9880	.15539	1.	.25312	1.	.09208	1.
.9875	.15861	1.	.25824	1.	.09575	1.
.9870	.16177	1.	.26325	1.	.09941	1.
.9865	.16488	1.	.26816	1.	.10306	1.
.9860	.16792	1.	.27298	1.	.10669	1.
.9855	.17092	1.	.27771	1.	.11031	1.
.9850	.17386	1.	.28235	1.	.11392	1.
.9845	.17676	1.	.28691	1.	.11751	1.
.9840	.17961	1.	.29138	1.	.12110	1.
.9835	.18242	1.	.29579	1.	.12467	1.
.9830	.18518	1.	.30012	1.	.12822	1.
.9825	.18791	1.	.30439	1.	.13177	1.
.9820	.19060	1.	.30859	1.	.13530	1.
.9815	.19325	1.	.31273	1.	.13882	1.
.9810	.19587	1.	.31680	1.	.14233	1.
.9805	.19846	1.	.32082	1.	.14583	1.
.9800	.20101	1.	.32478	1.	.14931	1.
.9795	.20353	1.	.32869	1.	.15278	1.
.9790	.20603	1.	.33255	1.	.15624	1.
.9785	.20849	1.	.33636	1.	.15969	1.
.9780	.21093	1.	.34011	1.	.16313	1.
.9775	.21334	1.	.34382	1.	.16655	1.
.9770	.21573	1.	.34749	1.	.16996	1.
.9765	.21809	1.	.35111	1.	.17336	1.
.9760	.22042	1.	.35469	1.	.17675	1.
.9755	.22273	1.	.35823	1.	.18013	1.

TABLE I. Isentropic Exponent = 1.00

p/p_t	M	T/T_t	GAMMA	ISOTHERMAL T_t/T_t*	RAYLEIGH T_t/T_t*	SHOCK p_t1/p_t2
.9750	.22502	1.	.36173	1.	.18349	1.
.9745	.22729	1.	.36519	1.	.18684	1.
.9740	.22954	1.	.36861	1.	.19018	1.
.9735	.23176	1.	.37199	1.	.19351	1.
.9730	.23397	1.	.37534	1.	.19683	1.
.9725	.23616	1.	.37865	1.	.20014	1.
.9720	.23833	1.	.38193	1.	.20343	1.
.9715	.24047	1.	.38518	1.	.20671	1.
.9710	.24261	1.	.38839	1.	.20999	1.
.9705	.24472	1.	.39157	1.	.21325	1.
.9700	.24682	1.	.39472	1.	.21649	1.
.9695	.24890	1.	.39785	1.	.21973	1.
.9690	.25096	1.	.40094	1.	.22296	1.
.9685	.25301	1.	.40400	1.	.22617	1.
.9680	.25504	1.	.40704	1.	.22938	1.
.9675	.25706	1.	.41005	1.	.23257	1.
.9670	.25906	1.	.41303	1.	.23575	1.
.9665	.26105	1.	.41598	1.	.23892	1.
.9660	.26303	1.	.41891	1.	.24208	1.
.9655	.26499	1.	.42182	1.	.24523	1.
.9650	.26694	1.	.42470	1.	.24836	1.
.9645	.26887	1.	.42755	1.	.25149	1.
.9640	.27079	1.	.43039	1.	.25460	1.
.9635	.27270	1.	.43320	1.	.25771	1.
.9630	.27460	1.	.43598	1.	.26080	1.
.9625	.27648	1.	.43875	1.	.26388	1.
.9620	.27836	1.	.44149	1.	.26696	1.
.9615	.28022	1.	.44421	1.	.27002	1.
.9610	.28207	1.	.44691	1.	.27307	1.
.9605	.28391	1.	.44959	1.	.27611	1.
.9600	.28573	1.	.45225	1.	.27914	1.
.9595	.28755	1.	.45489	1.	.28215	1.
.9590	.28936	1.	.45751	1.	.28516	1.
.9585	.29116	1.	.46011	1.	.28816	1.
.9580	.29294	1.	.46270	1.	.29115	1.
.9575	.29472	1.	.46526	1.	.29412	1.
.9570	.29649	1.	.46780	1.	.29709	1.
.9565	.29824	1.	.47033	1.	.30005	1.
.9560	.29999	1.	.47284	1.	.30299	1.
.9555	.30173	1.	.47533	1.	.30593	1.
.9550	.30346	1.	.47781	1.	.30885	1.
.9545	.30518	1.	.48026	1.	.31176	1.
.9540	.30689	1.	.48271	1.	.31467	1.
.9535	.30860	1.	.48513	1.	.31756	1.
.9530	.31029	1.	.48754	1.	.32045	1.
.9525	.31198	1.	.48993	1.	.32332	1.
.9520	.31366	1.	.49231	1.	.32619	1.
.9515	.31533	1.	.49467	1.	.32904	1.
.9510	.31699	1.	.49702	1.	.33188	1.
.9505	.31864	1.	.49935	1.	.33472	1.

TABLE I. Isentropic Exponent = 1.00

p/p_t	M	T/T_t	GAMMA	ISOTHERMAL T_t/T_t^*	RAYLEIGH T_t/T_t^*	SHOCK p_{t1}/p_{t2}
.950	.32029	1.	.50167	1.	.33754	1.
.949	.32356	1.	.50626	1.	.34316	1.
.948	.32681	1.	.51079	1.	.34874	1.
.947	.33002	1.	.51527	1.	.35428	1.
.946	.33320	1.	.51970	1.	.35978	1.
.945	.33636	1.	.52407	1.	.36524	1.
.944	.33950	1.	.52839	1.	.37066	1.
.943	.34260	1.	.53266	1.	.37605	1.
.942	.34569	1.	.53689	1.	.38140	1.
.941	.34875	1.	.54106	1.	.38671	1.
.940	.35178	1.	.54519	1.	.39198	1.
.939	.35480	1.	.54928	1.	.39722	1.
.938	.35779	1.	.55332	1.	.40242	1.
.937	.36075	1.	.55731	1.	.40758	1.
.936	.36370	1.	.56127	1.	.41271	1.
.935	.36663	1.	.56518	1.	.41780	1.
.934	.36954	1.	.56905	1.	.42286	1.
.933	.37242	1.	.57289	1.	.42788	1.
.932	.37529	1.	.57668	1.	.43286	1.
.931	.37814	1.	.58043	1.	.43781	1.
.930	.38097	1.	.58415	1.	.44272	1.
.929	.38379	1.	.58783	1.	.44760	1.
.928	.38658	1.	.59148	1.	.45245	1.
.927	.38936	1.	.59509	1.	.45726	1.
.926	.39213	1.	.59866	1.	.46204	1.
.925	.39487	1.	.60220	1.	.46678	1.
.924	.39760	1.	.60571	1.	.47149	1.
.923	.40032	1.	.60919	1.	.47617	1.
.922	.40301	1.	.61263	1.	.48081	1.
.921	.40570	1.	.61604	1.	.48542	1.
.920	.40837	1.	.61942	1.	.49000	1.
.919	.41102	1.	.62277	1.	.49454	1.
.918	.41366	1.	.62609	1.	.49906	1.
.917	.41629	1.	.62938	1.	.50354	1.
.916	.41890	1.	.63264	1.	.50799	1.
.915	.42150	1.	.63587	1.	.51241	1.
.914	.42409	1.	.63907	1.	.51679	1.
.913	.42666	1.	.64224	1.	.52115	1.
.912	.42922	1.	.64539	1.	.52547	1.
.911	.43177	1.	.64851	1.	.52977	1.
.910	.43431	1.	.65160	1.	.53403	1.
.909	.43683	1.	.65467	1.	.53826	1.
.908	.43934	1.	.65771	1.	.54246	1.
.907	.44184	1.	.66073	1.	.54663	1.
.906	.44433	1.	.66372	1.	.55078	1.
.905	.44681	1.	.66668	1.	.55489	1.
.904	.44928	1.	.66963	1.	.55897	1.
.903	.45174	1.	.67254	1.	.56303	1.
.902	.45418	1.	.67544	1.	.56705	1.
.901	.45662	1.	.67831	1.	.57105	1.

TABLE I.   Isentropic Exponent = 1.00

p/p_t	M	T/T_t	GAMMA	ISOTHERMAL T_t/T_t^*	RAYLEIGH T_t/T_t^*	SHOCK p_{t1}/p_{t2}
.900	.45904	1.	.68115	1.	.57502	1.
.899	.46146	1.	.68397	1.	.57895	1.
.898	.46386	1.	.68678	1.	.58287	1.
.897	.46626	1.	.68955	1.	.58675	1.
.896	.46865	1.	.69231	1.	.59060	1.
.895	.47102	1.	.69504	1.	.59443	1.
.894	.47339	1.	.69776	1.	.59823	1.
.893	.47575	1.	.70045	1.	.60200	1.
.892	.47810	1.	.70312	1.	.60574	1.
.891	.48044	1.	.70577	1.	.60946	1.
.890	.48277	1.	.70840	1.	.61315	1.
.889	.48509	1.	.71101	1.	.61682	1.
.888	.48741	1.	.71360	1.	.62045	1.
.887	.48971	1.	.71617	1.	.62406	1.
.886	.49201	1.	.71872	1.	.62765	1.
.885	.49430	1.	.72125	1.	.63121	1.
.884	.49658	1.	.72376	1.	.63474	1.
.883	.49886	1.	.72625	1.	.63825	1.
.882	.50113	1.	.72872	1.	.64173	1.
.881	.50338	1.	.73118	1.	.64518	1.
.880	.50564	1.	.73361	1.	.64861	1.
.879	.50788	1.	.73603	1.	.65202	1.
.878	.51012	1.	.73843	1.	.65540	1.
.877	.51234	1.	.74081	1.	.65875	1.
.876	.51457	1.	.74318	1.	.66209	1.
.875	.51678	1.	.74552	1.	.66539	1.
.874	.51899	1.	.74785	1.	.66867	1.
.873	.52119	1.	.75017	1.	.67193	1.
.872	.52338	1.	.75246	1.	.67517	1.
.871	.52557	1.	.75474	1.	.67838	1.
.870	.52775	1.	.75700	1.	.68156	1.
.869	.52993	1.	.75925	1.	.68472	1.
.868	.53210	1.	.76148	1.	.68786	1.
.867	.53426	1.	.76369	1.	.69098	1.
.866	.53641	1.	.76589	1.	.69407	1.
.865	.53856	1.	.76807	1.	.69714	1.
.864	.54071	1.	.77024	1.	.70019	1.
.863	.54285	1.	.77239	1.	.70321	1.
.862	.54498	1.	.77452	1.	.70621	1.
.861	.54710	1.	.77664	1.	.70919	1.
.860	.54922	1.	.77874	1.	.71215	1.
.859	.55134	1.	.78083	1.	.71509	1.
.858	.55345	1.	.78291	1.	.71800	1.
.857	.55555	1.	.78497	1.	.72089	1.
.856	.55765	1.	.78701	1.	.72376	1.
.855	.55974	1.	.78904	1.	.72660	1.
.854	.56183	1.	.79106	1.	.72943	1.
.853	.56391	1.	.79306	1.	.73223	1.
.852	.56598	1.	.79504	1.	.73502	1.
.851	.56806	1.	.79702	1.	.73778	1.

TABLE I. Isentropic Exponent = 1.00

p/p$_t$	M	T/T$_t$	GAMMA	ISOTHERMAL T$_t$/T*	RAYLEIGH T$_t$/T$_t^*$	SHOCK p$_{t1}$/p$_{t2}$
.850	.57012	1.	.79898	1.	.74052	1.
.849	.57218	1.	.80092	1.	.74324	1.
.848	.57424	1.	.80285	1.	.74594	1.
.847	.57629	1.	.80477	1.	.74862	1.
.846	.57834	1.	.80667	1.	.75128	1.
.845	.58038	1.	.80856	1.	.75392	1.
.844	.58241	1.	.81044	1.	.75653	1.
.843	.58445	1.	.81230	1.	.75913	1.
.842	.58647	1.	.81416	1.	.76171	1.
.841	.58850	1.	.81599	1.	.76427	1.
.840	.59051	1.	.81782	1.	.76681	1.
.839	.59253	1.	.81963	1.	.76932	1.
.838	.59454	1.	.82143	1.	.77182	1.
.837	.59654	1.	.82322	1.	.77430	1.
.836	.59854	1.	.82499	1.	.77676	1.
.835	.60054	1.	.82675	1.	.77920	1.
.834	.60253	1.	.82850	1.	.78163	1.
.833	.60452	1.	.83024	1.	.78403	1.
.832	.60650	1.	.83196	1.	.78641	1.
.831	.60848	1.	.83367	1.	.78878	1.
.830	.61046	1.	.83537	1.	.79113	1.
.829	.61243	1.	.83706	1.	.79346	1.
.828	.61440	1.	.83874	1.	.79577	1.
.827	.61636	1.	.84040	1.	.79806	1.
.826	.61832	1.	.84206	1.	.80033	1.
.825	.62028	1.	.84370	1.	.80259	1.
.824	.62223	1.	.84533	1.	.80483	1.
.823	.62418	1.	.84695	1.	.80705	1.
.822	.62612	1.	.84855	1.	.80925	1.
.821	.62806	1.	.85015	1.	.81143	1.
.820	.63000	1.	.85173	1.	.81360	1.
.819	.63194	1.	.85330	1.	.81575	1.
.818	.63387	1.	.85487	1.	.81788	1.
.817	.63579	1.	.85642	1.	.82000	1.
.816	.63772	1.	.85796	1.	.82210	1.
.815	.63964	1.	.85948	1.	.82418	1.
.814	.64155	1.	.86100	1.	.82624	1.
.813	.64347	1.	.86251	1.	.82829	1.
.812	.64538	1.	.86400	1.	.83032	1.
.811	.64728	1.	.86549	1.	.83234	1.
.810	.64919	1.	.86696	1.	.83433	1.
.809	.65109	1.	.86843	1.	.83632	1.
.808	.65298	1.	.86988	1.	.83828	1.
.807	.65488	1.	.87133	1.	.84023	1.
.806	.65677	1.	.87276	1.	.84216	1.
.805	.65866	1.	.87418	1.	.84408	1.
.804	.66054	1.	.87559	1.	.84598	1.
.803	.66242	1.	.87699	1.	.84786	1.
.802	.66430	1.	.87839	1.	.84973	1.
.801	.66618	1.	.87977	1.	.85159	1.

TABLE I. Isentropic Exponent = 1.00

p/p_t	M	T/T_t	GAMMA	ISOTHERMAL T_t/T_t^*	RAYLEIGH T_t/T_t^*	SHOCK p_{t1}/p_{t2}
.800	.66805	1.	.88114	1.	.85342	1.
.799	.66992	1.	.88250	1.	.85525	1.
.798	.67178	1.	.88385	1.	.85705	1.
.797	.67365	1.	.88519	1.	.85885	1.
.796	.67551	1.	.88653	1.	.86062	1.
.795	.67737	1.	.88785	1.	.86238	1.
.794	.67922	1.	.88916	1.	.86413	1.
.793	.68108	1.	.89046	1.	.86586	1.
.792	.68293	1.	.89176	1.	.86758	1.
.791	.68477	1.	.89304	1.	.86928	1.
.790	.68662	1.	.89431	1.	.87097	1.
.789	.68846	1.	.89558	1.	.87264	1.
.788	.69030	1.	.89683	1.	.87430	1.
.787	.69214	1.	.89808	1.	.87594	1.
.786	.69397	1.	.89931	1.	.87757	1.
.785	.69580	1.	.90054	1.	.87919	1.
.784	.69763	1.	.90176	1.	.88079	1.
.783	.69946	1.	.90297	1.	.88238	1.
.782	.70129	1.	.90417	1.	.88395	1.
.781	.70311	1.	.90536	1.	.88551	1.
.780	.70493	1.	.90654	1.	.88705	1.
.779	.70675	1.	.90771	1.	.88859	1.
.778	.70856	1.	.90887	1.	.89010	1.
.777	.71037	1.	.91003	1.	.89161	1.
.776	.71218	1.	.91117	1.	.89310	1.
.775	.71399	1.	.91231	1.	.89457	1.
.774	.71580	1.	.91344	1.	.89604	1.
.773	.71760	1.	.91456	1.	.89749	1.
.772	.71940	1.	.91567	1.	.89893	1.
.771	.72120	1.	.91677	1.	.90035	1.
.770	.72300	1.	.91786	1.	.90176	1.
.769	.72480	1.	.91894	1.	.90316	1.
.768	.72659	1.	.92002	1.	.90454	1.
.767	.72838	1.	.92109	1.	.90592	1.
.766	.73017	1.	.92215	1.	.90728	1.
.765	.73196	1.	.92319	1.	.90862	1.
.764	.73374	1.	.92424	1.	.90996	1.
.763	.73552	1.	.92527	1.	.91128	1.
.762	.73730	1.	.92629	1.	.91259	1.
.761	.73908	1.	.92731	1.	.91388	1.
.760	.74086	1.	.92832	1.	.91517	1.
.759	.74264	1.	.92932	1.	.91644	1.
.758	.74441	1.	.93031	1.	.91770	1.
.757	.74618	1.	.93129	1.	.91895	1.
.756	.74795	1.	.93227	1.	.92018	1.
.755	.74972	1.	.93324	1.	.92141	1.
.754	.75148	1.	.93419	1.	.92262	1.
.753	.75325	1.	.93515	1.	.92382	1.
.752	.75501	1.	.93609	1.	.92500	1.
.751	.75677	1.	.93702	1.	.92618	1.

TABLE I. Isentropic Exponent = 1.00

p/p_t	M	T/T_t	GAMMA	ISOTHERMAL T_t/T_t^*	RAYLEIGH T_t/T_t^*	SHOCK p_{t1}/p_{t2}
.750	.75853	1.	.93795	1.	.92734	1.
.749	.76028	1.	.93887	1.	.92850	1.
.748	.76204	1.	.93978	1.	.92964	1.
.747	.76379	1.	.94068	1.	.93077	1.
.746	.76555	1.	.94158	1.	.93189	1.
.745	.76730	1.	.94247	1.	.93299	1.
.744	.76904	1.	.94335	1.	.93409	1.
.743	.77079	1.	.94422	1.	.93517	1.
.742	.77254	1.	.94508	1.	.93625	1.
.741	.77428	1.	.94594	1.	.93731	1.
.740	.77602	1.	.94679	1.	.93836	1.
.739	.77776	1.	.94763	1.	.93940	1.
.738	.77950	1.	.94846	1.	.94043	1.
.737	.78124	1.	.94929	1.	.94145	1.
.736	.78298	1.	.95011	1.	.94245	1.
.735	.78471	1.	.95092	1.	.94345	1.
.734	.78644	1.	.95172	1.	.94444	1.
.733	.78817	1.	.95252	1.	.94541	1.
.732	.78990	1.	.95331	1.	.94638	1.
.731	.79163	1.	.95409	1.	.94733	1.
.730	.79336	1.	.95486	1.	.94828	1.
.729	.79509	1.	.95563	1.	.94921	1.
.728	.79681	1.	.95639	1.	.95013	1.
.727	.79853	1.	.95714	1.	.95105	1.
.726	.80026	1.	.95788	1.	.95195	1.
.725	.80198	1.	.95862	1.	.95284	1.
.724	.80370	1.	.95935	1.	.95372	1.
.723	.80541	1.	.96007	1.	.95460	1.
.722	.80713	1.	.96079	1.	.95546	1.
.721	.80885	1.	.96150	1.	.95631	1.
.720	.81056	1.	.96220	1.	.95715	1.
.719	.81227	1.	.96289	1.	.95799	1.
.718	.81399	1.	.96358	1.	.95881	1.
.717	.81570	1.	.96426	1.	.95962	1.
.716	.81740	1.	.96493	1.	.96043	1.
.715	.81911	1.	.96560	1.	.96122	1.
.714	.82082	1.	.96626	1.	.96200	1.
.713	.82253	1.	.96691	1.	.96278	1.
.712	.82423	1.	.96755	1.	.96354	1.
.711	.82593	1.	.96819	1.	.96430	1.
.710	.82764	1.	.96882	1.	.96505	1.
.709	.82934	1.	.96945	1.	.96578	1.
.708	.83104	1.	.97007	1.	.96651	1.
.707	.83274	1.	.97068	1.	.96723	1.
.706	.83443	1.	.97128	1.	.96794	1.
.705	.83613	1.	.97188	1.	.96864	1.
.704	.83783	1.	.97247	1.	.96933	1.
.703	.83952	1.	.97305	1.	.97002	1.
.702	.84122	1.	.97363	1.	.97069	1.
.701	.84291	1.	.97419	1.	.97135	1.

TABLE I. Isentropic Exponent = 1.00

p/p_t	M	T/T_t	GAMMA	ISOTHERMAL T_t/T_t^*	RAYLEIGH T_t/T_t^*	SHOCK p_{t1}/p_{t2}
.700	.84460	1.	.97476	1.	.97201	1.
.699	.84629	1.	.97531	1.	.97266	1.
.698	.84798	1.	.97586	1.	.97329	1.
.697	.84967	1.	.97641	1.	.97392	1.
.696	.85136	1.	.97694	1.	.97454	1.
.695	.85305	1.	.97747	1.	.97516	1.
.694	.85473	1.	.97800	1.	.97576	1.
.693	.85642	1.	.97851	1.	.97636	1.
.692	.85810	1.	.97902	1.	.97694	1.
.691	.85979	1.	.97952	1.	.97752	1.
.690	.86147	1.	.98002	1.	.97809	1.
.689	.86315	1.	.98051	1.	.97865	1.
.688	.86483	1.	.98100	1.	.97920	1.
.687	.86651	1.	.98147	1.	.97975	1.
.686	.86819	1.	.98194	1.	.98029	1.
.685	.86987	1.	.98241	1.	.98081	1.
.684	.87155	1.	.98287	1.	.98133	1.
.683	.87322	1.	.98332	1.	.98185	1.
.682	.87490	1.	.98376	1.	.98235	1.
.681	.87658	1.	.98420	1.	.98285	1.
.680	.87825	1.	.98463	1.	.98333	1.
.679	.87993	1.	.98506	1.	.98381	1.
.678	.88160	1.	.98548	1.	.98429	1.
.677	.88327	1.	.98589	1.	.98475	1.
.676	.88494	1.	.98630	1.	.98521	1.
.675	.88661	1.	.98670	1.	.98566	1.
.674	.88829	1.	.98710	1.	.98610	1.
.673	.88996	1.	.98748	1.	.98653	1.
.672	.89162	1.	.98787	1.	.98696	1.
.671	.89329	1.	.98824	1.	.98737	1.
.670	.89496	1.	.98861	1.	.98778	1.
.669	.89663	1.	.98898	1.	.98819	1.
.668	.89830	1.	.98933	1.	.98858	1.
.667	.89996	1.	.98968	1.	.98897	1.
.666	.90163	1.	.99003	1.	.98935	1.
.665	.90329	1.	.99037	1.	.98973	1.
.664	.90496	1.	.99070	1.	.99009	1.
.663	.90662	1.	.99103	1.	.99045	1.
.662	.90828	1.	.99135	1.	.99080	1.
.661	.90995	1.	.99166	1.	.99115	1.
.660	.91161	1.	.99197	1.	.99148	1.
.659	.91327	1.	.99228	1.	.99181	1.
.658	.91493	1.	.99257	1.	.99214	1.
.657	.91659	1.	.99286	1.	.99245	1.
.656	.91825	1.	.99315	1.	.99276	1.
.655	.91991	1.	.99343	1.	.99306	1.
.654	.92157	1.	.99370	1.	.99336	1.
.653	.92323	1.	.99396	1.	.99365	1.
.652	.92489	1.	.99423	1.	.99393	1.
.651	.92655	1.	.99448	1.	.99420	1.

TABLE I. Isentropic Exponent = 1.00

p/p_t	M	T/T_t	GAMMA	ISOTHERMAL T_t/T_t^*	RAYLEIGH T_t/T_t^*	SHOCK p_{t1}/p_{t2}
.650	.92821	1.	.99473	1.	.99447	1.
.649	.92986	1.	.99497	1.	.99473	1.
.648	.93152	1.	.99521	1.	.99498	1.
.647	.93318	1.	.99544	1.	.99523	1.
.646	.93483	1.	.99567	1.	.99547	1.
.645	.93649	1.	.99589	1.	.99571	1.
.644	.93814	1.	.99610	1.	.99593	1.
.643	.93980	1.	.99631	1.	.99615	1.
.642	.94145	1.	.99651	1.	.99637	1.
.641	.94311	1.	.99670	1.	.99658	1.
.640	.94476	1.	.99689	1.	.99678	1.
.639	.94642	1.	.99708	1.	.99697	1.
.638	.94807	1.	.99726	1.	.99716	1.
.637	.94972	1.	.99743	1.	.99734	1.
.636	.95137	1.	.99760	1.	.99752	1.
.635	.95303	1.	.99776	1.	.99769	1.
.634	.95468	1.	.99792	1.	.99785	1.
.633	.95633	1.	.99807	1.	.99801	1.
.632	.95798	1.	.99821	1.	.99816	1.
.631	.95963	1.	.99835	1.	.99830	1.
.630	.96129	1.	.99848	1.	.99844	1.
.629	.96294	1.	.99861	1.	.99858	1.
.628	.96459	1.	.99873	1.	.99870	1.
.627	.96624	1.	.99885	1.	.99882	1.
.626	.96789	1.	.99896	1.	.99894	1.
.625	.96954	1.	.99906	1.	.99904	1.
.624	.97119	1.	.99916	1.	.99915	1.
.623	.97284	1.	.99926	1.	.99924	1.
.622	.97449	1.	.99934	1.	.99933	1.
.621	.97614	1.	.99943	1.	.99942	1.
.620	.97779	1.	.99950	1.	.99950	1.
.619	.97944	1.	.99957	1.	.99957	1.
.618	.98109	1.	.99964	1.	.99964	1.
.617	.98274	1.	.99970	1.	.99970	1.
.616	.98439	1.	.99975	1.	.99975	1.
.615	.98604	1.	.99980	1.	.99980	1.
.614	.98768	1.	.99985	1.	.99985	1.
.613	.98933	1.	.99989	1.	.99989	1.
.612	.99098	1.	.99992	1.	.99992	1.
.611	.99263	1.	.99995	1.	.99995	1.
.610	.99428	1.	.99997	1.	.99997	1.
.609	.99593	1.	.99998	1.	.99998	1.
.608	.99758	1.	.99999	1.	.99999	1.
.607	.99923	1.	.00000	1.	.00000	1.
.606	1.00088	1.	.00000	1.	.00000	1.00000
.605	1.00252	1.	.99999	1.	.99999	1.00000
.604	1.00417	1.	.99998	1.	.99998	1.00000
.603	1.00582	1.	.99997	1.	.99997	1.00000
.602	1.00747	1.	.99994	1.	.99994	1.00000
.601	1.00912	1.	.99992	1.	.99992	1.00000

TABLE I. Isentropic Exponent = 1.00

p/p_t	M	T/T_t	GAMMA	ISOTHERMAL T_t/T_t*	RAYLEIGH T_t/T_t*	SHOCK p_t1/p_t2
.600	1.01077	1.	.99988	1.	.99989	1.00000
.599	1.01242	1.	.99985	1.	.99985	1.00000
.598	1.01407	1.	.99980	1.	.99980	1.00000
.597	1.01571	1.	.99975	1.	.99976	1.00001
.596	1.01736	1.	.99970	1.	.99970	1.00001
.595	1.01901	1.	.99964	1.	.99965	1.00001
.594	1.02066	1.	.99958	1.	.99958	1.00001
.593	1.02231	1.	.99951	1.	.99951	1.00001
.592	1.02396	1.	.99943	1.	.99944	1.00002
.591	1.02561	1.	.99935	1.	.99936	1.00002
.590	1.02726	1.	.99926	1.	.99928	1.00003
.589	1.02891	1.	.99917	1.	.99919	1.00003
.588	1.03056	1.	.99908	1.	.99909	1.00004
.587	1.03221	1.	.99897	1.	.99900	1.00004
.586	1.03386	1.	.99887	1.	.99889	1.00005
.585	1.03551	1.	.99875	1.	.99878	1.00006
.584	1.03716	1.	.99864	1.	.99867	1.00006
.583	1.03881	1.	.99851	1.	.99855	1.00007
.582	1.04047	1.	.99839	1.	.99843	1.00008
.581	1.04212	1.	.99825	1.	.99830	1.00009
.580	1.04377	1.	.99811	1.	.99817	1.00010
.579	1.04542	1.	.99797	1.	.99803	1.00012
.578	1.04707	1.	.99782	1.	.99789	1.00013
.577	1.04873	1.	.99767	1.	.99774	1.00014
.576	1.05038	1.	.99751	1.	.99759	1.00016
.575	1.05203	1.	.99734	1.	.99743	1.00017
.574	1.05368	1.	.99717	1.	.99727	1.00019
.573	1.05534	1.	.99700	1.	.99710	1.00021
.572	1.05699	1.	.99682	1.	.99693	1.00023
.571	1.05865	1.	.99663	1.	.99676	1.00025
.570	1.06030	1.	.99644	1.	.99658	1.00027
.569	1.06196	1.	.99624	1.	.99640	1.00029
.568	1.06361	1.	.99604	1.	.99621	1.00031
.567	1.06527	1.	.99584	1.	.99601	1.00034
.566	1.06692	1.	.99563	1.	.99582	1.00036
.565	1.06858	1.	.99541	1.	.99561	1.00039
.564	1.07023	1.	.99519	1.	.99541	1.00042
.563	1.07189	1.	.99496	1.	.99520	1.00045
.562	1.07355	1.	.99473	1.	.99498	1.00048
.561	1.07521	1.	.99449	1.	.99476	1.00051
.560	1.07686	1.	.99425	1.	.99454	1.00054
.559	1.07852	1.	.99400	1.	.99431	1.00058
.558	1.08018	1.	.99375	1.	.99407	1.00061
.557	1.08184	1.	.99350	1.	.99384	1.00065
.556	1.08350	1.	.99323	1.	.99360	1.00069
.555	1.08516	1.	.99297	1.	.99335	1.00073
.554	1.08682	1.	.99269	1.	.99310	1.00077
.553	1.08848	1.	.99242	1.	.99285	1.00081
.552	1.09014	1.	.99213	1.	.99259	1.00086
.551	1.09181	1.	.99185	1.	.99232	1.00091

TABLE I. Isentropic Exponent = 1.00

p/p_t	M	T/T_t	GAMMA	ISOTHERMAL T_t/T_t^*	RAYLEIGH T_t/T_t^*	SHOCK p_{t1}/p_{t2}
.550	1.09347	1.	.99155	1.	.99206	1.00095
.549	1.09513	1.	.99126	1.	.99179	1.00100
.548	1.09680	1.	.99095	1.	.99151	1.00105
.547	1.09846	1.	.99065	1.	.99123	1.00111
.546	1.10012	1.	.99033	1.	.99095	1.00116
.545	1.10179	1.	.99002	1.	.99066	1.00122
.544	1.10345	1.	.98969	1.	.99037	1.00128
.543	1.10512	1.	.98937	1.	.99008	1.00134
.542	1.10679	1.	.98903	1.	.98978	1.00140
.541	1.10845	1.	.98870	1.	.98947	1.00146
.540	1.11012	1.	.98835	1.	.98916	1.00152
.539	1.11179	1.	.98801	1.	.98885	1.00159
.538	1.11346	1.	.98765	1.	.98854	1.00166
.537	1.11513	1.	.98729	1.	.98822	1.00173
.536	1.11680	1.	.98693	1.	.98790	1.00180
.535	1.11847	1.	.98656	1.	.98757	1.00188
.534	1.12014	1.	.98619	1.	.98724	1.00195
.533	1.12181	1.	.98582	1.	.98690	1.00203
.532	1.12349	1.	.98543	1.	.98656	1.00211
.531	1.12516	1.	.98505	1.	.98622	1.00220
.530	1.12683	1.	.98465	1.	.98588	1.00228
.529	1.12851	1.	.98426	1.	.98552	1.00237
.528	1.13019	1.	.98385	1.	.98517	1.00245
.527	1.13186	1.	.98345	1.	.98481	1.00254
.526	1.13354	1.	.98304	1.	.98445	1.00264
.525	1.13522	1.	.98262	1.	.98409	1.00273
.524	1.13689	1.	.98220	1.	.98372	1.00283
.523	1.13857	1.	.98177	1.	.98335	1.00293
.522	1.14025	1.	.98134	1.	.98297	1.00303
.521	1.14193	1.	.98090	1.	.98259	1.00313
.520	1.14361	1.	.98046	1.	.98221	1.00324
.519	1.14530	1.	.98001	1.	.98182	1.00335
.518	1.14698	1.	.97956	1.	.98143	1.00346
.517	1.14866	1.	.97911	1.	.98103	1.00357
.516	1.15035	1.	.97865	1.	.98064	1.00368
.515	1.15203	1.	.97818	1.	.98023	1.00380
.514	1.15372	1.	.97771	1.	.97983	1.00392
.513	1.15540	1.	.97723	1.	.97942	1.00404
.512	1.15709	1.	.97675	1.	.97901	1.00417
.511	1.15878	1.	.97627	1.	.97859	1.00430
.510	1.16047	1.	.97578	1.	.97817	1.00442
.509	1.16216	1.	.97528	1.	.97775	1.00456
.508	1.16385	1.	.97478	1.	.97733	1.00469
.507	1.16554	1.	.97428	1.	.97690	1.00483
.506	1.16724	1.	.97377	1.	.97646	1.00497
.505	1.16893	1.	.97325	1.	.97603	1.00511
.504	1.17062	1.	.97274	1.	.97559	1.00525
.503	1.17232	1.	.97221	1.	.97514	1.00540
.502	1.17401	1.	.97168	1.	.97470	1.00555
.501	1.17571	1.	.97115	1.	.97425	1.00570

TABLE I. Isentropic Exponent = 1.00

p/p_t	M	T/T_t	GAMMA	ISOTHERMAL T_t/T_t^*	RAYLEIGH T_t/T_t^*	SHOCK p_{t1}/p_{t2}
.500	1.17741	1.	.97061	1.	.97379	1.00586
.499	1.17911	1.	.97007	1.	.97334	1.00601
.498	1.18081	1.	.96952	1.	.97288	1.00617
.497	1.18251	1.	.96897	1.	.97242	1.00634
.496	1.18421	1.	.96841	1.	.97195	1.00650
.495	1.18592	1.	.96785	1.	.97148	1.00667
.494	1.18762	1.	.96728	1.	.97101	1.00684
.493	1.18932	1.	.96671	1.	.97053	1.00702
.492	1.19103	1.	.96613	1.	.97005	1.00719
.491	1.19274	1.	.96555	1.	.96957	1.00737
.490	1.19445	1.	.96496	1.	.96908	1.00756
.489	1.19615	1.	.96437	1.	.96859	1.00774
.488	1.19786	1.	.96377	1.	.96810	1.00793
.487	1.19958	1.	.96317	1.	.96761	1.00812
.486	1.20129	1.	.96257	1.	.96711	1.00831
.485	1.20300	1.	.96196	1.	.96660	1.00851
.484	1.20472	1.	.96134	1.	.96610	1.00871
.483	1.20643	1.	.96072	1.	.96559	1.00891
.482	1.20815	1.	.96010	1.	.96508	1.00912
.481	1.20987	1.	.95947	1.	.96457	1.00933
.480	1.21159	1.	.95883	1.	.96405	1.00954
.479	1.21331	1.	.95819	1.	.96353	1.00976
.478	1.21503	1.	.95755	1.	.96301	1.00998
.477	1.21675	1.	.95690	1.	.96248	1.01020
.476	1.21847	1.	.95625	1.	.96195	1.01042
.475	1.22020	1.	.95559	1.	.96142	1.01065
.474	1.22192	1.	.95493	1.	.96088	1.01088
.473	1.22365	1.	.95426	1.	.96034	1.01111
.472	1.22538	1.	.95359	1.	.95980	1.01135
.471	1.22711	1.	.95291	1.	.95926	1.01159
.470	1.22884	1.	.95223	1.	.95871	1.01184
.469	1.23057	1.	.95154	1.	.95816	1.01208
.468	1.23230	1.	.95085	1.	.95761	1.01233
.467	1.23404	1.	.95015	1.	.95705	1.01259
.466	1.23577	1.	.94945	1.	.95649	1.01285
.465	1.23751	1.	.94875	1.	.95593	1.01311
.464	1.23925	1.	.94803	1.	.95536	1.01337
.463	1.24099	1.	.94732	1.	.95479	1.01364
.462	1.24273	1.	.94660	1.	.95422	1.01391
.461	1.24447	1.	.94588	1.	.95365	1.01418
.460	1.24622	1.	.94515	1.	.95307	1.01446
.459	1.24796	1.	.94441	1.	.95249	1.01474
.458	1.24971	1.	.94367	1.	.95191	1.01503
.457	1.25146	1.	.94293	1.	.95133	1.01532
.456	1.25321	1.	.94218	1.	.95074	1.01561
.455	1.25496	1.	.94143	1.	.95015	1.01590
.454	1.25671	1.	.94067	1.	.94955	1.01620
.453	1.25846	1.	.93991	1.	.94896	1.01651
.452	1.26022	1.	.93914	1.	.94836	1.01681
.451	1.26197	1.	.93837	1.	.94776	1.01712

TABLE I. Isentropic Exponent $= 1.00$

p/p_t	M	T/T_t	GAMMA	ISOTHERMAL T_t/T_t	RAYLEIGH T_t/T_t^*	SHOCK p_{t1}/p_{t2}
.450	1.26373	1.	.93759	1.	.94715	
.449	1.26549	1.	.93681	1.	.94655	1.01744
.448	1.26725	1.	.93603	1.	.94594	1.01776
.447	1.26901	1.	.93523	1.	.94532	1.01808
.446	1.27078	1.	.93444	1.	.94471	1.01840
						1.01873
.445	1.27254	1.	.93364	1.	.94409	
.444	1.27431	1.	.93283	1.	.94347	1.01906
.443	1.27608	1.	.93203	1.	.94284	1.01940
.442	1.27785	1.	.93121	1.	.94222	1.01974
.441	1.27962	1.	.93039	1.	.94159	1.02009
						1.02044
.440	1.28139	1.	.92957	1.	.94096	
.439	1.28316	1.	.92874	1.	.94032	1.02079
.438	1.28494	1.	.92791	1.	.93969	1.02115
.437	1.28672	1.	.92707	1.	.93905	1.02151
.436	1.28850	1.	.92623	1.	.93840	1.02187
						1.02224
.435	1.29028	1.	.92538	1.	.93776	
.434	1.29206	1.	.92453	1.	.93711	1.02261
.433	1.29385	1.	.92367	1.	.93646	1.02299
.432	1.29563	1.	.92281	1.	.93581	1.02337
.431	1.29742	1.	.92194	1.	.93515	1.02376
						1.02414
.430	1.29921	1.	.92107	1.	.93450	
.429	1.30100	1.	.92020	1.	.93384	1.02454
.428	1.30279	1.	.91932	1.	.93317	1.02494
.427	1.30459	1.	.91843	1.	.93251	1.02534
.426	1.30638	1.	.91754	1.	.93184	1.02574
						1.02616
.425	1.30818	1.	.91665	1.	.93117	
.424	1.30998	1.	.91575	1.	.93050	1.02657
.423	1.31178	1.	.91485	1.	.92982	1.02699
.422	1.31358	1.	.91394	1.	.92914	1.02741
.421	1.31539	1.	.91303	1.	.92846	1.02784
						1.02828
.420	1.31719	1.	.91211	1.	.92778	
.419	1.31900	1.	.91119	1.	.92709	1.02871
.418	1.32081	1.	.91026	1.	.92641	1.02915
.417	1.32263	1.	.90933	1.	.92572	1.02960
.416	1.32444	1.	.90839	1.	.92502	1.03005
						1.03051
.415	1.32626	1.	.90745	1.	.92433	
.414	1.32807	1.	.90650	1.	.92363	1.03097
.413	1.32989	1.	.90555	1.	.92293	1.03143
.412	1.33171	1.	.90460	1.	.92223	1.03190
.411	1.33354	1.	.90364	1.	.92152	1.03238
						1.03286
.410	1.33536	1.	.90267	1.	.92081	
.409	1.33719	1.	.90170	1.	.92010	1.03334
.408	1.33902	1.	.90073	1.	.91939	1.03383
.407	1.34085	1.	.89975	1.	.91868	1.03432
.406	1.34269	1.	.89877	1.	.91796	1.03482
						1.03532
.405	1.34452	1.	.89778	1.	.91724	
.404	1.34636	1.	.89679	1.	.91652	1.03583
.403	1.34820	1.	.89579	1.	.91579	1.03634
.402	1.35004	1.	.89479	1.	.91507	1.03686
.401	1.35188	1.	.89378	1.	.91434	1.03739
						1.03791

TABLE I. Isentropic Exponent = 1.00

p/p_t	M	T/T_t	GAMMA	ISOTHERMAL T_t/T_t^*	RAYLEIGH T_t/T_t^*	SHOCK p_{t1}/p_{t2}
.400	1.35373	1.	.89277	1.	.91361	1.03845
.399	1.35558	1.	.89175	1.	.91287	1.03899
.398	1.35743	1.	.89073	1.	.91213	1.03953
.397	1.35928	1.	.88971	1.	.91140	1.04008
.396	1.36113	1.	.88867	1.	.91066	1.04063
.395	1.36299	1.	.88764	1.	.90991	1.04119
.394	1.36485	1.	.88660	1.	.90917	1.04176
.393	1.36671	1.	.88556	1.	.90842	1.04233
.392	1.36857	1.	.88451	1.	.90767	1.04290
.391	1.37044	1.	.88345	1.	.90692	1.04348
.390	1.37230	1.	.88239	1.	.90616	1.04407
.389	1.37417	1.	.88133	1.	.90540	1.04466
.388	1.37605	1.	.88026	1.	.90465	1.04526
.387	1.37792	1.	.87919	1.	.90388	1.04586
.386	1.37980	1.	.87811	1.	.90312	1.04647
.385	1.38167	1.	.87703	1.	.90235	1.04708
.384	1.38356	1.	.87594	1.	.90159	1.04770
.383	1.38544	1.	.87485	1.	.90081	1.04833
.382	1.38732	1.	.87375	1.	.90004	1.04896
.381	1.38921	1.	.87265	1.	.89927	1.04959
.380	1.39110	1.	.87155	1.	.89849	1.05024
.379	1.39300	1.	.87043	1.	.89771	1.05088
.378	1.39489	1.	.86932	1.	.89693	1.05154
.377	1.39679	1.	.86820	1.	.89614	1.05220
.376	1.39869	1.	.86707	1.	.89536	1.05286
.375	1.40059	1.	.86594	1.	.89457	1.05354
.374	1.40250	1.	.86481	1.	.89378	1.05421
.373	1.40441	1.	.86367	1.	.89298	1.05490
.372	1.40632	1.	.86253	1.	.89219	1.05559
.371	1.40823	1.	.86138	1.	.89139	1.05629
.370	1.41014	1.	.86023	1.	.89059	1.05699
.369	1.41206	1.	.85907	1.	.88979	1.05770
.368	1.41398	1.	.85790	1.	.88899	1.05841
.367	1.41591	1.	.85674	1.	.88818	1.05914
.366	1.41783	1.	.85556	1.	.88737	1.05986
.365	1.41976	1.	.85439	1.	.88656	1.06060
.364	1.42169	1.	.85321	1.	.88575	1.06134
.363	1.42362	1.	.85202	1.	.88493	1.06209
.362	1.42556	1.	.85083	1.	.88412	1.06284
.361	1.42750	1.	.84963	1.	.88330	1.06360
.360	1.42944	1.	.84843	1.	.88247	1.06437
.359	1.43139	1.	.84722	1.	.88165	1.06514
.358	1.43333	1.	.84601	1.	.88083	1.06592
.357	1.43528	1.	.84480	1.	.88000	1.06671
.356	1.43724	1.	.84358	1.	.87917	1.06751
.355	1.43919	1.	.84235	1.	.87834	1.06831
.354	1.44115	1.	.84112	1.	.87750	1.06912
.353	1.44311	1.	.83989	1.	.87666	1.06993
.352	1.44508	1.	.83865	1.	.87583	1.07075
.351	1.44704	1.	.83741	1.	.87498	1.07158

TABLE I. Isentropic Exponent = 1.00

p/p_t	M	T/T_t	GAMMA	ISOTHERMAL T_t/T_t^*	RAYLEIGH T_t/T_t^*	SHOCK p_{t1}/p_{t2}
.350	1.44902	1.	.83616	1.	.87414	1.07242
.349	1.45099	1.	.83490	1.	.87330	1.07327
.348	1.45296	1.	.83365	1.	.87245	1.07412
.347	1.45494	1.	.83238	1.	.87160	1.07498
.346	1.45693	1.	.83111	1.	.87075	1.07584
.345	1.45891	1.	.82984	1.	.86990	1.07671
.344	1.46090	1.	.82856	1.	.86904	1.07760
.343	1.46289	1.	.82728	1.	.86818	1.07848
.342	1.46489	1.	.82599	1.	.86732	1.07938
.341	1.46688	1.	.82470	1.	.86646	1.08028
.340	1.46888	1.	.82340	1.	.86560	1.08120
.339	1.47089	1.	.82210	1.	.86473	1.08211
.338	1.47289	1.	.82080	1.	.86386	1.08304
.337	1.47491	1.	.81949	1.	.86299	1.08398
.336	1.47692	1.	.81817	1.	.86212	1.08492
.335	1.47894	1.	.81685	1.	.86124	1.08587
.334	1.48096	1.	.81552	1.	.86037	1.08683
.333	1.48298	1.	.81419	1.	.85949	1.08780
.332	1.48501	1.	.81286	1.	.85861	1.08877
.331	1.48704	1.	.81151	1.	.85772	1.08976
.330	1.48907	1.	.81017	1.	.85684	1.09075
.329	1.49111	1.	.80882	1.	.85595	1.09175
.328	1.49315	1.	.80746	1.	.85506	1.09276
.327	1.49519	1.	.80610	1.	.85417	1.09378
.326	1.49724	1.	.80474	1.	.85328	1.09480
.325	1.49929	1.	.80337	1.	.85238	1.09584
.324	1.50134	1.	.80199	1.	.85148	1.09688
.323	1.50340	1.	.80061	1.	.85059	1.09793
.322	1.50546	1.	.79923	1.	.84968	1.09899
.321	1.50752	1.	.79784	1.	.84878	1.10006
.320	1.50959	1.	.79645	1.	.84787	1.10114
.319	1.51166	1.	.79505	1.	.84697	1.10223
.318	1.51374	1.	.79364	1.	.84606	1.10333
.317	1.51582	1.	.79223	1.	.84514	1.10443
.316	1.51790	1.	.79082	1.	.84423	1.10555
.315	1.51999	1.	.78940	1.	.84331	1.10668
.314	1.52208	1.	.78798	1.	.84239	1.10781
.313	1.52417	1.	.78655	1.	.84147	1.10896
.312	1.52627	1.	.78512	1.	.84055	1.11011
.311	1.52837	1.	.78368	1.	.83963	1.11127
.310	1.53048	1.	.78223	1.	.83870	1.11245
.309	1.53259	1.	.78078	1.	.83777	1.11363
.308	1.53470	1.	.77933	1.	.83684	1.11482
.307	1.53682	1.	.77787	1.	.83591	1.11603
.306	1.53894	1.	.77641	1.	.83497	1.11724
.305	1.54107	1.	.77494	1.	.83404	1.11847
.304	1.54320	1.	.77347	1.	.83310	1.11970
.303	1.54533	1.	.77199	1.	.83215	1.12094
.302	1.54747	1.	.77051	1.	.83121	1.12220
.301	1.54961	1.	.76902	1.	.83027	1.12347

TABLE I. Isentropic Exponent = 1.00

p/p_t	M	T/T_t	GAMMA	ISOTHERMAL T_t/T_t^*	RAYLEIGH T_t/T_t^*	SHOCK p_{t1}/p_{t2}
.300	1.55176	1.	.76752	1.	.82932	1.12474
.299	1.55391	1.	.76603	1.	.82837	1.12603
.298	1.55606	1.	.76452	1.	.82742	1.12733
.297	1.55822	1.	.76301	1.	.82646	1.12864
.296	1.56038	1.	.76150	1.	.82551	1.12996
.295	1.56255	1.	.75998	1.	.82455	1.13129
.294	1.56472	1.	.75846	1.	.82359	1.13263
.293	1.56690	1.	.75693	1.	.82263	1.13398
.292	1.56908	1.	.75540	1.	.82166	1.13535
.291	1.57126	1.	.75386	1.	.82070	1.13673
.290	1.57345	1.	.75231	1.	.81973	1.13811
.289	1.57565	1.	.75076	1.	.81876	1.13951
.288	1.57784	1.	.74921	1.	.81778	1.14093
.287	1.58005	1.	.74765	1.	.81681	1.14235
.286	1.58225	1.	.74609	1.	.81583	1.14379
.285	1.58447	1.	.74452	1.	.81485	1.14524
.284	1.58668	1.	.74294	1.	.81387	1.14670
.283	1.58890	1.	.74136	1.	.81289	1.14817
.282	1.59113	1.	.73978	1.	.81190	1.14966
.281	1.59336	1.	.73819	1.	.81092	1.15115
.280	1.59560	1.	.73659	1.	.80993	1.15267
.279	1.59784	1.	.73499	1.	.80894	1.15419
.278	1.60008	1.	.73339	1.	.80794	1.15573
.277	1.60233	1.	.73178	1.	.80695	1.15728
.276	1.60459	1.	.73016	1.	.80595	1.15884
.275	1.60685	1.	.72854	1.	.80495	1.16042
.274	1.60912	1.	.72692	1.	.80394	1.16201
.273	1.61139	1.	.72529	1.	.80294	1.16361
.272	1.61366	1.	.72365	1.	.80193	1.16523
.271	1.61594	1.	.72201	1.	.80092	1.16687
.270	1.61823	1.	.72036	1.	.79991	1.16851
.269	1.62052	1.	.71871	1.	.79890	1.17017
.268	1.62282	1.	.71705	1.	.79788	1.17185
.267	1.62512	1.	.71539	1.	.79687	1.17354
.266	1.62743	1.	.71372	1.	.79585	1.17524
.265	1.62974	1.	.71205	1.	.79483	1.17696
.264	1.63206	1.	.71037	1.	.79380	1.17870
.263	1.63438	1.	.70869	1.	.79278	1.18044
.262	1.63671	1.	.70700	1.	.79175	1.18221
.261	1.63905	1.	.70531	1.	.79072	1.18399
.260	1.64139	1.	.70361	1.	.78968	1.18578
.259	1.64373	1.	.70190	1.	.78865	1.18760
.258	1.64608	1.	.70019	1.	.78761	1.18942
.257	1.64844	1.	.69848	1.	.78657	1.19127
.256	1.65080	1.	.69676	1.	.78553	1.19313
.255	1.65317	1.	.69503	1.	.78449	1.19500
.254	1.65555	1.	.69330	1.	.78344	1.19689
.253	1.65793	1.	.69157	1.	.78239	1.19880
.252	1.66032	1.	.68982	1.	.78134	1.20073
.251	1.66271	1.	.68808	1.	.78029	1.20267

TABLE I.   Isentropic Exponent = 1.00

p/p_t	M	T/T_t	GAMMA	ISOTHERMAL T_t/T*	RAYLEIGH T_t/T*	SHOCK p_tt/p_t2
.250	1.66511	1.	.68633	1.	.77923	1.20463
.249	1.66751	1.	.68457	1.	.77817	1.20661
.248	1.66993	1.	.68280	1.	.77711	1.20861
.247	1.67234	1.	.68104	1.	.77605	1.21062
.246	1.67477	1.	.67926	1.	.77499	1.21265
.245	1.67720	1.	.67748	1.	.77392	1.21470
.244	1.67964	1.	.67570	1.	.77285	1.21677
.243	1.68208	1.	.67391	1.	.77178	1.21886
.242	1.68453	1.	.67211	1.	.77071	1.22097
.241	1.68698	1.	.67031	1.	.76963	1.22309
.240	1.68945	1.	.66850	1.	.76855	1.22524
.239	1.69192	1.	.66669	1.	.76747	1.22740
.238	1.69439	1.	.66487	1.	.76639	1.22959
.237	1.69688	1.	.66305	1.	.76530	1.23179
.236	1.69937	1.	.66122	1.	.76422	1.23401
.235	1.70186	1.	.65939	1.	.76313	1.23626
.234	1.70437	1.	.65755	1.	.76203	1.23853
.233	1.70688	1.	.65570	1.	.76094	1.24081
.232	1.70940	1.	.65385	1.	.75984	1.24312
.231	1.71192	1.	.65199	1.	.75874	1.24545
.230	1.71445	1.	.65013	1.	.75764	1.24780
.229	1.71699	1.	.64826	1.	.75653	1.25017
.228	1.71954	1.	.64639	1.	.75543	1.25257
.227	1.72210	1.	.64451	1.	.75432	1.25498
.226	1.72466	1.	.64263	1.	.75321	1.25742
.225	1.72723	1.	.64074	1.	.75209	1.25989
.224	1.72980	1.	.63884	1.	.75097	1.26237
.223	1.73239	1.	.63694	1.	.74985	1.26488
.222	1.73498	1.	.63503	1.	.74873	1.26742
.221	1.73758	1.	.63312	1.	.74761	1.26998
.220	1.74019	1.	.63120	1.	.74648	1.27256
.219	1.74280	1.	.62927	1.	.74535	1.27517
.218	1.74543	1.	.62734	1.	.74422	1.27780
.217	1.74806	1.	.62541	1.	.74308	1.28046
.216	1.75070	1.	.62347	1.	.74195	1.28314
.215	1.75335	1.	.62152	1.	.74081	1.28585
.214	1.75601	1.	.61957	1.	.73966	1.28859
.213	1.75867	1.	.61761	1.	.73852	1.29135
.212	1.76135	1.	.61564	1.	.73737	1.29414
.211	1.76403	1.	.61367	1.	.73622	1.29695
.210	1.76672	1.	.61169	1.	.73507	1.29980
.209	1.76942	1.	.60971	1.	.73391	1.30267
.208	1.77213	1.	.60772	1.	.73275	1.30557
.207	1.77484	1.	.60573	1.	.73159	1.30850
.206	1.77757	1.	.60373	1.	.73043	1.31146
.205	1.78031	1.	.60172	1.	.72926	1.31445
.204	1.78305	1.	.59971	1.	.72809	1.31747
.203	1.78581	1.	.59769	1.	.72692	1.32052
.202	1.78857	1.	.59567	1.	.72575	1.32360
.201	1.79134	1.	.59364	1.	.72457	1.32671

TABLE I. Isentropic Exponent = 1.00

p/p_t	M	T/T_t	GAMMA	ISOTHERMAL T_t/T_t^*	RAYLEIGH T_t/T_t^*	SHOCK p_{t1}/p_{t2}
.200	1.79412	1.	.59160	1.	.72339	1.32986
.199	1.79691	1.	.58956	1.	.72220	1.33303
.198	1.79972	1.	.58751	1.	.72102	1.33624
.197	1.80253	1.	.58546	1.	.71983	1.33948
.196	1.80535	1.	.58340	1.	.71864	1.34276
.195	1.80818	1.	.58133	1.	.71744	1.34607
.194	1.81102	1.	.57926	1.	.71624	1.34941
.193	1.81387	1.	.57718	1.	.71504	1.35279
.192	1.81673	1.	.57510	1.	.71384	1.35620
.191	1.81961	1.	.57300	1.	.71263	1.35965
.190	1.82249	1.	.57091	1.	.71142	1.36314
.189	1.82538	1.	.56880	1.	.71021	1.36666
.188	1.82829	1.	.56669	1.	.70899	1.37022
.187	1.83120	1.	.56458	1.	.70778	1.37382
.186	1.83413	1.	.56246	1.	.70655	1.37746
.185	1.83706	1.	.56033	1.	.70533	1.38114
.184	1.84001	1.	.55819	1.	.70410	1.38485
.183	1.84297	1.	.55605	1.	.70287	1.38861
.182	1.84594	1.	.55391	1.	.70164	1.39241
.181	1.84892	1.	.55175	1.	.70040	1.39625
.180	1.85192	1.	.54959	1.	.69916	1.40013
.179	1.85492	1.	.54743	1.	.69791	1.40406
.178	1.85794	1.	.54525	1.	.69667	1.40803
.177	1.86097	1.	.54308	1.	.69542	1.41204
.176	1.86401	1.	.54089	1.	.69416	1.41610
.175	1.86707	1.	.53870	1.	.69290	1.42020
.174	1.87013	1.	.53650	1.	.69164	1.42435
.173	1.87321	1.	.53429	1.	.69038	1.42855
.172	1.87631	1.	.53208	1.	.68911	1.43280
.171	1.87941	1.	.52986	1.	.68784	1.43709
.170	1.88253	1.	.52764	1.	.68657	1.44143
.169	1.88566	1.	.52541	1.	.68529	1.44583
.168	1.88880	1.	.52317	1.	.68401	1.45027
.167	1.89196	1.	.52093	1.	.68272	1.45477
.166	1.89513	1.	.51868	1.	.68143	1.45932
.165	1.89832	1.	.51642	1.	.68014	1.46393
.164	1.90152	1.	.51415	1.	.67885	1.46858
.163	1.90473	1.	.51188	1.	.67755	1.47330
.162	1.90796	1.	.50960	1.	.67624	1.47807
.161	1.91120	1.	.50732	1.	.67494	1.48290
.160	1.91446	1.	.50503	1.	.67363	1.48779
.159	1.91773	1.	.50273	1.	.67231	1.49273
.158	1.92102	1.	.50042	1.	.67099	1.49774
.157	1.92432	1.	.49811	1.	.66967	1.50281
.156	1.92764	1.	.49579	1.	.66835	1.50794
.155	1.93097	1.	.49346	1.	.66702	1.51314
.154	1.93432	1.	.49113	1.	.66568	1.51840
.153	1.93769	1.	.48879	1.	.66434	1.52372
.152	1.94107	1.	.48644	1.	.66300	1.52912
.151	1.94447	1.	.48409	1.	.66166	1.53458

TABLE I. Isentropic Exponent = 1.00

p/p_t	M	T/T_t	GAMMA	ISOTHERMAL T_t/T_t^*	RAYLEIGH T_t/T_t^*	SHOCK p_{t1}/p_{t2}
.150	1.94788	1.	.48173	1.	.66031	1.54012
.149	1.95131	1.	.47936	1.	.65895	1.54572
.148	1.95476	1.	.47698	1.	.65759	1.55140
.147	1.95823	1.	.47460	1.	.65623	1.55715
.146	1.96171	1.	.47221	1.	.65486	1.56297
.145	1.96521	1.	.46981	1.	.65349	1.56888
.144	1.96873	1.	.46741	1.	.65211	1.57486
.143	1.97226	1.	.46499	1.	.65073	1.58092
.142	1.97582	1.	.46258	1.	.64935	1.58706
.141	1.97939	1.	.46015	1.	.64796	1.59329
.140	1.98298	1.	.45771	1.	.64657	1.59960
.139	1.98660	1.	.45527	1.	.64517	1.60600
.138	1.99023	1.	.45282	1.	.64376	1.61248
.137	1.99388	1.	.45037	1.	.64235	1.61906
.136	1.99755	1.	.44790	1.	.64094	1.62572
.135	2.00124	1.	.44543	1.	.63952	1.63248
.134	2.00495	1.	.44295	1.	.63810	1.63934
.133	2.00868	1.	.44046	1.	.63667	1.64629
.132	2.01244	1.	.43797	1.	.63524	1.65334
.131	2.01621	1.	.43547	1.	.63380	1.66050
.130	2.02001	1.	.43296	1.	.63236	1.66776
.129	2.02383	1.	.43044	1.	.63091	1.67512
.128	2.02767	1.	.42791	1.	.62946	1.68260
.127	2.03154	1.	.42538	1.	.62800	1.69018
.126	2.03542	1.	.42284	1.	.62653	1.69788
.125	2.03933	1.	.42029	1.	.62507	1.70569
.124	2.04327	1.	.41773	1.	.62359	1.71362
.123	2.04723	1.	.41516	1.	.62211	1.72168
.122	2.05121	1.	.41259	1.	.62062	1.72985
.121	2.05522	1.	.41001	1.	.61913	1.73816
.120	2.05925	1.	.40742	1.	.61763	1.74659
.119	2.06331	1.	.40482	1.	.61613	1.75516
.118	2.06740	1.	.40221	1.	.61462	1.76386
.117	2.07151	1.	.39960	1.	.61310	1.77270
.116	2.07565	1.	.39697	1.	.61158	1.78169
.115	2.07982	1.	.39434	1.	.61005	1.79082
.114	2.08401	1.	.39170	1.	.60852	1.80010
.113	2.08824	1.	.38905	1.	.60698	1.80953
.112	2.09249	1.	.38639	1.	.60543	1.81912
.111	2.09677	1.	.38373	1.	.60387	1.82887
.110	2.10108	1.	.38105	1.	.60231	1.83879
.109	2.10543	1.	.37837	1.	.60074	1.84888
.108	2.10980	1.	.37567	1.	.59917	1.85914
.107	2.11420	1.	.37297	1.	.59759	1.86957
.106	2.11864	1.	.37026	1.	.59600	1.88020
.105	2.12311	1.	.36754	1.	.59440	1.89100
.104	2.12761	1.	.36482	1.	.59280	1.90201
.103	2.13215	1.	.36208	1.	.59119	1.91321
.102	2.13672	1.	.35933	1.	.58957	1.92461
.101	2.14132	1.	.35658	1.	.58794	1.93623

TABLE I. Isentropic Exponent = 1.00

p/p_t	M	T/T_t	GAMMA	ISOTHERMAL T_t/T_t^*	RAYLEIGH T_t/T_t^*	SHOCK p_{t1}/p_{t2}
.100	2.14597	1.	.35381	1.	.58631	1.94806
.099	2.15064	1.	.35104	1.	.58467	1.96011
.098	2.15536	1.	.34825	1.	.58302	1.97238
.097	2.16011	1.	.34546	1.	.58136	1.98490
.096	2.16491	1.	.34266	1.	.57970	1.99765
.095	2.16974	1.	.33984	1.	.57802	2.01065
.094	2.17461	1.	.33702	1.	.57634	2.02391
.093	2.17952	1.	.33419	1.	.57464	2.03743
.092	2.18448	1.	.33135	1.	.57294	2.05122
.091	2.18947	1.	.32849	1.	.57123	2.06529
.090	2.19451	1.	.32563	1.	.56951	2.07965
.089	2.19960	1.	.32276	1.	.56778	2.09431
.088	2.20473	1.	.31988	1.	.56605	2.10928
.087	2.20991	1.	.31699	1.	.56430	2.12456
.086	2.21513	1.	.31408	1.	.56254	2.14017
.085	2.22041	1.	.31117	1.	.56077	2.15611
.084	2.22573	1.	.30825	1.	.55899	2.17241
.083	2.23110	1.	.30531	1.	.55720	2.18907
.082	2.23653	1.	.30237	1.	.55540	2.20610
.081	2.24201	1.	.29941	1.	.55359	2.22351
.080	2.24754	1.	.29645	1.	.55177	2.24133
.079	2.25313	1.	.29347	1.	.54993	2.25956
.078	2.25878	1.	.29048	1.	.54809	2.27822
.077	2.26449	1.	.28748	1.	.54623	2.29733
.076	2.27025	1.	.28447	1.	.54436	2.31690
.075	2.27608	1.	.28145	1.	.54248	2.33694
.074	2.28197	1.	.27841	1.	.54058	2.35749
.073	2.28792	1.	.27537	1.	.53867	2.37854
.072	2.29394	1.	.27231	1.	.53675	2.40014
.071	2.30003	1.	.26924	1.	.53482	2.42229
.070	2.30619	1.	.26616	1.	.53287	2.44502
.069	2.31242	1.	.26307	1.	.53090	2.46836
.068	2.31873	1.	.25996	1.	.52893	2.49232
.067	2.32511	1.	.25684	1.	.52693	2.51694
.066	2.33157	1.	.25371	1.	.52492	2.54224
.065	2.33811	1.	.25057	1.	.52290	2.56825
.064	2.34473	1.	.24741	1.	.52086	2.59500
.063	2.35143	1.	.24424	1.	.51880	2.62254
.062	2.35823	1.	.24106	1.	.51673	2.65088
.061	2.36511	1.	.23786	1.	.51463	2.68007
.060	2.37209	1.	.23466	1.	.51252	2.71016
.059	2.37917	1.	.23143	1.	.51039	2.74117
.058	2.38634	1.	.22820	1.	.50824	2.77316
.057	2.39362	1.	.22495	1.	.50608	2.80618
.056	2.40100	1.	.22168	1.	.50389	2.84027
.055	2.40849	1.	.21840	1.	.50168	2.87549
.054	2.41610	1.	.21511	1.	.49945	2.91191
.053	2.42382	1.	.21180	1.	.49719	2.94958
.052	2.43167	1.	.20848	1.	.49492	2.98857
.051	2.43964	1.	.20514	1.	.49262	3.02896

TABLE I. Isentropic Exponent = 1.00

p/p_t	M	T/T_t	GAMMA	ISOTHERMAL T_t/T_t^*	RAYLEIGH T_t/T_t^*	SHOCK p_{t1}/p_{t2}
.050	2.44775	1.	.20178	1.	.49029	3.07082
.049	2.45599	1.	.19841	1.	.48794	3.11424
.048	2.46437	1.	.19503	1.	.48557	3.15931
.047	2.47290	1.	.19162	1.	.48317	3.20613
.046	2.48158	1.	.18821	1.	.48073	3.25481
.045	2.49042	1.	.18477	1.	.47827	3.30546
.044	2.49943	1.	.18132	1.	.47578	3.35821
.043	2.50861	1.	.17785	1.	.47326	3.41319
.042	2.51797	1.	.17436	1.	.47070	3.47057
.041	2.52752	1.	.17085	1.	.46812	3.53049
.040	2.53727	1.	.16733	1.	.46549	3.59315
.039	2.54723	1.	.16379	1.	.46283	3.65874
.038	2.55741	1.	.16023	1.	.46013	3.72748
.037	2.56782	1.	.15664	1.	.45739	3.79960
.036	2.57846	1.	.15304	1.	.45460	3.87538
.035	2.58937	1.	.14942	1.	.45178	3.95511
.034	2.60054	1.	.14578	1.	.44890	4.03911
.033	2.61199	1.	.14211	1.	.44598	4.12776
.032	2.62375	1.	.13843	1.	.44300	4.22147
.031	2.63582	1.	.13472	1.	.43997	4.32068
.030	2.64823	1.	.13099	1.	.43689	4.42593
.029	2.66100	1.	.12723	1.	.43374	4.53781
.028	2.67415	1.	.12345	1.	.43053	4.65697
.027	2.68772	1.	.11965	1.	.42725	4.78419
.026	2.70173	1.	.11581	1.	.42389	4.92034
.025	2.71620	1.	.11196	1.	.42046	5.06644
.024	2.73119	1.	.10807	1.	.41695	5.22365
.023	2.74673	1.	.10416	1.	.41335	5.39335
.022	2.76287	1.	.10021	1.	.40965	5.57713
.021	2.77965	1.	.09624	1.	.40585	5.77691
.020	2.79715	1.	.09223	1.	.40193	5.99494
.019	2.81543	1.	.08820	1.	.39790	6.23395
.018	2.83457	1.	.08412	1.	.39373	6.49723
.017	2.85466	1.	.08001	1.	.38942	6.78884
.016	2.87582	1.	.07586	1.	.38494	7.11379
.015	2.89817	1.	.07167	1.	.38028	7.47838
.014	2.92188	1.	.06744	1.	.37543	7.89064
.013	2.94714	1.	.06317	1.	.37034	8.36095
.012	2.97417	1.	.05884	1.	.36501	8.90305
.011	3.00329	1.	.05447	1.	.35937	9.53542
.010	3.03485	1.	.05004	1.	.35339	10.28367
.009	3.06938	1.	.04554	1.	.34701	11.18431
.008	3.10751	1.	.04099	1.	.34013	12.29131
.007	3.15019	1.	.03636	1.	.33266	13.68824
.006	3.19875	1.	.03164	1.	.32442	15.51197
.005	3.25525	1.	.02683	1.	.31518	18.00408
.004	3.32309	1.	.02192	1.	.30457	21.63681
.003	3.40856	1.	.01686	1.	.29188	27.48199
.002	3.52551	1.	.01163	1.	.27568	38.64187
.001	3.71693	1.	.00613	1.	.25176	69.81049

TABLE II. Isentropic Exponent = 1.10

p/p_t	M	T/T_t	GAMMA	ISOTHERMAL T_t/T_t*	RAYLEIGH T_t/T_t*	SHOCK p_t1/p_t2
1.0000	.00000	1.00000	.00000	.95652	.00000	1.
.9995	.03015	.99995	.05031	.95657	.00381	1.
.9990	.04265	.99991	.07112	.95661	.00761	1.
.9985	.05224	.99986	.08708	.95665	.01140	1.
.9980	.06034	.99982	.10051	.95670	.01517	1.
.9975	.06746	.99977	.11234	.95674	.01893	1.
.9970	.07391	.99973	.12302	.95678	.02268	1.
.9965	.07985	.99968	.13283	.95683	.02641	1.
.9960	.08537	.99964	.14195	.95687	.03014	1.
.9955	.09056	.99959	.15051	.95691	.03385	1.
.9950	.09548	.99954	.15860	.95696	.03755	1.
.9945	.10015	.99950	.16628	.95700	.04123	1.
.9940	.10462	.99945	.17362	.95705	.04491	1.
.9935	.10890	.99941	.18065	.95709	.04857	1.
.9930	.11303	.99936	.18740	.95713	.05222	1.
.9925	.11701	.99932	.19391	.95718	.05585	1.
.9920	.12087	.99927	.20020	.95722	.05948	1.
.9915	.12461	.99922	.20629	.95726	.06309	1.
.9910	.12824	.99918	.21220	.95731	.06669	1.
.9905	.13177	.99913	.21794	.95735	.07028	1.
.9900	.13521	.99909	.22353	.95740	.07385	1.
.9895	.13857	.99904	.22897	.95744	.07742	1.
.9890	.14185	.99899	.23427	.95748	.08097	1.
.9885	.14506	.99895	.23946	.95753	.08451	1.
.9880	.14820	.99890	.24452	.95757	.08804	1.
.9875	.15127	.99886	.24948	.95762	.09155	1.
.9870	.15429	.99881	.25433	.95766	.09506	1.
.9865	.15725	.99877	.25909	.95770	.09855	1.
.9860	.16016	.99872	.26375	.95775	.10203	1.
.9855	.16302	.99867	.26833	.95779	.10550	1.
.9850	.16583	.99863	.27282	.95784	.10896	1.
.9845	.16859	.99858	.27723	.95788	.11241	1.
.9840	.17131	.99853	.28157	.95793	.11584	1.
.9835	.17399	.99849	.28584	.95797	.11926	1.
.9830	.17663	.99844	.29004	.95801	.12268	1.
.9825	.17924	.99840	.29417	.95806	.12608	1.
.9820	.18180	.99835	.29824	.95810	.12946	1.
.9815	.18434	.99830	.30225	.95815	.13284	1.
.9810	.18684	.99826	.30620	.95819	.13621	1.
.9805	.18931	.99821	.31009	.95824	.13956	1.
.9800	.19174	.99817	.31393	.95828	.14291	1.
.9795	.19415	.99812	.31772	.95832	.14624	1.
.9790	.19653	.99807	.32146	.95837	.14956	1.
.9785	.19889	.99803	.32515	.95841	.15287	1.
.9780	.20121	.99798	.32880	.95846	.15617	1.
.9775	.20352	.99793	.33239	.95850	.15946	1.
.9770	.20579	.99789	.33595	.95855	.16274	1.
.9765	.20805	.99784	.33946	.95859	.16600	1.
.9760	.21028	.99779	.34294	.95864	.16926	1.
.9755	.21249	.99775	.34637	.95868	.17250	1.

TABLE II. Isentropic Exponent = 1.10

p/p_t	M	T/T_t	GAMMA	ISOTHERMAL T_t/T_t^*	RAYLEIGH T_t/T_t^*	SHOCK p_{t1}/p_{t2}
.9750	.21468	.99770	.34976	.95873	.17574	1.
.9745	.21684	.99765	.35312	.95877	.17896	1.
.9740	.21899	.99761	.35644	.95882	.18217	1.
.9735	.22111	.99756	.35972	.95886	.18537	1.
.9730	.22322	.99751	.36298	.95890	.18856	1.
.9725	.22531	.99747	.36619	.95895	.19174	1.
.9720	.22738	.99742	.36938	.95899	.19491	1.
.9715	.22943	.99737	.37253	.95904	.19807	1.
.9710	.23147	.99733	.37565	.95908	.20122	1.
.9705	.23349	.99728	.37874	.95913	.20435	1.
.9700	.23549	.99723	.38180	.95917	.20748	1.
.9695	.23748	.99719	.38484	.95922	.21060	1.
.9690	.23945	.99714	.38784	.95926	.21370	1.
.9685	.24141	.99709	.39082	.95931	.21680	1.
.9680	.24335	.99705	.39377	.95935	.21988	1.
.9675	.24528	.99700	.39669	.95940	.22296	1.
.9670	.24720	.99695	.39959	.95944	.22602	1.
.9665	.24910	.99691	.40247	.95949	.22908	1.
.9660	.25098	.99686	.40531	.95953	.23212	1.
.9655	.25286	.99681	.40814	.95958	.23515	1.
.9650	.25472	.99677	.41094	.95962	.23818	1.
.9645	.25657	.99672	.41372	.95967	.24119	1.
.9640	.25840	.99667	.41648	.95972	.24419	1.
.9635	.26023	.99663	.41921	.95976	.24719	1.
.9630	.26204	.99658	.42192	.95981	.25017	1.
.9625	.26384	.99653	.42461	.95985	.25314	1.
.9620	.26564	.99648	.42728	.95990	.25611	1.
.9615	.26741	.99644	.42993	.95994	.25906	1.
.9610	.26918	.99639	.43256	.95999	.26200	1.
.9605	.27094	.99634	.43517	.96003	.26494	1.
.9600	.27269	.99630	.43776	.96008	.26786	1.
.9595	.27443	.99625	.44033	.96012	.27077	1.
.9590	.27616	.99620	.44288	.96017	.27368	1.
.9585	.27787	.99615	.44541	.96021	.27657	1.
.9580	.27958	.99611	.44793	.96026	.27946	1.
.9575	.28128	.99606	.45043	.96031	.28233	1.
.9570	.28297	.99601	.45291	.96035	.28520	1.
.9565	.28465	.99597	.45537	.96040	.28805	1.
.9560	.28632	.99592	.45782	.96044	.28805	1.
.9555	.28799	.99587	.46024	.96049	.29090	1.
.9550	.28964	.99582	.46266	.96053	.29656	1.
.9545	.29129	.99578	.46505	.96058	.29938	1.
.9540	.29292	.99573	.46744	.96063	.30219	1.
.9535	.29455	.99568	.46980	.96067	.30499	1.
.9530	.29618	.99563	.47215	.96072	.30778	1.
.9525	.29779	.99559	.47448	.96076	.31056	1.
.9520	.29939	.99554	.47680	.96081	.31333	1.
.9515	.30099	.99549	.47911	.96085	.31609	1.
.9510	.30258	.99544	.48140	.96090	.31884	1.
.9505	.30417	.99540	.48367	.96095	.32158	1.

TABLE II. Isentropic Exponent = 1.10

p/p_t	M	T/T_t	GAMMA	ISOTHERMAL T_t/T_t*	RAYLEIGH T_t/T_t*	SHOCK p_t1/p_t2
.950	.30574	.99535	.48594	.96099	.32432	1.
.949	.30887	.99525	.49042	.96108	.32976	1.
.948	.31197	.99516	.49485	.96118	.33516	1.
.947	.31505	.99506	.49922	.96127	.34053	1.
.946	.31810	.99497	.50354	.96136	.34586	1.
.945	.32112	.99487	.50782	.96145	.35115	1.
.944	.32412	.99477	.51204	.96155	.35641	1.
.943	.32710	.99468	.51622	.96164	.36164	1.
.942	.33005	.99458	.52035	.96173	.36683	1.
.941	.33298	.99449	.52444	.96182	.37198	1.
.940	.33588	.99439	.52848	.96192	.37710	1.
.939	.33877	.99429	.53247	.96201	.38219	1.
.938	.34163	.99420	.53643	.96210	.38724	1.
.937	.34448	.99410	.54034	.96220	.39226	1.
.936	.34730	.99401	.54422	.96229	.39724	1.
.935	.35010	.99391	.54805	.96238	.40219	1.
.934	.35289	.99381	.55185	.96248	.40711	1.
.933	.35565	.99372	.55560	.96257	.41199	1.
.932	.35840	.99362	.55932	.96267	.41684	1.
.931	.36113	.99352	.56301	.96276	.42166	1.
.930	.36384	.99342	.56666	.96285	.42645	1.
.929	.36654	.99333	.57027	.96295	.43120	1.
.928	.36922	.99323	.57385	.96304	.43592	1.
.927	.37188	.99313	.57739	.96314	.44061	1.
.926	.37453	.99304	.58090	.96323	.44527	1.
.925	.37716	.99294	.58438	.96333	.44990	1.
.924	.37978	.99284	.58783	.96342	.45449	1.
.923	.38238	.99274	.59125	.96351	.45905	1.
.922	.38497	.99264	.59463	.96361	.46359	1.
.921	.38754	.99255	.59799	.96370	.46809	1.
.920	.39010	.99245	.60131	.96380	.47256	1.
.919	.39265	.99235	.60461	.96390	.47700	1.
.918	.39518	.99225	.60787	.96399	.48142	1.
.917	.39770	.99215	.61111	.96409	.48580	1.
.916	.40020	.99206	.61432	.96418	.49015	1.
.915	.40270	.99196	.61751	.96428	.49447	1.
.914	.40518	.99186	.62066	.96437	.49877	1.
.913	.40765	.99176	.62379	.96447	.50303	1.
.912	.41010	.99166	.62690	.96457	.50727	1.
.911	.41255	.99156	.62997	.96466	.51147	1.
.910	.41498	.99146	.63303	.96476	.51565	1.
.909	.41741	.99136	.63605	.96485	.51980	1.
.908	.41982	.99126	.63906	.96495	.52392	1.
.907	.42222	.99117	.64203	.96505	.52801	1.
.906	.42461	.99107	.64499	.96514	.53208	1.
.905	.42699	.99097	.64792	.96524	.53612	1.
.904	.42936	.99087	.65083	.96534	.54013	1.
.903	.43171	.99077	.65371	.96544	.54411	1.
.902	.43406	.99067	.65657	.96553	.54806	1.
.901	.43640	.99057	.65941	.96563	.55199	1.

TABLE II. Isentropic Exponent = 1.10

p/p_t	M	T/T_t	GAMMA	ISOTHERMAL T_t/T_t^*	RAYLEIGH T_t/T_t^*	SHOCK p_{t1}/p_{t2}
.900	.43873	.99047	.66223	.96573	.55589	1.
.899	.44105	.99037	.66502	.96583	.55977	1.
.898	.44336	.99027	.66780	.96592	.56362	1.
.897	.44566	.99017	.67055	.96602	.56744	1.
.896	.44795	.99007	.67328	.96612	.57123	1.
.895	.45024	.98997	.67599	.96622	.57500	1.
.894	.45251	.98987	.67868	.96631	.57875	1.
.893	.45478	.98976	.68135	.96641	.58246	1.
.892	.45704	.98966	.68400	.96651	.58616	1.
.891	.45928	.98956	.68663	.96661	.58982	1.
.890	.46153	.98946	.68925	.96671	.59347	1.
.889	.46376	.98936	.69184	.96681	.59708	1.
.888	.46598	.98926	.69441	.96691	.60067	1.
.887	.46820	.98916	.69696	.96701	.60424	1.
.886	.47041	.98906	.69950	.96710	.60778	1.
.885	.47261	.98896	.70201	.96720	.61130	1.
.884	.47480	.98885	.70451	.96730	.61479	1.
.883	.47699	.98875	.70699	.96740	.61826	1.
.882	.47917	.98865	.70946	.96750	.62171	1.
.881	.48134	.98855	.71190	.96760	.62513	1.
.880	.48351	.98845	.71433	.96770	.62853	1.
.879	.48567	.98834	.71674	.96780	.63190	1.
.878	.48782	.98824	.71913	.96790	.63525	1.
.877	.48996	.98814	.72151	.96800	.63858	1.
.876	.49210	.98804	.72387	.96810	.64188	1.
.875	.49423	.98793	.72621	.96820	.64516	1.
.874	.49636	.98783	.72853	.96830	.64842	1.
.873	.49847	.98773	.73084	.96841	.65166	1.
.872	.50059	.98763	.73314	.96851	.65487	1.
.871	.50269	.98752	.73542	.96861	.65806	1.
.870	.50479	.98742	.73768	.96871	.66123	1.
.869	.50688	.98732	.73992	.96881	.66437	1.
.868	.50897	.98721	.74216	.96891	.66749	1.
.867	.51105	.98711	.74437	.96901	.67060	1.
.866	.51313	.98701	.74657	.96911	.67368	1.
.865	.51520	.98690	.74876	.96922	.67673	1.
.864	.51726	.98680	.75093	.96932	.67977	1.
.863	.51932	.98669	.75308	.96942	.68278	1.
.862	.52137	.98659	.75522	.96952	.68578	1.
.861	.52342	.98649	.75735	.96962	.68875	1.
.860	.52546	.98638	.75946	.96973	.69170	1.
.859	.52750	.98628	.76156	.96983	.69463	1.
.858	.52953	.98617	.76364	.96993	.69754	1.
.857	.53156	.98607	.76571	.97004	.70043	1.
.856	.53358	.98596	.76777	.97014	.70330	1.
.855	.53560	.98586	.76981	.97024	.70615	1.
.854	.53761	.98575	.77184	.97034	.70897	1.
.853	.53961	.98565	.77385	.97045	.71178	1.
.852	.54161	.98554	.77585	.97055	.71457	1.
.851	.54361	.98544	.77784	.97065	.71734	1.

TABLE II. Isentropic Exponent = 1.10

p/p_t	M	T/T_t	GAMMA	ISOTHERMAL T_t/T_t*	RAYLEIGH T_t/T_t*	SHOCK p_t1/p_t2
.850	.54560	.98533	.77981	.97076	.72008	1.
.849	.54759	.98523	.78178	.97086	.72281	1.
.848	.54957	.98512	.78372	.97097	.72552	1.
.847	.55155	.98502	.78566	.97107	.72821	1.
.846	.55352	.98491	.78758	.97118	.73088	1.
.845	.55549	.98481	.78949	.97128	.73353	1.
.844	.55746	.98470	.79139	.97138	.73616	1.
.843	.55942	.98459	.79327	.97149	.73877	1.
.842	.56137	.98449	.79514	.97159	.74136	1.
.841	.56332	.98438	.79700	.97170	.74394	1.
.840	.56527	.98427	.79885	.97180	.74649	1.
.839	.56721	.98417	.80069	.97191	.74903	1.
.838	.56915	.98406	.80251	.97201	.75155	1.
.837	.57109	.98395	.80432	.97212	.75405	1.
.836	.57302	.98385	.80612	.97223	.75653	1.
.835	.57495	.98374	.80791	.97233	.75899	1.
.834	.57687	.98363	.80968	.97244	.76144	1.
.833	.57879	.98353	.81145	.97254	.76386	1.
.832	.58070	.98342	.81320	.97265	.76627	1.
.831	.58261	.98331	.81494	.97276	.76867	1.
.830	.58452	.98320	.81667	.97286	.77104	1.
.829	.58643	.98310	.81839	.97297	.77340	1.
.828	.58833	.98299	.82009	.97308	.77574	1.
.827	.59022	.98288	.82179	.97318	.77806	1.
.826	.59212	.98277	.82347	.97329	.78036	1.
.825	.59401	.98266	.82514	.97340	.78265	1.
.824	.59589	.98256	.82681	.97350	.78492	1.
.823	.59777	.98245	.82846	.97361	.78717	1.
.822	.59965	.98234	.83010	.97372	.78941	1.
.821	.60153	.98223	.83173	.97383	.79163	1.
.820	.60340	.98212	.83335	.97393	.79383	1.
.819	.60527	.98201	.83496	.97404	.79602	1.
.818	.60714	.98190	.83655	.97415	.79819	1.
.817	.60900	.98179	.83814	.97426	.80035	1.
.816	.61086	.98168	.83972	.97437	.80248	1.
.815	.61272	.98157	.84128	.97448	.80460	1.
.814	.61457	.98147	.84284	.97459	.80671	1.
.813	.61642	.98136	.84439	.97469	.80880	1.
.812	.61827	.98125	.84592	.97480	.81087	1.
.811	.62011	.98114	.84745	.97491	.81293	1.
.810	.62195	.98103	.84896	.97502	.81497	1.
.809	.62379	.98092	.85047	.97513	.81700	1.
.808	.62562	.98081	.85196	.97524	.81901	1.
.807	.62746	.98069	.85345	.97535	.82101	1.
.806	.62929	.98058	.85492	.97546	.82299	1.
.805	.63111	.98047	.85639	.97557	.82495	1.
.804	.63294	.98036	.85785	.97568	.82690	1.
.803	.63476	.98025	.85929	.97579	.82884	1.
.802	.63657	.98014	.86073	.97590	.83075	1.
.801	.63839	.98003	.86216	.97601	.83266	1.

TABLE II. Isentropic Exponent = 1.10

p/p_t	M	T/T_t	GAMMA	ISOTHERMAL T_t/T_t^*	RAYLEIGH T_t/T_t^*	SHOCK p_{t1}/p_{t2}
.800	.64020	.97992	.86357	.97612	.83455	
.799	.64201	.97981	.86498	.97623	.83642	1.
.798	.64382	.97970	.86638	.97635	.83828	1.
.797	.64562	.97958	.86777	.97646	.84013	1.
.796	.64743	.97947	.86915	.97657	.84196	1.
						1.
.795	.64923	.97936	.87052	.97668	.84378	
.794	.65102	.97925	.87188	.97679	.84558	1.
.793	.65282	.97914	.87324	.97690	.84737	1.
.792	.65461	.97902	.87458	.97702	.84914	1.
.791	.65640	.97891	.87591	.97713	.85090	1.
						1.
.790	.65819	.97880	.87724	.97724	.85265	
.789	.65997	.97869	.87856	.97735	.85438	1.
.788	.66176	.97857	.87986	.97747	.85610	1.
.787	.66354	.97846	.88116	.97758	.85780	1.
.786	.66531	.97835	.88245	.97769	.85949	1.
						1.
.785	.66709	.97823	.88373	.97780	.86117	
.784	.66886	.97812	.88500	.97792	.86283	1.
.783	.67063	.97801	.88627	.97803	.86448	1.
.782	.67240	.97789	.88752	.97815	.86612	1.
.781	.67417	.97778	.88877	.97826	.86774	1.
						1.
.780	.67594	.97767	.89000	.97837	.86935	
.779	.67770	.97755	.89123	.97849	.87095	1.
.778	.67946	.97744	.89245	.97860	.87253	1.
.777	.68122	.97732	.89367	.97872	.87410	1.
.776	.68297	.97721	.89487	.97883	.87566	1.
						1.
.775	.68473	.97709	.89606	.97895	.87720	
.774	.68648	.97698	.89725	.97906	.87874	1.
.773	.68823	.97686	.89843	.97918	.88025	1.
.772	.68998	.97675	.89960	.97929	.88176	1.
.771	.69172	.97663	.90076	.97941	.88325	1.
						1.
.770	.69347	.97652	.90191	.97952	.88474	
.769	.69521	.97640	.90306	.97964	.88620	1.
.768	.69695	.97629	.90420	.97975	.88766	1.
.767	.69869	.97617	.90533	.97987	.88910	1.
.766	.70043	.97606	.90645	.97999	.89054	1.
						1.
.765	.70216	.97594	.90756	.98010	.89196	
.764	.70390	.97583	.90866	.98022	.89336	1.
.763	.70563	.97571	.90976	.98033	.89476	1.
.762	.70736	.97559	.91085	.98045	.89614	1.
.761	.70909	.97548	.91193	.98057	.89751	1.
						1.
.760	.71081	.97536	.91300	.98069	.89887	
.759	.71254	.97524	.91407	.98080	.90022	1.
.758	.71426	.97513	.91513	.98092	.90156	1.
.757	.71598	.97501	.91618	.98104	.90288	1.
.756	.71770	.97489	.91722	.98116	.90419	1.
						1.
.755	.71942	.97477	.91825	.98127	.90549	
.754	.72113	.97466	.91928	.98139	.90678	1.
.753	.72285	.97454	.92030	.98151	.90806	1.
.752	.72456	.97442	.92131	.98163	.90933	1.
.751	.72627	.97430	.92232	.98175	.91058	1.
						1.

TABLE II. Isentropic Exponent = 1.10

p/p_t	M	T/T_t	GAMMA	ISOTHERMAL T_t/T_t^*	RAYLEIGH T_t/T_t^*	SHOCK p_{t1}/p_{t2}
.750	.72798	.97419	.92331	.98187	.91183	1.
.749	.72969	.97407	.92430	.98199	.91306	1.
.748	.73140	.97395	.92528	.98211	.91428	1.
.747	.73310	.97383	.92626	.98223	.91549	1.
.746	.73481	.97371	.92722	.98235	.91669	1.
.745	.73651	.97359	.92818	.98246	.91788	1.
.744	.73821	.97348	.92913	.98258	.91906	1.
.743	.73991	.97336	.93008	.98270	.92023	1.
.742	.74161	.97324	.93102	.98283	.92138	1.
.741	.74330	.97312	.93194	.98295	.92253	1.
.740	.74500	.97300	.93287	.98307	.92366	1.
.739	.74669	.97288	.93378	.98319	.92479	1.
.738	.74839	.97276	.93469	.98331	.92590	1.
.737	.75008	.97264	.93559	.98343	.92701	1.
.736	.75177	.97252	.93648	.98355	.92810	1.
.735	.75346	.97240	.93737	.98367	.92918	1.
.734	.75515	.97228	.93825	.98379	.93025	1.
.733	.75683	.97216	.93912	.98392	.93132	1.
.732	.75852	.97204	.93999	.98404	.93237	1.
.731	.76020	.97192	.94085	.98416	.93341	1.
.730	.76188	.97180	.94170	.98428	.93444	1.
.729	.76356	.97167	.94254	.98441	.93546	1.
.728	.76524	.97155	.94338	.98453	.93647	1.
.727	.76692	.97143	.94421	.98465	.93748	1.
.726	.76860	.97131	.94503	.98477	.93847	1.
.725	.77028	.97119	.94585	.98490	.93945	1.
.724	.77195	.97107	.94666	.98502	.94042	1.
.723	.77363	.97094	.94746	.98515	.94138	1.
.722	.77530	.97082	.94826	.98527	.94234	1.
.721	.77697	.97070	.94905	.98539	.94328	1.
.720	.77865	.97058	.94983	.98552	.94421	1.
.719	.78032	.97045	.95060	.98564	.94514	1.
.718	.78198	.97033	.95137	.98577	.94605	1.
.717	.78365	.97021	.95213	.98589	.94696	1.
.716	.78532	.97009	.95289	.98602	.94785	1.
.715	.78699	.96996	.95364	.98614	.94874	1.
.714	.78865	.96984	.95438	.98627	.94962	1.
.713	.79032	.96972	.95511	.98639	.95049	1.
.712	.79198	.96959	.95584	.98652	.95134	1.
.711	.79364	.96947	.95656	.98665	.95219	1.
.710	.79530	.96934	.95728	.98677	.95304	1.
.709	.79696	.96922	.95799	.98690	.95387	1.
.708	.79862	.96910	.95869	.98703	.95469	1.
.707	.80028	.96897	.95939	.98715	.95550	1.
.706	.80194	.96885	.96008	.98728	.95631	1.
.705	.80360	.96872	.96076	.98741	.95710	1.
.704	.80525	.96860	.96144	.98753	.95789	1.
.703	.80691	.96847	.96211	.98766	.95867	1.
.702	.80856	.96835	.96277	.98779	.95944	1.
.701	.81021	.96822	.96343	.98792	.96020	1.

TABLE II. Isentropic Exponent = 1.10

p/p_t	M	T/T_t	GAMMA	ISOTHERMAL T_t/T^*	RAYLEIGH T_t/T^*	SHOCK p_{t1}/p_{t2}
.700	.81187	.96810	.96408	.98805	.96095	1.
.699	.81352	.96797	.96472	.98817	.96170	1.
.698	.81517	.96784	.96536	.98830	.96243	1.
.697	.81682	.96772	.96599	.98843	.96316	1.
.696	.81847	.96759	.96662	.98856	.96388	1.
.695	.82012	.96746	.96724	.98869	.96459	1.
.694	.82177	.96734	.96785	.98882	.96529	1.
.693	.82342	.96721	.96845	.98895	.96598	1.
.692	.82506	.96708	.96905	.98908	.96667	1.
.691	.82671	.96696	.96965	.98921	.96735	1.
.690	.82835	.96683	.97024	.98934	.96802	1.
.689	.83000	.96670	.97082	.98947	.96868	1.
.688	.83164	.96657	.97139	.98960	.96933	1.
.687	.83329	.96645	.97196	.98973	.96997	1.
.686	.83493	.96632	.97253	.98986	.97061	1.
.685	.83657	.96619	.97308	.98999	.97124	1.
.684	.83821	.96606	.97363	.99012	.97186	1.
.683	.83985	.96593	.97418	.99026	.97247	1.
.682	.84149	.96581	.97472	.99039	.97308	1.
.681	.84313	.96568	.97525	.99052	.97367	1.
.680	.84477	.96555	.97578	.99065	.97426	1.
.679	.84641	.96542	.97630	.99078	.97485	1.
.678	.84805	.96529	.97681	.99092	.97542	1.
.677	.84969	.96516	.97732	.99105	.97599	1.
.676	.85132	.96503	.97782	.99118	.97655	1.
.675	.85296	.96490	.97832	.99132	.97710	1.
.674	.85460	.96477	.97881	.99145	.97764	1.
.673	.85623	.96464	.97929	.99158	.97818	1.
.672	.85787	.96451	.97977	.99172	.97871	1.
.671	.85950	.96438	.98024	.99185	.97923	1.
.670	.86114	.96425	.98071	.99199	.97974	1.
.669	.86277	.96412	.98117	.99212	.98025	1.
.668	.85440	.96399	.98163	.99226	.98075	1.
.667	.86604	.96385	.98208	.99239	.98124	1.
.666	.86767	.96372	.98252	.99253	.98172	1.
.665	.86930	.96359	.98296	.99266	.98220	1.
.664	.87093	.96346	.98339	.99280	.98267	1.
.663	.87257	.96333	.98381	.99294	.98314	1.
.662	.87420	.96320	.98423	.99307	.98359	1.
.661	.87583	.96306	.98465	.99321	.98404	1.
.660	.87746	.96293	.98506	.99334	.98448	1.
.659	.87909	.96280	.98546	.99348	.98492	1.
.658	.88072	.96266	.98585	.99362	.98535	1.
.657	.88235	.96253	.98625	.99376	.98577	1.
.656	.88398	.96240	.98663	.99389	.98618	1.
.655	.88561	.96226	.98701	.99403	.98659	1.
.654	.88723	.96213	.98738	.99417	.98699	1.
.653	.88886	.96200	.98775	.99431	.98739	1.
.652	.89049	.96186	.98812	.99445	.98777	1.
.651	.89212	.96173	.98847	.99459	.98815	1.

TABLE II. Isentropic Exponent = 1.10

p/p_t	M	T/T_t	GAMMA	ISOTHERMAL T_t/T_t^*	RAYLEIGH T_t/T_t^*	SHOCK p_{t1}/p_{t2}
.650	.89375	.96159	.98882	.99472	.98853	1.
.649	.89537	.96146	.98917	.99486	.98890	1.
.648	.89700	.96133	.98951	.99500	.98926	1.
.647	.89863	.96119	.98984	.99514	.98961	1.
.646	.90025	.96106	.99017	.99528	.98996	1.
.645	.90188	.96092	.99049	.99542	.99030	1.
.644	.90351	.96078	.99081	.99556	.99063	1.
.643	.90513	.96065	.99112	.99570	.99096	1.
.642	.90676	.96051	.99143	.99584	.99128	1.
.641	.90838	.96038	.99173	.99599	.99160	1.
.640	.91001	.96024	.99203	.99613	.99191	1.
.639	.91163	.96010	.99232	.99627	.99221	1.
.638	.91326	.95997	.99260	.99641	.99251	1.
.637	.91489	.95983	.99288	.99655	.99280	1.
.636	.91651	.95969	.99315	.99670	.99308	1.
.635	.91814	.95956	.99342	.99684	.99336	1.
.634	.91976	.95942	.99368	.99698	.99363	1.
.633	.92139	.95928	.99394	.99712	.99390	1.
.632	.92301	.95914	.99419	.99727	.99416	1.
.631	.92464	.95900	.99444	.99741	.99441	1.
.630	.92626	.95887	.99468	.99755	.99466	1.
.629	.92788	.95873	.99491	.99770	.99490	1.
.628	.92951	.95859	.99514	.99784	.99514	1.
.627	.93113	.95845	.99536	.99799	.99537	1.
.626	.93276	.95831	.99558	.99813	.99559	1.
.625	.93438	.95817	.99580	.99828	.99581	1.
.624	.93601	.95803	.99600	.99842	.99602	1.
.623	.93763	.95789	.99621	.99857	.99623	1.
.622	.93926	.95775	.99640	.99871	.99643	1.
.621	.94088	.95761	.99660	.99886	.99662	1.
.620	.94251	.95747	.99678	.99901	.99681	1.
.619	.94413	.95733	.99696	.99915	.99699	1.
.618	.94576	.95719	.99714	.99930	.99717	1.
.617	.94738	.95705	.99731	.99945	.99734	1.
.616	.94901	.95691	.99748	.99959	.99751	1.
.615	.95063	.95677	.99763	.99974	.99767	1.
.614	.95226	.95663	.99779	.99989	.99782	1.
.613	.95388	.95649	.99794	1.00004	.99797	1.
.612	.95551	.95634	.99808	1.00019	.99812	1.
.611	.95713	.95620	.99822	1.00034	.99825	1.
.610	.95876	.95606	.99835	1.00048	.99839	1.
.609	.96039	.95592	.99848	1.00063	.99851	1.
.608	.96201	.95577	.99861	1.00078	.99864	1.
.607	.96364	.95563	.99872	1.00093	.99875	1.
.606	.96526	.95549	.99884	1.00108	.99886	1.
.605	.96689	.95534	.99894	1.00123	.99897	1.
.604	.96852	.95520	.99904	1.00138	.99907	1.
.603	.97014	.95506	.99914	1.00153	.99916	1.
.602	.97177	.95491	.99923	1.00169	.99925	1.
.601	.97340	.95477	.99932	1.00184	.99934	1.

TABLE II. Isentropic Exponent = 1.10

p/p_t	M	T/T_t	GAMMA	ISOTHERMAL T_t/T^*	RAYLEIGH T_t/T^*	SHOCK p_{t1}/p_{t2}
.600	.97503	.95462	.99940	1.00199	.99942	1.
.599	.97665	.95448	.99948	1.00214	.99949	1.
.598	.97828	.95433	.99955	1.00229	.99956	1.
.597	.97991	.95419	.99961	1.00245	.99963	1.
.596	.98154	.95404	.99967	1.00260	.99968	1.
.595	.98317	.95390	.99973	1.00275	.99974	1.
.594	.98480	.95375	.99978	1.00290	.99979	1.
.593	.98643	.95361	.99982	1.00306	.99983	1.
.592	.98806	.95346	.99986	1.00321	.99987	1.
.591	.98969	.95331	.99990	1.00337	.99990	1.
.590	.99132	.95317	.99993	1.00352	.99993	1.
.589	.99295	.95302	.99995	1.00368	.99995	1.
.588	.99458	.95287	.99997	1.00383	.99997	1.
.587	.99621	.95272	.99999	1.00399	.99999	1.
.586	.99784	.95258	.00000	1.00414	.00000	1.
.585	.99948	.95243	.00000	1.00430	.00000	1.
.584	1.00111	.95228	.00000	1.00445	.00000	1.00000
.583	1.00274	.95213	.99999	1.00461	.99999	1.00000
.582	1.00438	.95198	.99998	1.00477	.99998	1.00000
.581	1.00601	.95183	.99997	1.00492	.99997	1.00000
.580	1.00764	.95169	.99994	1.00508	.99995	1.00000
.579	1.00928	.95154	.99992	1.00524	.99992	1.00000
.578	1.01091	.95139	.99989	1.00540	.99989	1.00000
.577	1.01255	.95124	.99985	1.00556	.99986	1.00000
.576	1.01419	.95109	.99981	1.00571	.99982	1.00000
.575	1.01582	.95094	.99976	1.00587	.99978	1.00000
.574	1.01746	.95079	.99971	1.00603	.99973	1.00001
.573	1.01910	.95064	.99966	1.00619	.99968	1.00001
.572	1.02073	.95048	.99959	1.00635	.99962	1.00001
.571	1.02237	.95033	.99953	1.00651	.99956	1.00001
.570	1.02401	.95018	.99946	1.00667	.99949	1.00002
.569	1.02565	.95003	.99938	1.00683	.99942	1.00002
.568	1.02729	.94988	.99930	1.00699	.99934	1.00003
.567	1.02893	.94973	.99921	1.00716	.99926	1.00003
.566	1.03057	.94957	.99912	1.00732	.99918	1.00004
.565	1.03221	.94942	.99902	1.00748	.99909	1.00004
.564	1.03386	.94927	.99892	1.00764	.99900	1.00005
.563	1.03550	.94912	.99882	1.00780	.99890	1.00006
.562	1.03714	.94896	.99871	1.00797	.99880	1.00006
.561	1.03879	.94881	.99859	1.00813	.99869	1.00007
.560	1.04043	.94865	.99847	1.00829	.99858	1.00008
.559	1.04208	.94850	.99834	1.00846	.99847	1.00009
.558	1.04372	.94835	.99821	1.00862	.99835	1.00010
.557	1.04537	.94819	.99808	1.00879	.99822	1.00012
.556	1.04701	.94804	.99794	1.00895	.99810	1.00013
.555	1.04866	.94788	.99779	1.00912	.99796	1.00014
.554	1.05031	.94773	.99764	1.00928	.99783	1.00016
.553	1.05196	.94757	.99748	1.00945	.99769	1.00017
.552	1.05361	.94741	.99732	1.00961	.99754	1.00019
.551	1.05526	.94726	.99716	1.00978	.99739	1.00021

TABLE II.   Isentropic Exponent = 1.10

p/p_t	M	T/T_t	GAMMA	ISOTHERMAL T_t/T_t^*	RAYLEIGH T_t/T_t^*	SHOCK p_{t1}/p_{t2}
.550	1.05691	.94710	.99699	1.00995	.99724	1.00022
.549	1.05856	.94694	.99681	1.01011	.99708	1.00024
.548	1.06021	.94679	.99663	1.01028	.99692	1.00026
.547	1.06187	.94663	.99645	1.01045	.99676	1.00029
.546	1.06352	.94647	.99626	1.01062	.99659	1.00031
.545	1.06517	.94632	.99606	1.01079	.99642	1.00033
.544	1.06683	.94616	.99586	1.01095	.99624	1.00036
.543	1.06849	.94600	.99566	1.01112	.99606	1.00038
.542	1.07014	.94584	.99545	1.01129	.99587	1.00041
.541	1.07180	.94568	.99524	1.01146	.99568	1.00044
.540	1.07346	.94552	.99502	1.01163	.99549	1.00047
.539	1.07512	.94536	.99479	1.01180	.99529	1.00050
.538	1.07678	.94520	.99456	1.01197	.99509	1.00053
.537	1.07844	.94504	.99433	1.01214	.99488	1.00057
.536	1.08010	.94488	.99409	1.01232	.99468	1.00060
.535	1.08176	.94472	.99385	1.01249	.99446	1.00064
.534	1.08342	.94456	.99360	1.01266	.99425	1.00068
.533	1.08509	.94440	.99335	1.01283	.99402	1.00072
.532	1.08675	.94424	.99309	1.01301	.99380	1.00076
.531	1.08842	.94408	.99283	1.01318	.99357	1.00080
.530	1.09009	.94392	.99256	1.01335	.99334	1.00084
.529	1.09175	.94376	.99229	1.01353	.99310	1.00089
.528	1.09342	.94359	.99201	1.01370	.99286	1.00094
.527	1.09509	.94343	.99173	1.01388	.99262	1.00099
.526	1.09676	.94327	.99144	1.01405	.99237	1.00104
.525	1.09843	.94310	.99115	1.01423	.99212	1.00109
.524	1.10010	.94294	.99086	1.01440	.99187	1.00114
.523	1.10178	.94278	.99055	1.01458	.99161	1.00120
.522	1.10345	.94261	.99025	1.01475	.99135	1.00125
.521	1.10513	.94245	.98994	1.01493	.99108	1.00131
.520	1.10680	.94228	.98962	1.01511	.99081	1.00137
.519	1.10848	.94212	.98930	1.01529	.99054	1.00144
.518	1.11016	.94195	.98898	1.01546	.99026	1.00150
.517	1.11183	.94179	.98865	1.01564	.98998	1.00157
.516	1.11351	.94162	.98831	1.01582	.98970	1.00163
.515	1.11520	.94146	.98797	1.01600	.98941	1.00170
.514	1.11688	.94129	.98763	1.01618	.98912	1.00177
.513	1.11856	.94112	.98728	1.01636	.98883	1.00185
.512	1.12024	.94096	.98693	1.01654	.98853	1.00192
.511	1.12193	.94079	.98657	1.01672	.98823	1.00200
.510	1.12361	.94062	.98621	1.01690	.98792	1.00208
.509	1.12530	.94045	.98584	1.01708	.98762	1.00216
.508	1.12699	.94029	.98547	1.01727	.98730	1.00224
.507	1.12868	.94012	.98509	1.01745	.98699	1.00233
.506	1.13037	.93995	.98471	1.01763	.98667	1.00242
.505	1.13206	.93978	.98432	1.01781	.98635	1.00250
.504	1.13375	.93961	.98393	1.01800	.98602	1.00260
.503	1.13545	.93944	.98353	1.01818	.98569	1.00269
.502	1.13714	.93927	.98313	1.01837	.98536	1.00279
.501	1.13884	.93910	.98272	1.01855	.98503	1.00288

TABLE II. Isentropic Exponent = 1.10

p/p_t	M	T/T_t	GAMMA	ISOTHERMAL T_t/T*	RAYLEIGH T_t/T*	SHOCK p_t1/p_t2
.500	1.14054	.93893	.98231	1.01873	.98469	1.00298
.499	1.14223	.93876	.98190	1.01892	.98434	1.00309
.498	1.14393	.93859	.98148	1.01911	.98400	1.00319
.497	1.14563	.93842	.98105	1.01929	.98365	1.00330
.496	1.14734	.93825	.98062	1.01948	.98330	1.00341
.495	1.14904	.93807	.98019	1.01967	.98294	1.00352
.494	1.15074	.93790	.97975	1.01985	.98258	1.00363
.493	1.15245	.93773	.97931	1.02004	.98222	1.00375
.492	1.15416	.93756	.97886	1.02023	.98186	1.00386
.491	1.15587	.93738	.97840	1.02042	.98149	1.00398
.490	1.15757	.93721	.97795	1.02061	.98112	1.00411
.489	1.15929	.93703	.97748	1.02080	.98074	1.00423
.488	1.16100	.93686	.97702	1.02099	.98036	1.00436
.487	1.16271	.93668	.97654	1.02118	.97998	1.00449
.486	1.16443	.93651	.97607	1.02137	.97960	1.00462
.485	1.16614	.93633	.97559	1.02156	.97921	1.00476
.484	1.16786	.93616	.97510	1.02175	.97882	1.00490
.483	1.16958	.93598	.97461	1.02194	.97842	1.00504
.482	1.17130	.93581	.97411	1.02214	.97803	1.00518
.481	1.17302	.93563	.97361	1.02233	.97763	1.00532
.480	1.17474	.93545	.97311	1.02252	.97722	1.00547
.479	1.17647	.93528	.97260	1.02272	.97682	1.00562
.478	1.17819	.93510	.97208	1.02291	.97641	1.00578
.477	1.17992	.93492	.97156	1.02311	.97600	1.00593
.476	1.18165	.93474	.97104	1.02330	.97558	1.00609
.475	1.18338	.93456	.97051	1.02350	.97516	1.00625
.474	1.18511	.93438	.96998	1.02369	.97474	1.00642
.473	1.18684	.93420	.96944	1.02389	.97431	1.00658
.472	1.18857	.93402	.96890	1.02409	.97389	1.00675
.471	1.19031	.93384	.96835	1.02428	.97346	1.00693
.470	1.19205	.93366	.96780	1.02448	.97302	1.00710
.469	1.19379	.93348	.96724	1.02468	.97259	1.00728
.468	1.19553	.93330	.96668	1.02488	.97215	1.00746
.467	1.19727	.93312	.96611	1.02508	.97170	1.00764
.466	1.19901	.93294	.96554	1.02528	.97126	1.00783
.465	1.20076	.93276	.96496	1.02548	.97081	1.00802
.464	1.20250	.93257	.96438	1.02568	.97036	1.00821
.463	1.20425	.93239	.96380	1.02588	.96991	1.00841
.462	1.20600	.93221	.96321	1.02608	.96945	1.00861
.461	1.20775	.93202	.96261	1.02628	.96899	1.00881
.460	1.20950	.93184	.96201	1.02649	.96853	1.00901
.459	1.21126	.93166	.96141	1.02669	.96806	1.00922
.458	1.21301	.93147	.96080	1.02689	.96759	1.00943
.457	1.21477	.93129	.96019	1.02710	.96712	1.00964
.456	1.21653	.93110	.95957	1.02730	.96665	1.00986
.455	1.21829	.93092	.95895	1.02751	.96617	1.01008
.454	1.22005	.93073	.95832	1.02771	.96569	1.01030
.453	1.22182	.93054	.95769	1.02792	.96521	1.01053
.452	1.22358	.93036	.95705	1.02812	.96472	1.01076
.451	1.22535	.93017	.95641	1.02833	.96423	1.01099

TABLE II. Isentropic Exponent = 1.10

p/p_t	M	T/T_t	GAMMA	ISOTHERMAL T_t/T_t^*	RAYLEIGH T_t/T_t^*	SHOCK p_{t1}/p_{t2}
.450	1.22712	.92998	.95576	1.02854	.96374	1.01123
.449	1.22889	.92979	.95511	1.02875	.96325	1.01147
.448	1.23067	.92960	.95446	1.02896	.96275	1.01171
.447	1.23244	.92942	.95380	1.02917	.96225	1.01196
.446	1.23422	.92923	.95313	1.02937	.96175	1.01221
.445	1.23599	.92904	.95246	1.02958	.96124	1.01246
.444	1.23777	.92885	.95179	1.02980	.96074	1.01271
.443	1.23956	.92866	.95111	1.03001	.96023	1.01297
.442	1.24134	.92847	.95042	1.03022	.95971	1.01324
.441	1.24313	.92827	.94974	1.03043	.95920	1.01350
.440	1.24491	.92808	.94904	1.03064	.95868	1.01377
.439	1.24670	.92789	.94834	1.03086	.95816	1.01405
.438	1.24849	.92770	.94764	1.03107	.95764	1.01432
.437	1.25029	.92751	.94693	1.03128	.95711	1.01460
.436	1.25208	.92731	.94622	1.03150	.95658	1.01489
.435	1.25388	.92712	.94551	1.03171	.95605	1.01518
.434	1.25568	.92692	.94478	1.03193	.95552	1.01547
.433	1.25748	.92673	.94406	1.03215	.95498	1.01576
.432	1.25928	.92654	.94333	1.03236	.95444	1.01606
.431	1.26108	.92634	.94259	1.03258	.95390	1.01636
.430	1.26289	.92614	.94185	1.03280	.95335	1.01667
.429	1.26470	.92595	.94111	1.03302	.95281	1.01698
.428	1.26651	.92575	.94036	1.03324	.95226	1.01729
.427	1.26832	.92556	.93960	1.03346	.95170	1.01761
.426	1.27014	.92536	.93884	1.03368	.95115	1.01793
.425	1.27195	.92516	.93808	1.03390	.95059	1.01825
.424	1.27377	.92496	.93731	1.03412	.95003	1.01858
.423	1.27559	.92476	.93654	1.03434	.94947	1.01892
.422	1.27742	.92457	.93576	1.03456	.94890	1.01925
.421	1.27924	.92437	.93498	1.03479	.94834	1.01959
.420	1.28107	.92417	.93419	1.03501	.94777	1.01994
.419	1.28290	.92397	.93340	1.03524	.94719	1.02029
.418	1.28473	.92376	.93260	1.03546	.94662	1.02064
.417	1.28656	.92356	.93180	1.03569	.94604	1.02100
.416	1.28840	.92336	.93099	1.03591	.94546	1.02136
.415	1.29024	.92316	.93018	1.03614	.94488	1.02172
.414	1.29208	.92296	.92936	1.03637	.94429	1.02209
.413	1.29392	.92275	.92854	1.03659	.94370	1.02246
.412	1.29576	.92255	.92772	1.03682	.94311	1.02284
.411	1.29761	.92235	.92689	1.03705	.94252	1.02322
.410	1.29946	.92214	.92605	1.03728	.94193	1.02361
.409	1.30131	.92194	.92521	1.03751	.94133	1.02400
.408	1.30317	.92173	.92437	1.03774	.94073	1.02439
.407	1.30502	.92153	.92352	1.03797	.94013	1.02479
.406	1.30688	.92132	.92266	1.03821	.93952	1.02519
.405	1.30874	.92112	.92181	1.03844	.93892	1.02560
.404	1.31060	.92091	.92094	1.03867	.93831	1.02601
.403	1.31247	.92070	.92007	1.03891	.93770	1.02643
.402	1.31434	.92049	.91920	1.03914	.93708	1.02685
.401	1.31621	.92028	.91832	1.03938	.93646	1.02728

TABLE II. Isentropic Exponent $= 1.10$

p/p_t	M	T/T_t	GAMMA	ISOTHERMAL T_t/T_t^*	RAYLEIGH T_t/T_t^*	SHOCK p_{t1}/p_{t2}
.400	1.31808	.92008	.91744	1.03961	.93585	1.02770
.399	1.31996	.91987	.91655	1.03985	.93522	1.02814
.398	1.32183	.91966	.91566	1.04009	.93460	1.02858
.397	1.32371	.91945	.91476	1.04032	.93397	1.02902
.396	1.32560	.91924	.91386	1.04056	.93335	1.02947
.395	1.32748	.91902	.91296	1.04080	.93272	1.02992
.394	1.32937	.91881	.91205	1.04104	.93208	1.03038
.393	1.33126	.91860	.91113	1.04128	.93145	1.03084
.392	1.33315	.91839	.91021	1.04152	.93081	1.03131
.391	1.33505	.91817	.90928	1.04176	.93017	1.03178
.390	1.33695	.91796	.90835	1.04201	.92953	1.03226
.389	1.33885	.91775	.90742	1.04225	.92888	1.03274
.388	1.34075	.91753	.90648	1.04249	.92823	1.03323
.387	1.34266	.91732	.90553	1.04274	.92758	1.03372
.386	1.34457	.91710	.90458	1.04298	.92693	1.03421
.385	1.34648	.91688	.90363	1.04323	.92628	1.03472
.384	1.34839	.91667	.90267	1.04348	.92562	1.03522
.383	1.35031	.91645	.90171	1.04372	.92496	1.03573
.382	1.35223	.91623	.90074	1.04397	.92430	1.03625
.381	1.35415	.91601	.89976	1.04422	.92364	1.03677
.380	1.35607	.91580	.89878	1.04447	.92298	1.03730
.379	1.35800	.91558	.89780	1.04472	.92231	1.03783
.378	1.35993	.91536	.89681	1.04497	.92164	1.03837
.377	1.36187	.91514	.89582	1.04522	.92097	1.03891
.376	1.36380	.91491	.89482	1.04548	.92029	1.03946
.375	1.36574	.91469	.89382	1.04573	.91961	1.04001
.374	1.36768	.91447	.89281	1.04598	.91894	1.04057
.373	1.36963	.91425	.89180	1.04624	.91825	1.04114
.372	1.37158	.91403	.89078	1.04649	.91757	1.04171
.371	1.37353	.91380	.88976	1.04675	.91689	1.04228
.370	1.37548	.91358	.88873	1.04701	.91620	1.04286
.369	1.37744	.91335	.88770	1.04726	.91551	1.04345
.368	1.37940	.91313	.88666	1.04752	.91482	1.04404
.367	1.38136	.91290	.88562	1.04778	.91412	1.04464
.366	1.38333	.91268	.88458	1.04804	.91342	1.04524
.365	1.38529	.91245	.88352	1.04830	.91273	1.04585
.364	1.38727	.91222	.88247	1.04856	.91202	1.04646
.363	1.38924	.91199	.88141	1.04883	.91132	1.04708
.362	1.39122	.91176	.88034	1.04909	.91062	1.04771
.361	1.39320	.91154	.87927	1.04935	.90991	1.04834
.360	1.39519	.91131	.87819	1.04962	.90920	1.04898
.359	1.39717	.91107	.87711	1.04988	.90849	1.04963
.358	1.39916	.91084	.87603	1.05015	.90777	1.05028
.357	1.40116	.91061	.87494	1.05042	.90706	1.05093
.356	1.40316	.91038	.87384	1.05068	.90634	1.05160
.355	1.40516	.91015	.87274	1.05095	.90562	1.05226
.354	1.40716	.90991	.87163	1.05122	.90489	1.05294
.353	1.40917	.90968	.87052	1.05149	.90417	1.05362
.352	1.41118	.90945	.86941	1.05176	.90344	1.05431
.351	1.41319	.90921	.86829	1.05204	.90271	1.05500

TABLE II.   Isentropic Exponent = 1.10

p/p_t	M	T/T_t	GAMMA	ISOTHERMAL T_t/T_t^*	RAYLEIGH T_t/T_t^*	SHOCK p_{t1}/p_{t2}
.350	1.41521	.90897	.86716	1.05231	.90198	1.05570
.349	1.41723	.90874	.86603	1.05258	.90125	1.05641
.348	1.41926	.90850	.86490	1.05286	.90051	1.05712
.347	1.42128	.90826	.86376	1.05313	.89978	1.05784
.346	1.42331	.90803	.86261	1.05341	.89904	1.05857
.345	1.42535	.90779	.86146	1.05369	.89829	1.05930
.344	1.42739	.90755	.86031	1.05396	.89755	1.06004
.343	1.42943	.90731	.85915	1.05424	.89680	1.06079
.342	1.43147	.90707	.85798	1.05452	.89605	1.06154
.341	1.43352	.90682	.85681	1.05480	.89530	1.06230
.340	1.43558	.90658	.85564	1.05509	.89455	1.06307
.339	1.43763	.90634	.85446	1.05537	.89380	1.06384
.338	1.43969	.90610	.85327	1.05565	.89304	1.06462
.337	1.44176	.90585	.85208	1.05594	.89228	1.06541
.336	1.44382	.90561	.85089	1.05622	.89152	1.06620
.335	1.44589	.90536	.84969	1.05651	.89076	1.06700
.334	1.44797	.90512	.84848	1.05679	.88999	1.06781
.333	1.45005	.90487	.84727	1.05708	.88922	1.06863
.332	1.45213	.90462	.84605	1.05737	.88845	1.06945
.331	1.45422	.90437	.84483	1.05766	.88768	1.07029
.330	1.45631	.90413	.84361	1.05795	.88691	1.07112
.329	1.45840	.90388	.84238	1.05824	.88613	1.07197
.328	1.46050	.90363	.84114	1.05854	.88535	1.07282
.327	1.46260	.90337	.83990	1.05883	.88457	1.07368
.326	1.46471	.90312	.83865	1.05913	.88379	1.07455
.325	1.46682	.90287	.83740	1.05942	.88301	1.07543
.324	1.46893	.90262	.83615	1.05972	.88222	1.07631
.323	1.47105	.90236	.83489	1.06002	.88143	1.07721
.322	1.47317	.90211	.83362	1.06032	.88064	1.07811
.321	1.47530	.90186	.83235	1.06062	.87985	1.07901
.320	1.47743	.90160	.83107	1.06092	.87905	1.07993
.319	1.47957	.90134	.82979	1.06122	.87826	1.08085
.318	1.48170	.90109	.82850	1.06152	.87746	1.08179
.317	1.48385	.90083	.82721	1.06183	.87666	1.08273
.316	1.48600	.90057	.82591	1.06213	.87585	1.08367
.315	1.48815	.90031	.82461	1.06244	.87505	1.08463
.314	1.49030	.90005	.82330	1.06274	.87424	1.08560
.313	1.49246	.89979	.82199	1.06305	.87343	1.08657
.312	1.49463	.89953	.82067	1.06336	.87262	1.08755
.311	1.49680	.89926	.81935	1.06367	.87181	1.08855
.310	1.49897	.89900	.81802	1.06398	.87099	1.08955
.309	1.50115	.89874	.81668	1.06430	.87018	1.09055
.308	1.50333	.89847	.81534	1.06461	.86936	1.09157
.307	1.50552	.89821	.81400	1.06492	.86853	1.09260
.306	1.50771	.89794	.81265	1.06524	.86771	1.09363
.305	1.50991	.89767	.81130	1.06556	.86688	1.09468
.304	1.51211	.89741	.80994	1.06588	.86606	1.09573
.303	1.51432	.89714	.80857	1.06619	.86523	1.09679
.302	1.51653	.89687	.80720	1.06651	.86439	1.09787
.301	1.51874	.89660	.80582	1.06684	.86356	1.09895

TABLE II. Isentropic Exponent = 1.10

p/p_t	M	T/T_t	GAMMA	ISOTHERMAL T_t/T_t^*	RAYLEIGH T_t/T_t^*	SHOCK p_{t1}/p_{t2}
.300	1.52096	.89633	.80444	1.06716	.86272	1.10004
.299	1.52319	.89605	.80306	1.06748	.86188	1.10114
.298	1.52542	.89578	.80166	1.06781	.86104	1.10225
.297	1.52765	.89551	.80027	1.06813	.86020	1.10337
.296	1.52989	.89523	.79886	1.06846	.85936	1.10450
.295	1.53214	.89496	.79746	1.06879	.85851	1.10564
.294	1.53439	.89468	.79604	1.06912	.85766	1.10679
.293	1.53664	.89440	.79462	1.06945	.85681	1.10795
.292	1.53890	.89413	.79320	1.06978	.85596	1.10912
.291	1.54117	.89385	.79177	1.07012	.85510	1.11030
.290	1.54344	.89357	.79034	1.07045	.85424	1.11149
.289	1.54572	.89329	.78890	1.07079	.85338	1.11269
.288	1.54800	.89300	.78745	1.07113	.85252	1.11390
.287	1.55028	.89272	.78600	1.07147	.85166	1.11513
.286	1.55257	.89244	.78454	1.07181	.85079	1.11636
.285	1.55487	.89216	.78308	1.07215	.84993	1.11760
.284	1.55717	.89187	.78161	1.07249	.84906	1.11886
.283	1.55948	.89158	.78014	1.07283	.84818	1.12012
.282	1.56179	.89130	.77866	1.07318	.84731	1.12140
.281	1.56411	.89101	.77718	1.07353	.84643	1.12269
.280	1.56644	.89072	.77569	1.07387	.84556	1.12399
.279	1.56877	.89043	.77419	1.07422	.84468	1.12530
.278	1.57110	.89014	.77269	1.07457	.84379	1.12662
.277	1.57344	.88985	.77119	1.07493	.84291	1.12796
.276	1.57579	.88956	.76968	1.07528	.84202	1.12930
.275	1.57814	.88926	.76816	1.07563	.84113	1.13066
.274	1.58050	.88897	.76664	1.07599	.84024	1.13203
.273	1.58287	.88867	.76511	1.07635	.83935	1.13341
.272	1.58524	.88838	.76358	1.07671	.83845	1.13480
.271	1.58761	.88808	.76204	1.07707	.83755	1.13621
.270	1.58999	.88778	.76049	1.07743	.83666	1.13763
.269	1.59238	.88748	.75894	1.07779	.83575	1.13906
.268	1.59478	.88718	.75738	1.07816	.83485	1.14050
.267	1.59718	.88688	.75582	1.07852	.83394	1.14196
.266	1.59958	.88658	.75425	1.07889	.83304	1.14343
.265	1.60200	.88627	.75268	1.07926	.83212	1.14491
.264	1.60442	.88597	.75110	1.07963	.83121	1.14641
.263	1.60684	.88566	.74952	1.08001	.83030	1.14792
.262	1.60927	.88536	.74793	1.08038	.82938	1.14944
.261	1.61171	.88505	.74633	1.08076	.82846	1.15098
.260	1.61416	.88474	.74473	1.08113	.82754	1.15253
.259	1.61661	.88443	.74312	1.08151	.82662	1.15409
.258	1.61907	.88412	.74151	1.08189	.82569	1.15567
.257	1.62153	.88381	.73989	1.08227	.82476	1.15726
.256	1.62400	.88349	.73827	1.08266	.82383	1.15887
.255	1.62648	.88318	.73664	1.08304	.82290	1.16049
.254	1.62897	.88286	.73500	1.08343	.82197	1.16213
.253	1.63146	.88255	.73336	1.08382	.82103	1.16378
.252	1.63396	.88223	.73171	1.08421	.82009	1.16545
.251	1.63647	.88191	.73006	1.08460	.81915	1.16713

TABLE II. Isentropic Exponent $=1.10$

p/p_t	M	T/T_t	GAMMA	ISOTHERMAL T_t/T_t^*	RAYLEIGH T_t/T_t^*	SHOCK p_{t1}/p_{t2}
.250	1.63898	.88159	.72840	1.08499	.81821	1.16882
.249	1.64150	.88127	.72673	1.08539	.81726	1.17053
.248	1.64403	.88095	.72506	1.08579	.81631	1.17226
.247	1.64656	.88062	.72338	1.08619	.81536	1.17400
.246	1.64910	.88030	.72170	1.08659	.81441	1.17576
.245	1.65165	.87997	.72001	1.08699	.81346	1.17754
.244	1.65421	.87965	.71831	1.08739	.81250	1.17933
.243	1.65677	.87932	.71661	1.08780	.81154	1.18113
.242	1.65934	.87899	.71490	1.08821	.81058	1.18296
.241	1.66192	.87866	.71319	1.08862	.80961	1.18480
.240	1.66451	.87833	.71147	1.08903	.80865	1.18666
.239	1.66711	.87799	.70975	1.08944	.80768	1.18853
.238	1.66971	.87766	.70802	1.08986	.80671	1.19042
.237	1.67232	.87732	.70628	1.09027	.80574	1.19233
.236	1.67494	.87698	.70453	1.09069	.80476	1.19426
.235	1.67756	.87665	.70278	1.09111	.80378	1.19621
.234	1.68020	.87631	.70103	1.09196	.80280	1.19817
.233	1.68284	.87597	.69927	1.09196	.80182	1.20015
.232	1.68549	.87562	.69750	1.09239	.80084	1.20215
.231	1.68815	.87528	.69573	1.09282	.79985	1.20417
.230	1.69082	.87493	.69394	1.09325	.79886	1.20621
.229	1.69350	.87459	.69216	1.09368	.79787	1.20827
.228	1.69618	.87424	.69037	1.09412	.79687	1.21035
.227	1.69887	.87389	.68857	1.09456	.79588	1.21245
.226	1.70157	.87354	.68676	1.09500	.79488	1.21456
.225	1.70429	.87319	.68495	1.09544	.79388	1.21670
.224	1.70701	.87283	.68313	1.09588	.79287	1.21886
.223	1.70973	.87248	.68131	1.09633	.79187	1.22103
.222	1.71247	.87212	.67948	1.09677	.79086	1.22323
.221	1.71522	.87176	.67764	1.09722	.78985	1.22545
.220	1.71797	.87141	.67580	1.09768	.78883	1.22770
.219	1.72074	.87104	.67395	1.09813	.78782	1.22996
.218	1.72351	.87068	.67209	1.09859	.78680	1.23224
.217	1.72630	.87032	.67023	1.09905	.78578	1.23455
.216	1.72909	.86995	.66836	1.09951	.78475	1.23688
.215	1.73189	.86959	.66649	1.09997	.78373	1.23923
.214	1.73470	.86922	.66460	1.10044	.78270	1.24161
.213	1.73753	.86885	.66271	1.10091	.78166	1.24401
.212	1.74036	.86848	.66082	1.10138	.78063	1.24643
.211	1.74320	.86810	.65892	1.10185	.77959	1.24888
.210	1.74605	.86773	.65701	1.10233	.77856	1.25135
.209	1.74892	.86735	.65510	1.10281	.77751	1.25385
.208	1.75179	.86697	.65317	1.10329	.77647	1.25637
.207	1.75467	.86659	.65125	1.10377	.77542	1.25891
.206	1.75756	.86621	.64931	1.10426	.77437	1.26148
.205	1.76047	.86583	.64737	1.10475	.77332	1.26408
.204	1.76338	.86544	.64542	1.10524	.77226	1.26670
.203	1.76631	.86506	.64347	1.10573	.77121	1.26935
.202	1.76924	.86467	.64151	1.10623	.77015	1.27203
.201	1.77219	.86428	.63954	1.10673	.76908	1.27473

TABLE II. Isentropic Exponent = 1.10

p/p_t	M	T/T_t	GAMMA	ISOTHERMAL T_t/T_t*	RAYLEIGH T_t/T_t*	SHOCK p_t1/p_t2
.200	1.77515	.86389	.63756	1.10723	.76802	1.27747
.199	1.77812	.86349	.63558	1.10773	.76695	1.28023
.198	1.78110	.86310	.63359	1.10824	.76588	1.28301
.197	1.78409	.86270	.63160	1.10875	.76480	1.28583
.196	1.78710	.86230	.62960	1.10926	.76373	1.28868
.195	1.79011	.86190	.62759	1.10978	.76265	1.29155
.194	1.79314	.86150	.62557	1.11030	.76157	1.29446
.193	1.79618	.86109	.62355	1.11082	.76048	1.29739
.192	1.79923	.86069	.62152	1.11135	.75939	1.30036
.191	1.80230	.86028	.61948	1.11187	.75830	1.30336
.190	1.80537	.85987	.61744	1.11240	.75721	1.30639
.189	1.80846	.85946	.61538	1.11294	.75611	1.30945
.188	1.81156	.85904	.61333	1.11348	.75501	1.31255
.187	1.81467	.85863	.61126	1.11402	.75391	1.31567
.186	1.81780	.85821	.60919	1.11456	.75280	1.31884
.185	1.82094	.85779	.60711	1.11510	.75169	1.32203
.184	1.82409	.85736	.60502	1.11565	.75058	1.32526
.183	1.82726	.85694	.60293	1.11621	.74947	1.32853
.182	1.83044	.85651	.60082	1.11676	.74835	1.33183
.181	1.83363	.85608	.59871	1.11732	.74723	1.33517
.180	1.83684	.85565	.59660	1.11789	.74610	1.33854
.179	1.84006	.85522	.59447	1.11845	.74498	1.34195
.178	1.84329	.85478	.59234	1.11902	.74385	1.34540
.177	1.84654	.85435	.59020	1.11959	.74271	1.34889
.176	1.84980	.85391	.58806	1.12017	.74158	1.35241
.175	1.85308	.85346	.58591	1.12075	.74044	1.35598
.174	1.85637	.85302	.58374	1.12134	.73929	1.35959
.173	1.85968	.85257	.58158	1.12192	.73815	1.36323
.172	1.86300	.85212	.57940	1.12252	.73700	1.36692
.171	1.86634	.85167	.57722	1.12311	.73585	1.37065
.170	1.86969	.85122	.57502	1.12371	.73469	1.37443
.169	1.87306	.85076	.57282	1.12431	.73353	1.37825
.168	1.87644	.85030	.57062	1.12492	.73237	1.38211
.167	1.87984	.84984	.56840	1.12553	.73120	1.38601
.166	1.88326	.84938	.56618	1.12614	.73003	1.38997
.165	1.88669	.84891	.56395	1.12676	.72886	1.39397
.164	1.89014	.84844	.56171	1.12739	.72768	1.39801
.163	1.89360	.84797	.55947	1.12801	.72650	1.40211
.162	1.89708	.84750	.55721	1.12864	.72532	1.40625
.161	1.90058	.84702	.55495	1.12928	.72413	1.41044
.160	1.90410	.84654	.55268	1.12992	.72294	1.41469
.159	1.90763	.84606	.55040	1.13056	.72174	1.41898
.158	1.91118	.84557	.54812	1.13121	.72055	1.42333
.157	1.91475	.84508	.54582	1.13187	.71934	1.42773
.156	1.91834	.84459	.54352	1.13252	.71814	1.43218
.155	1.92195	.84410	.54121	1.13319	.71693	1.43669
.154	1.92557	.84360	.53889	1.13385	.71572	1.44126
.153	1.92922	.84310	.53656	1.13452	.71450	1.44588
.152	1.93288	.84260	.53423	1.13520	.71328	1.45056
.151	1.93656	.84210	.53188	1.13588	.71205	1.45530

TABLE II.   Isentropic Exponent = 1.10

p/p_t	M	T/T_t	GAMMA	ISOTHERMAL T_t/T_t^*	RAYLEIGH T_t/T_t^*	SHOCK p_{t1}/p_{t2}
.150	1.94026	.84159	.52953	1.13657	.71082	1.46010
.149	1.94398	.84108	.52717	1.13726	.70959	1.46496
.148	1.94772	.84056	.52480	1.13796	.70836	1.46989
.147	1.95149	.84004	.52242	1.13866	.70711	1.47487
.146	1.95527	.83952	.52003	1.13936	.70587	1.47993
.145	1.95907	.83900	.51764	1.14008	.70462	1.48505
.144	1.96290	.83847	.51523	1.14079	.70337	1.49023
.143	1.96675	.83794	.51282	1.14152	.70211	1.49549
.142	1.97061	.83740	.51040	1.14225	.70085	1.50081
.141	1.97450	.83687	.50797	1.14298	.69958	1.50621
.140	1.97842	.83633	.50553	1.14372	.69831	1.51167
.139	1.98235	.83578	.50308	1.14447	.69704	1.51722
.138	1.98631	.83523	.50062	1.14522	.69576	1.52283
.137	1.99030	.83468	.49816	1.14597	.69447	1.52853
.136	1.99430	.83412	.49568	1.14674	.69319	1.53430
.135	1.99833	.83356	.49320	1.14751	.69189	1.54016
.134	2.00239	.83300	.49070	1.14828	.69059	1.54609
.133	2.00647	.83243	.48820	1.14907	.68929	1.55211
.132	2.01057	.83186	.48569	1.14985	.68799	1.55822
.131	2.01470	.83129	.48316	1.15065	.68667	1.56441
.130	2.01886	.83071	.48063	1.15145	.68536	1.57069
.129	2.02304	.83013	.47809	1.15226	.68403	1.57706
.128	2.02725	.82954	.47554	1.15308	.68271	1.58352
.127	2.03149	.82895	.47298	1.15390	.68138	1.59008
.126	2.03575	.82835	.47041	1.15473	.68004	1.59674
.125	2.04005	.82775	.46783	1.15556	.67870	1.60349
.124	2.04437	.82715	.46524	1.15641	.67735	1.61035
.123	2.04872	.82654	.46264	1.15726	.67600	1.61731
.122	2.05310	.82593	.46003	1.15812	.67464	1.62438
.121	2.05751	.82531	.45741	1.15899	.67328	1.63155
.120	2.06195	.82469	.45478	1.15986	.67191	1.63883
.119	2.06642	.82406	.45214	1.16074	.67053	1.64623
.118	2.07092	.82343	.44949	1.16163	.66915	1.65374
.117	2.07545	.82279	.44683	1.16253	.66777	1.66138
.116	2.08002	.82215	.44416	1.16344	.66638	1.66913
.115	2.08462	.82150	.44148	1.16436	.66498	1.67701
.114	2.08925	.82085	.43878	1.16528	.66358	1.68501
.113	2.09392	.82019	.43608	1.16621	.66217	1.69315
.112	2.09862	.81953	.43337	1.16716	.66075	1.70142
.111	2.10336	.81886	.43064	1.16811	.65933	1.70982
.110	2.10813	.81819	.42791	1.16907	.65790	1.71837
.109	2.11294	.81751	.42516	1.17004	.65647	1.72706
.108	2.11779	.81683	.42241	1.17102	.65503	1.73590
.107	2.12267	.81614	.41964	1.17201	.65358	1.74489
.106	2.12760	.81544	.41686	1.17301	.65212	1.75403
.105	2.13256	.81474	.41407	1.17403	.65066	1.76333
.104	2.13756	.81403	.41126	1.17505	.64920	1.77280
.103	2.14261	.81331	.40845	1.17608	.64772	1.78244
.102	2.14769	.81259	.40562	1.17712	.64624	1.79224
.101	2.15282	.81186	.40279	1.17818	.64475	1.80223

TABLE II. Isentropic Exponent = 1.10

p/p_t	M	T/T_t	GAMMA	ISOTHERMAL T_t/T^*	RAYLEIGH T_t/T_t^*	SHOCK p_{t1}/p_{t2}
.100	2.15799	.81113	.39994	1.17924	.64325	1.81240
.099	2.16321	.81039	.39708	1.18032	.64175	1.82275
.098	2.16847	.80964	.39420	1.18141	.64024	1.83330
.097	2.17378	.80889	.39132	1.18251	.63872	1.84404
.096	2.17913	.80813	.38842	1.18363	.63719	1.85499
.095	2.18453	.80736	.38551	1.18476	.63566	1.86615
.094	2.18998	.80658	.38259	1.18590	.63412	1.87752
.093	2.19548	.80580	.37965	1.18705	.63256	1.88912
.092	2.20103	.80501	.37671	1.18822	.63101	1.90095
.091	2.20664	.80421	.37375	1.18940	.62944	1.91301
.090	2.21229	.80340	.37077	1.19059	.62786	1.92531
.089	2.21800	.80258	.36779	1.19180	.62628	1.93786
.088	2.22377	.80176	.36479	1.19303	.62468	1.95068
.087	2.22959	.80093	.36177	1.19427	.62308	1.96376
.086	2.23547	.80009	.35875	1.19552	.62146	1.97711
.085	2.24141	.79923	.35571	1.19680	.61984	1.99075
.084	2.24741	.79838	.35265	1.19808	.61821	2.00468
.083	2.25348	.79751	.34958	1.19939	.61657	2.01892
.082	2.25960	.79663	.34650	1.20071	.61492	2.03347
.081	2.26580	.79574	.34341	1.20205	.61325	2.04834
.080	2.27205	.79484	.34030	1.20341	.61158	2.06355
.079	2.27838	.79393	.33717	1.20479	.60990	2.07910
.078	2.28478	.79301	.33403	1.20618	.60820	2.09502
.077	2.29125	.79209	.33088	1.20760	.60650	2.11130
.076	2.29779	.79114	.32771	1.20904	.60478	2.12798
.075	2.30441	.79019	.32452	1.21049	.60305	2.14505
.074	2.31110	.78923	.32132	1.21197	.60131	2.16254
.073	2.31788	.78825	.31810	1.21347	.59956	2.18046
.072	2.32474	.78727	.31487	1.21499	.59779	2.19882
.071	2.33168	.78627	.31162	1.21654	.59602	2.21765
.070	2.33870	.78525	.30836	1.21811	.59423	2.23697
.069	2.34582	.78423	.30508	1.21970	.59242	2.25679
.068	2.35303	.78319	.30178	1.22132	.59060	2.27712
.067	2.36033	.78213	.29847	1.22297	.58877	2.29801
.066	2.36773	.78106	.29514	1.22464	.58692	2.31946
.065	2.37523	.77998	.29179	1.22634	.58506	2.34150
.064	2.38283	.77888	.28842	1.22807	.58319	2.36417
.063	2.39054	.77777	.28504	1.22983	.58130	2.38747
.062	2.39835	.77664	.28163	1.23162	.57939	2.41145
.061	2.40628	.77549	.27821	1.23344	.57746	2.43613
.060	2.41433	.77432	.27477	1.23530	.57552	2.46155
.059	2.42249	.77314	.27131	1.23719	.57357	2.48774
.058	2.43078	.77194	.26784	1.23911	.57159	2.51474
.057	2.43920	.77072	.26434	1.24107	.56960	2.54258
.056	2.44775	.76948	.26082	1.24307	.56758	2.57132
.055	2.45644	.76822	.25728	1.24511	.56555	2.60098
.054	2.46527	.76694	.25373	1.24719	.56350	2.63163
.053	2.47425	.76564	.25015	1.24931	.56142	2.66331
.052	2.48338	.76432	.24655	1.25147	.55933	2.69608
.051	2.49267	.76297	.24292	1.25369	.55721	2.73000

TABLE II. Isentropic Exponent = 1.10

p/p_t	M	T/T_t	GAMMA	ISOTHERMAL T_t/T_t^*	RAYLEIGH T_t/T_t^*	SHOCK p_{t1}/p_{t2}
.050	2.50213	.76160	.23928	1.25594	.55508	2.76513
.049	2.51176	.76020	.23561	1.25825	.55291	2.80154
.048	2.52156	.75877	.23192	1.26061	.55073	2.83930
.047	2.53156	.75732	.22821	1.26303	.54851	2.87850
.046	2.54174	.75584	.22447	1.26550	.54628	2.91921
.045	2.55213	.75434	.22071	1.26803	.54401	2.96154
.044	2.56273	.75280	.21692	1.27062	.54172	3.00559
.043	2.57356	.75122	.21311	1.27328	.53940	3.05146
.042	2.58461	.74962	.20927	1.27601	.53704	3.09928
.041	2.59590	.74798	.20541	1.27881	.53466	3.14917
.040	2.60745	.74630	.20151	1.28168	.53224	3.20130
.039	2.61927	.74459	.19759	1.28464	.52979	3.25580
.038	2.63136	.74283	.19365	1.28767	.52731	3.31286
.037	2.64375	.74103	.18967	1.29080	.52478	3.37266
.036	2.65645	.73919	.18566	1.29402	.52222	3.43543
.035	2.66948	.73730	.18162	1.29734	.51961	3.50139
.034	2.68285	.73536	.17755	1.30076	.51697	3.57081
.033	2.69659	.73336	.17344	1.30429	.51427	3.64397
.032	2.71072	.73132	.16930	1.30795	.51153	3.72121
.031	2.72526	.72921	.16513	1.31173	.50874	3.80288
.030	2.74024	.72704	.16092	1.31564	.50590	3.88941
.029	2.75569	.72480	.15668	1.31970	.50300	3.98124
.028	2.77164	.72249	.15239	1.32392	.50004	4.07892
.027	2.78813	.72011	.14807	1.32831	.49702	4.18305
.026	2.80520	.71764	.14370	1.33287	.49394	4.29430
.025	2.82288	.71509	.13930	1.33763	.49078	4.41349
.024	2.84124	.71244	.13484	1.34261	.48754	4.54152
.023	2.86033	.70969	.13035	1.34781	.48422	4.67946
.022	2.88021	.70682	.12580	1.35327	.48082	4.82858
.021	2.90095	.70384	.12120	1.35900	.47732	4.99035
.020	2.92264	.70073	.11655	1.36504	.47372	5.16652
.019	2.94537	.69747	.11185	1.37142	.47000	5.35922
.018	2.96926	.69405	.10708	1.37818	.46616	5.57100
.017	2.99444	.69045	.10226	1.38536	.46219	5.80498
.016	3.02105	.68665	.09736	1.39302	.45807	6.06502
.015	3.04928	.68264	.09240	1.40121	.45379	6.35597
.014	3.07936	.67837	.08737	1.41003	.44932	6.68397
.013	3.11155	.67381	.08225	1.41956	.44465	7.05694
.012	3.14619	.66893	.07705	1.42993	.43974	7.48533
.011	3.18370	.66366	.07175	1.44129	.43456	7.98315
.010	3.22463	.65793	.06636	1.45383	.42906	8.56972
.009	3.26968	.65166	.06084	1.46782	.42320	9.27247
.008	3.31982	.64472	.05521	1.48362	.41689	10.13178
.007	3.37639	.63694	.04943	1.50174	.41003	11.20977
.006	3.44139	.62808	.04349	1.52293	.40248	12.60760
.005	3.51786	.61775	.03735	1.54839	.39402	14.50253
.004	3.61093	.60535	.03099	1.58012	.38431	17.23804
.003	3.73019	.58972	.02432	1.62199	.37272	21.58541
.002	3.89714	.56838	.01726	1.68289	.35796	29.74503
.001	4.18048	.53367	.00955	1.79235	.33627	51.90335

TABLE III. Isentropic Exponent = 1.20

p/p_t	M	T/T_t	GAMMA	ISOTHERMAL T_t/T_t^*	RAYLEIGH T_t/T_t^*	SHOCK p_{t1}/p_{t2}
1.0000	.00000	1.00000	.00000	.92308	.00000	1.
.9995	.02887	.99992	.04874	.92315	.00366	1.
.9990	.04084	.99983	.06891	.92323	.00731	1.
.9985	.05002	.99975	.08437	.92331	.01095	1.
.9980	.05777	.99967	.09740	.92338	.01457	1.
.9975	.06460	.99958	.10886	.92346	.01819	1.
.9970	.07077	.99950	.11921	.92354	.02179	1.
.9965	.07645	.99942	.12873	.92362	.02538	1.
.9960	.08175	.99933	.13757	.92369	.02896	1.
.9955	.08672	.99925	.14587	.92377	.03252	1.
.9950	.09142	.99916	.15371	.92385	.03608	1.
.9945	.09590	.99908	.16116	.92393	.03962	1.
.9940	.10018	.99900	.16828	.92400	.04315	1.
.9935	.10428	.99891	.17509	.92408	.04667	1.
.9930	.10823	.99883	.18165	.92416	.05018	1.
.9925	.11205	.99875	.18796	.92424	.05368	1.
.9920	.11574	.99866	.19407	.92431	.05717	1.
.9915	.11932	.99858	.19997	.92439	.06064	1.
.9910	.12280	.99849	.20571	.92447	.06411	1.
.9905	.12618	.99841	.21128	.92455	.06756	1.
.9900	.12948	.99833	.21670	.92462	.07100	1.
.9895	.13269	.99824	.22198	.92470	.07443	1.
.9890	.13584	.99816	.22713	.92478	.07785	1.
.9885	.13891	.99807	.23216	.92486	.08126	1.
.9880	.14192	.99799	.23708	.92494	.08466	1.
.9875	.14487	.99791	.24189	.92501	.08804	1.
.9870	.14776	.99782	.24661	.92509	.09142	1.
.9865	.15060	.99774	.25122	.92517	.09478	1.
.9860	.15338	.99765	.25575	.92525	.09814	1.
.9855	.15612	.99757	.26020	.92533	.10148	1.
.9850	.15881	.99748	.26456	.92541	.10481	1.
.9845	.15146	.99740	.26885	.92548	.10813	1.
.9840	.16407	.99732	.27306	.92556	.11144	1.
.9835	.16664	.99723	.27721	.92564	.11474	1.
.9830	.16917	.99715	.28129	.92572	.11803	1.
.9825	.17166	.99706	.28530	.92580	.12131	1.
.9820	.17412	.99698	.28926	.92588	.12458	1.
.9815	.17655	.99689	.29316	.92595	.12784	1.
.9810	.17895	.99681	.29700	.92603	.13108	1.
.9805	.18131	.99672	.30078	.92611	.13432	1.
.9800	.18365	.99664	.30452	.92619	.13754	1.
.9795	.18596	.99655	.30820	.92627	.14076	1.
.9790	.18824	.99647	.31184	.92635	.14396	1.
.9785	.19050	.99638	.31543	.92643	.14716	1.
.9780	.19273	.99630	.31897	.92651	.15034	1.
.9775	.19494	.99621	.32247	.92658	.15352	1.
.9770	.19712	.99613	.32593	.92666	.15668	1.
.9765	.19928	.99604	.32935	.92674	.15983	1.
.9760	.20142	.99596	.33273	.92682	.16298	1.
.9755	.20354	.99587	.33607	.92690	.16611	1.

TABLE III. Isentropic Exponent = 1.20

p/p_t	M	T/T_t	GAMMA	ISOTHERMAL T_t/T_t*	RAYLEIGH T_t/T_t*	SHOCK p_{t1}/p_{t2}
.9750	.20563	.99579	.33937	.92698	.16923	1.
.9745	.20771	.99570	.34264	.92706	.17235	1.
.9740	.20977	.99562	.34587	.92714	.17545	1.
.9735	.21181	.99553	.34907	.92722	.17854	1.
.9730	.21383	.99545	.35223	.92730	.18162	1.
.9725	.21583	.99536	.35536	.92738	.18470	1.
.9720	.21782	.99528	.35847	.92746	.18776	1.
.9715	.21979	.99519	.36154	.92754	.19081	1.
.9710	.22174	.99511	.36458	.92762	.19385	1.
.9705	.22368	.99502	.36759	.92770	.19689	1.
.9700	.22560	.99494	.37057	.92777	.19991	1.
.9695	.22750	.99485	.37352	.92785	.20292	1.
.9690	.22940	.99477	.37645	.92793	.20593	1.
.9685	.23127	.99468	.37935	.92801	.20892	1.
.9680	.23314	.99459	.38223	.92809	.21191	1.
.9675	.23499	.99451	.38508	.92817	.21488	1.
.9670	.23682	.99442	.38790	.92825	.21785	1.
.9665	.23865	.99434	.39070	.92833	.22080	1.
.9660	.24046	.99425	.39348	.92841	.22375	1.
.9655	.24225	.99417	.39624	.92849	.22669	1.
.9650	.24404	.99408	.39897	.92857	.22961	1.
.9645	.24581	.99399	.40168	.92865	.23253	1.
.9640	.24758	.99391	.40436	.92873	.23544	1.
.9635	.24933	.99382	.40703	.92882	.23834	1.
.9630	.25107	.99374	.40967	.92890	.24123	1.
.9625	.25280	.99365	.41230	.92898	.24411	1.
.9620	.25451	.99356	.41490	.92906	.24698	1.
.9615	.25622	.99348	.41749	.92914	.24984	1.
.9610	.25792	.99339	.42005	.92922	.25269	1.
.9605	.25961	.99331	.42260	.92930	.25553	1.
.9600	.26128	.99322	.42513	.92938	.25837	1.
.9595	.26295	.99313	.42764	.92946	.26119	1.
.9590	.26461	.99305	.43013	.92954	.26401	1.
.9585	.26626	.99296	.43260	.92962	.26681	1.
.9580	.26790	.99287	.43506	.92970	.26961	1.
.9575	.26953	.99279	.43750	.92978	.27240	1.
.9570	.27115	.99270	.43992	.92986	.27518	1.
.9565	.27276	.99261	.44232	.92994	.27795	1.
.9560	.27437	.99253	.44471	.93003	.28071	1.
.9555	.27596	.99244	.44709	.93011	.28346	1.
.9550	.27755	.99236	.44944	.93019	.28621	1.
.9545	.27913	.99227	.45178	.93027	.28894	1.
.9540	.28070	.99218	.45411	.93035	.29167	1.
.9535	.28227	.99210	.45642	.93043	.29438	1.
.9530	.28383	.99201	.45872	.93051	.29709	1.
.9525	.28537	.99192	.46100	.93059	.29979	1.
.9520	.28692	.99184	.46327	.93068	.30248	1.
.9515	.28845	.99175	.46552	.93076	.30516	1.
.9510	.28998	.99166	.46776	.93084	.30784	1.
.9505	.29150	.99157	.46999	.93092	.31050	1.

TABLE III. Isentropic Exponent = 1.20

p/p_t	M	T/T_t	GAMMA	ISOTHERMAL T_t/T_t*	RAYLEIGH T_t/T_t*	SHOCK p_{t1}/p_{t2}
.950	.29301	.99149	.47220	.93100	.31316	1.
.949	.29602	.99131	.47658	.93117	.31844	1.
.948	.29900	.99114	.48091	.93133	.32369	1.
.947	.30195	.99097	.48519	.93149	.32891	1.
.946	.30488	.99079	.48942	.93166	.33410	1.
.945	.30778	.99062	.49361	.93182	.33925	1.
.944	.31066	.99044	.49774	.93199	.34436	1.
.943	.31352	.99027	.50184	.93215	.34945	1.
.942	.31636	.99009	.50588	.93232	.35450	1.
.941	.31917	.98992	.50988	.93248	.35952	1.
.940	.32196	.98974	.51384	.93265	.36451	1.
.939	.32473	.98956	.51776	.93281	.36946	1.
.938	.32749	.98939	.52164	.93298	.37438	1.
.937	.33022	.98921	.52548	.93314	.37927	1.
.936	.33293	.98904	.52928	.93331	.38413	1.
.935	.33563	.98886	.53304	.93347	.38896	1.
.934	.33830	.98868	.53676	.93364	.39375	1.
.933	.34096	.98851	.54045	.93381	.39852	1.
.932	.34360	.98833	.54410	.93397	.40325	1.
.931	.34623	.98815	.54772	.93414	.40796	1.
.930	.34883	.98798	.55130	.93431	.41263	1.
.929	.35143	.98780	.55485	.93448	.41727	1.
.928	.35400	.98762	.55837	.93464	.42188	1.
.927	.35656	.98745	.56185	.93481	.42646	1.
.926	.35911	.98727	.56530	.93498	.43102	1.
.925	.36164	.98709	.56872	.93515	.43554	1.
.924	.36416	.98691	.57211	.93532	.44003	1.
.923	.36666	.98673	.57547	.93549	.44450	1.
.922	.36915	.98656	.57880	.93566	.44893	1.
.921	.37162	.98638	.58210	.93582	.45334	1.
.920	.37408	.98620	.58538	.93599	.45772	1.
.919	.37653	.98602	.58862	.93616	.46207	1.
.918	.37897	.98584	.59184	.93633	.46639	1.
.917	.38139	.98566	.59503	.93650	.47068	1.
.916	.38380	.98548	.59819	.93667	.47494	1.
.915	.38620	.98530	.60133	.93684	.47918	1.
.914	.38859	.98512	.60444	.93702	.48339	1.
.913	.39097	.98494	.60753	.93719	.48757	1.
.912	.39333	.98476	.61059	.93736	.49173	1.
.911	.39569	.98458	.61362	.93753	.49585	1.
.910	.39803	.98440	.61664	.93770	.49996	1.
.909	.40036	.98422	.61962	.93787	.50403	1.
.908	.40268	.98404	.62259	.93804	.50808	1.
.907	.40499	.98386	.62553	.93822	.51210	1.
.906	.40729	.98368	.62845	.93839	.51609	1.
.905	.40958	.98350	.63134	.93856	.52006	1.
.904	.41186	.98332	.63421	.93874	.52400	1.
.903	.41414	.98314	.63706	.93891	.52792	1.
.902	.41640	.98296	.63989	.93908	.53181	1.
.901	.41865	.98278	.64270	.93926	.53567	1.

TABLE III.   Isentropic Exponent = 1.20

p/p_t	M	T/T_t	GAMMA	ISOTHERMAL T_t/T_t^*	RAYLEIGH T_t/T_t^*	SHOCK p_{t1}/p_{t2}
.900	.42089	.98259	.64549	.93943	.53951	1.
.899	.42313	.98241	.64825	.93960	.54333	1.
.898	.42535	.98223	.65100	.93978	.54712	1.
.897	.42757	.98205	.65372	.93995	.55088	1.
.896	.42978	.98186	.65643	.94013	.55462	1.
.895	.43198	.98168	.65911	.94030	.55834	1.
.894	.43417	.98150	.66178	.94048	.56203	1.
.893	.43635	.98132	.66442	.94065	.56569	1.
.892	.43853	.98113	.66705	.94083	.56934	1.
.891	.44070	.98095	.66966	.94100	.57295	1.
.890	.44286	.98077	.67225	.94118	.57655	1.
.889	.44501	.98058	.67482	.94136	.58012	1.
.888	.44715	.98040	.67737	.94153	.58367	1.
.887	.44929	.98021	.67991	.94171	.58719	1.
.886	.45142	.98003	.68242	.94189	.59069	1.
.885	.45354	.97984	.68492	.94206	.59417	1.
.884	.45566	.97966	.68740	.94224	.59762	1.
.883	.45776	.97948	.68987	.94242	.60105	1.
.882	.45987	.97929	.69232	.94260	.60446	1.
.881	.46196	.97911	.69475	.94278	.60784	1.
.880	.46405	.97892	.69716	.94295	.61121	1.
.879	.46613	.97873	.69956	.94313	.61455	1.
.878	.46821	.97855	.70194	.94331	.61787	1.
.877	.47027	.97836	.70430	.94349	.62116	1.
.876	.47234	.97818	.70665	.94367	.62444	1.
.875	.47439	.97799	.70899	.94385	.62769	1.
.874	.47644	.97780	.71130	.94403	.63092	1.
.873	.47848	.97762	.71360	.94421	.63413	1.
.872	.48052	.97743	.71589	.94439	.63732	1.
.871	.48255	.97724	.71816	.94457	.64048	1.
.870	.48458	.97706	.72042	.94475	.64363	1.
.869	.48660	.97687	.72266	.94493	.64675	1.
.868	.48862	.97668	.72489	.94511	.64986	1.
.867	.49062	.97649	.72710	.94530	.65294	1.
.866	.49263	.97631	.72930	.94548	.65600	1.
.865	.49463	.97612	.73148	.94566	.65904	1.
.864	.49662	.97593	.73365	.94584	.66206	1.
.863	.49861	.97574	.73580	.94603	.66506	1.
.862	.50059	.97555	.73794	.94621	.66804	1.
.861	.50256	.97537	.74007	.94639	.67100	1.
.860	.50454	.97518	.74218	.94657	.67394	1.
.859	.50650	.97499	.74428	.94676	.67686	1.
.858	.50847	.97480	.74637	.94694	.67976	1.
.857	.51042	.97461	.74844	.94713	.68264	1.
.856	.51238	.97442	.75050	.94731	.68550	1.
.855	.51432	.97423	.75254	.94749	.68834	1.
.854	.51627	.97404	.75458	.94768	.69117	1.
.853	.51820	.97385	.75660	.94786	.69397	1.
.852	.52014	.97366	.75860	.94805	.69675	1.
.851	.52207	.97347	.76060	.94824	.69952	1.

TABLE III. Isentropic Exponent = 1.20

p/p_t	M	T/T_t	GAMMA	ISOTHERMAL T_t/T_t*	RAYLEIGH T_t/T_t*	SHOCK p_{t1}/p_{t2}
.850	.52399	.97328	.76258	.94842	.70226	1.
.849	.52591	.97309	.76455	.94861	.70499	1.
.848	.52783	.97289	.76650	.94879	.70770	1.
.847	.52974	.97270	.76845	.94898	.71039	1.
.846	.53165	.97251	.77038	.94917	.71306	1.
.845	.53355	.97232	.77230	.94935	.71571	1.
.844	.53545	.97213	.77421	.94954	.71835	1.
.843	.53734	.97194	.77610	.94973	.72097	1.
.842	.53923	.97174	.77799	.94992	.72356	1.
.841	.54112	.97155	.77986	.95011	.72615	1.
.840	.54300	.97136	.78172	.95029	.72871	1.
.839	.54488	.97117	.78357	.95048	.73125	1.
.838	.54676	.97097	.78541	.95067	.73378	1.
.837	.54863	.97078	.78723	.95086	.73629	1.
.836	.55050	.97059	.78905	.95105	.73878	1.
.835	.55236	.97039	.79085	.95124	.74126	1.
.834	.55422	.97020	.79264	.95143	.74372	1.
.833	.55608	.97001	.79442	.95162	.74616	1.
.832	.55793	.96981	.79619	.95181	.74858	1.
.831	.55978	.96962	.79795	.95200	.75099	1.
.830	.56162	.96942	.79970	.95219	.75338	1.
.829	.56347	.96923	.80144	.95238	.75575	1.
.828	.56531	.96903	.80316	.95258	.75811	1.
.827	.56714	.96884	.80488	.95277	.76045	1.
.826	.56897	.96864	.80658	.95296	.76277	1.
.825	.57080	.96845	.80828	.95315	.76508	1.
.824	.57263	.96825	.80996	.95334	.76737	1.
.823	.57445	.96805	.81164	.95354	.76965	1.
.822	.57627	.96786	.81330	.95373	.77191	1.
.821	.57809	.96766	.81495	.95392	.77415	1.
.820	.57990	.96747	.81660	.95412	.77638	1.
.819	.58171	.96727	.81823	.95431	.77859	1.
.818	.58352	.96707	.81985	.95451	.78078	1.
.817	.58532	.96688	.82147	.95470	.78296	1.
.816	.58712	.96668	.82307	.95490	.78513	1.
.815	.58892	.96648	.82466	.95509	.78728	1.
.814	.59071	.96628	.82624	.95529	.78941	1.
.813	.59251	.96608	.82782	.95548	.79153	1.
.812	.59429	.96589	.82938	.95568	.79363	1.
.811	.59608	.96569	.83094	.95587	.79572	1.
.810	.59786	.96549	.83248	.95607	.79779	1.
.809	.59965	.96529	.83401	.95627	.79985	1.
.808	.60142	.96509	.83554	.95647	.80189	1.
.807	.60320	.96489	.83706	.95666	.80392	1.
.806	.60497	.96469	.83856	.95686	.80593	1.
.805	.60674	.96449	.84006	.95706	.80793	1.
.804	.60851	.96429	.84155	.95726	.80991	1.
.803	.61028	.96409	.84303	.95746	.81188	1.
.802	.61204	.96389	.84450	.95765	.81384	1.
.801	.61380	.96369	.84596	.95785	.81578	1.

TABLE III. Isentropic Exponent = 1.20

p/p_t	M	T/T_t	GAMMA	ISOTHERMAL T_t/T_t*	RAYLEIGH T_t/T_t*	SHOCK p_{t1}/p_{t2}
.800	.61556	.96349	.84741	.95805	.81771	1.
.799	.61731	.96329	.84885	.95825	.81962	1.
.798	.61906	.96309	.85028	.95845	.82152	1.
.797	.62081	.96289	.85171	.95865	.82340	1.
.796	.62256	.96269	.85312	.95885	.82527	1.
.795	.62431	.96249	.85453	.95905	.82713	1.
.794	.62605	.96228	.85593	.95926	.82897	1.
.793	.62779	.96208	.85732	.95946	.83080	1.
.792	.62953	.96188	.85870	.95966	.83262	1.
.791	.63127	.96168	.86007	.95986	.83442	1.
.790	.63300	.96147	.86143	.96006	.83621	1.
.789	.63473	.96127	.86279	.96027	.83798	1.
.788	.63646	.96107	.86414	.96047	.83974	1.
.787	.63819	.96087	.86547	.96067	.84149	1.
.786	.63992	.96066	.86680	.96088	.84323	1.
.785	.64164	.96046	.86812	.96108	.84495	1.
.784	.64336	.96025	.86944	.96128	.84666	1.
.783	.64508	.96005	.87074	.96149	.84836	1.
.782	.64680	.95985	.87204	.96169	.85004	1.
.781	.64851	.95964	.87333	.96190	.85171	1.
.780	.65023	.95944	.87460	.96210	.85337	1.
.779	.65194	.95923	.87588	.96231	.85501	1.
.778	.65365	.95903	.87714	.96252	.85665	1.
.777	.65536	.95882	.87840	.96272	.85827	1.
.776	.65706	.95861	.87964	.96293	.85987	1.
.775	.65877	.95841	.88088	.96314	.86147	1.
.774	.66047	.95820	.88211	.96334	.86305	1.
.773	.66217	.95800	.88334	.96355	.86462	1.
.772	.66387	.95779	.88455	.96376	.86618	1.
.771	.66556	.95758	.88576	.96397	.86773	1.
.770	.66726	.95737	.88696	.96418	.86926	1.
.769	.66895	.95717	.88815	.96438	.87078	1.
.768	.67064	.95696	.88934	.96459	.87229	1.
.767	.67233	.95675	.89051	.96480	.87379	1.
.766	.67402	.95654	.89168	.96501	.87528	1.
.765	.67571	.95634	.89284	.96522	.87675	1.
.764	.67739	.95613	.89400	.96543	.87821	1.
.763	.67908	.95592	.89514	.96564	.87967	1.
.762	.68076	.95571	.89628	.96586	.88111	1.
.761	.68244	.95550	.89741	.96607	.88253	1.
.760	.68412	.95529	.89854	.96628	.88395	1.
.759	.68579	.95508	.89965	.96649	.88536	1.
.758	.68747	.95487	.90076	.96670	.88675	1.
.757	.68914	.95466	.90186	.96692	.88813	1.
.756	.69082	.95445	.90296	.96713	.88950	1.
.755	.69249	.95424	.90404	.96734	.89086	1.
.754	.69416	.95403	.90512	.96756	.89221	1.
.753	.69583	.95382	.90620	.96777	.89355	1.
.752	.69749	.95361	.90726	.96798	.89488	1.
.751	.69916	.95340	.90832	.96820	.89619	1.

TABLE III. Isentropic Exponent = 1.20

p/p_t	M	T/T_t	GAMMA	ISOTHERMAL T_t/T_t^*	RAYLEIGH T_t/T_t^*	SHOCK p_{t1}/p_{t2}
.750	.70082	.95318	.90937	.96841	.89750	1.
.749	.70248	.95297	.91041	.96863	.89880	1.
.748	.70415	.95276	.91145	.96884	.90008	1.
.747	.70581	.95255	.91248	.96906	.90135	1.
.746	.70746	.95234	.91350	.96928	.90261	1.
.745	.70912	.95212	.91451	.96949	.90387	1.
.744	.71078	.95191	.91552	.96971	.90511	1.
.743	.71243	.95170	.91652	.96993	.90634	1.
.742	.71409	.95148	.91752	.97015	.90756	1.
.741	.71574	.95127	.91850	.97036	.90877	1.
.740	.71739	.95105	.91948	.97058	.90997	1.
.739	.71904	.95084	.92046	.97080	.91116	1.
.738	.72069	.95063	.92142	.97102	.91234	1.
.737	.72234	.95041	.92238	.97124	.91351	1.
.736	.72398	.95020	.92333	.97146	.91467	1.
.735	.72563	.94998	.92428	.97168	.91582	1.
.734	.72727	.94976	.92522	.97190	.91696	1.
.733	.72891	.94955	.92615	.97212	.91809	1.
.732	.73056	.94933	.92708	.97234	.91921	1.
.731	.73220	.94912	.92800	.97256	.92032	1.
.730	.73384	.94890	.92891	.97279	.92142	1.
.729	.73548	.94868	.92981	.97301	.92251	1.
.728	.73711	.94847	.93071	.97323	.92359	1.
.727	.73875	.94825	.93160	.97345	.92466	1.
.726	.74039	.94803	.93249	.97368	.92572	1.
.725	.74202	.94781	.93337	.97390	.92677	1.
.724	.74366	.94760	.93424	.97413	.92781	1.
.723	.74529	.94738	.93511	.97435	.92885	1.
.722	.74692	.94716	.93597	.97457	.92987	1.
.721	.74855	.94694	.93682	.97480	.93088	1.
.720	.75018	.94672	.93766	.97503	.93189	1.
.719	.75181	.94650	.93850	.97525	.93289	1.
.718	.75344	.94628	.93934	.97548	.93387	1.
.717	.75507	.94606	.94017	.97570	.93485	1.
.716	.75669	.94584	.94099	.97593	.93582	1.
.715	.75832	.94562	.94180	.97616	.93678	1.
.714	.75994	.94540	.94261	.97639	.93773	1.
.713	.76157	.94518	.94341	.97661	.93867	1.
.712	.76319	.94496	.94421	.97684	.93960	1.
.711	.76481	.94474	.94499	.97707	.94052	1.
.710	.76644	.94452	.94578	.97730	.94144	1.
.709	.76806	.94430	.94655	.97753	.94235	1.
.708	.76968	.94407	.94732	.97776	.94324	1.
.707	.77130	.94385	.94809	.97799	.94413	1.
.706	.77292	.94363	.94885	.97822	.94501	1.
.705	.77453	.94341	.94960	.97845	.94588	1.
.704	.77615	.94318	.95034	.97868	.94675	1.
.703	.77777	.94296	.95108	.97892	.94760	1.
.702	.77938	.94273	.95182	.97915	.94845	1.
.701	.78100	.94251	.95254	.97938	.94928	1.

TABLE III. Isentropic Exponent = 1.20

p/p_t	M	T/T_t	GAMMA	ISOTHERMAL T_t/T_t^*	RAYLEIGH T_t/T_t^*	SHOCK p_{t1}/p_{t2}
.700	.78261	.94229	.95326	.97961	.95011	1.
.699	.78423	.94206	.95398	.97985	.95093	1.
.698	.78584	.94184	.95469	.98008	.95175	1.
.697	.78745	.94161	.95539	.98032	.95255	1.
.696	.78907	.94139	.95609	.98055	.95335	1.
.695	.79068	.94116	.95678	.98078	.95413	1.
.694	.79229	.94094	.95746	.98102	.95491	1.
.693	.79390	.94071	.95814	.98126	.95569	1.
.692	.79551	.94048	.95882	.98149	.95645	1.
.691	.79712	.94026	.95948	.98173	.95720	1.
.690	.79873	.94003	.96014	.98197	.95795	1.
.689	.80033	.93980	.96080	.98220	.95869	1.
.688	.80194	.93957	.96145	.98244	.95942	1.
.687	.80355	.93935	.96209	.98268	.96015	1.
.686	.80516	.93912	.96273	.98292	.96086	1.
.685	.80676	.93889	.96336	.98316	.96157	1.
.684	.80837	.93866	.96399	.98340	.96227	1.
.683	.80997	.93843	.96461	.98364	.96297	1.
.682	.81158	.93820	.96522	.98388	.96365	1.
.681	.81318	.93797	.96583	.98412	.96433	1.
.680	.81479	.93775	.96643	.98436	.96500	1.
.679	.81639	.93752	.96703	.98460	.96566	1.
.678	.81799	.93728	.96762	.98484	.96632	1.
.677	.81960	.93705	.96820	.98508	.96697	1.
.676	.82120	.93682	.96878	.98533	.96761	1.
.675	.82280	.93659	.96936	.98557	.96824	1.
.674	.82440	.93636	.96992	.98581	.96887	1.
.673	.82601	.93613	.97049	.98606	.96948	1.
.672	.82761	.93590	.97104	.98630	.97009	1.
.671	.82921	.93567	.97160	.98655	.97070	1.
.670	.83081	.93543	.97214	.98679	.97130	1.
.669	.83241	.93520	.97268	.98704	.97188	1.
.668	.83401	.93497	.97321	.98728	.97247	1.
.667	.83561	.93473	.97374	.98753	.97304	1.
.666	.83721	.93450	.97427	.98778	.97361	1.
.665	.83881	.93427	.97478	.98802	.97417	1.
.664	.84041	.93403	.97530	.98827	.97473	1.
.663	.84200	.93380	.97580	.98852	.97527	1.
.662	.84360	.93356	.97630	.98877	.97581	1.
.661	.84520	.93333	.97680	.98902	.97635	1.
.660	.84680	.93309	.97729	.98927	.97687	1.
.659	.84840	.93286	.97777	.98952	.97739	1.
.658	.84999	.93262	.97825	.98977	.97791	1.
.657	.85159	.93238	.97872	.99002	.97841	1.
.656	.85319	.93215	.97919	.99027	.97891	1.
.655	.85479	.93191	.97965	.99052	.97940	1.
.654	.85638	.93167	.98011	.99077	.97989	1.
.653	.85798	.93143	.98056	.99103	.98037	1.
.652	.85958	.93120	.98101	.99128	.98084	1.
.651	.86117	.93096	.98145	.99153	.98131	1.

TABLE III.   Isentropic Exponent = 1.20

p/p_t	M	T/T_t	GAMMA	ISOTHERMAL T_t/T_t^*	RAYLEIGH T_t/T_t^*	SHOCK p_{t1}/p_{t2}
.650	.86277	.93072	.98188	.99179	.98177	1.
.649	.86437	.93048	.98231	.99204	.98222	1.
.648	.86596	.93024	.98274	.99230	.98267	1.
.647	.86756	.93000	.98315	.99255	.98311	1.
.646	.86916	.92976	.98357	.99281	.98354	1.
.645	.87075	.92952	.98398	.99307	.98397	1.
.644	.87235	.92928	.98438	.99332	.98439	1.
.643	.87395	.92904	.98478	.99358	.98481	1.
.642	.87554	.92880	.98517	.99384	.98522	1.
.641	.87714	.92856	.98555	.99410	.98562	1.
.640	.87873	.92832	.98594	.99435	.98602	1.
.639	.88033	.92808	.98631	.99461	.98641	1.
.638	.88193	.92783	.98668	.99487	.98679	1.
.637	.88352	.92759	.98705	.99513	.98717	1.
.636	.88512	.92735	.98741	.99539	.98754	1.
.635	.88672	.92711	.98776	.99566	.98791	1.
.634	.88831	.92686	.98811	.99592	.98827	1.
.633	.88991	.92662	.98846	.99618	.98862	1.
.632	.89150	.92637	.98880	.99644	.98897	1.
.631	.89310	.92613	.98913	.99670	.98931	1.
.630	.89470	.92588	.98946	.99697	.98964	1.
.629	.89630	.92564	.98978	.99723	.98997	1.
.628	.89789	.92539	.99010	.99750	.99030	1.
.627	.89949	.92515	.99042	.99776	.99062	1.
.626	.90109	.92490	.99072	.99803	.99093	1.
.625	.90268	.92466	.99103	.99829	.99124	1.
.624	.90428	.92441	.99132	.99856	.99154	1.
.623	.90588	.92416	.99162	.99883	.99183	1.
.622	.90748	.92391	.99190	.99909	.99212	1.
.621	.90907	.92367	.99219	.99936	.99240	1.
.620	.91067	.92342	.99246	.99963	.99268	1.
.619	.91227	.92317	.99274	.99990	.99295	1.
.618	.91387	.92292	.99300	1.00017	.99322	1.
.617	.91547	.92267	.99326	1.00044	.99348	1.
.616	.91707	.92242	.99352	1.00071	.99374	1.
.615	.91867	.92217	.99377	1.00098	.99399	1.
.614	.92027	.92192	.99402	1.00125	.99423	1.
.613	.92187	.92167	.99426	1.00152	.99447	1.
.612	.92347	.92142	.99450	1.00180	.99471	1.
.611	.92507	.92117	.99473	1.00207	.99494	1.
.610	.92667	.92092	.99495	1.00234	.99516	1.
.609	.92827	.92067	.99517	1.00262	.99538	1.
.608	.92987	.92042	.99539	1.00289	.99559	1.
.607	.93147	.92016	.99560	1.00317	.99580	1.
.606	.93307	.91991	.99581	1.00344	.99600	1.
.605	.93468	.91966	.99601	1.00372	.99619	1.
.604	.93628	.91940	.99620	1.00400	.99638	1.
.603	.93788	.91915	.99640	1.00427	.99657	1.
.602	.93948	.91890	.99658	1.00455	.99675	1.
.601	.94109	.91864	.99676	1.00483	.99693	1.

TABLE III. Isentropic Exponent = 1.20

p/p_t	M	T/T_t	GAMMA	ISOTHERMAL T_t/T_t*	RAYLEIGH T_t/T_t*	SHOCK p_t1/p_t2
.600	.94269	.91839	.99694	1.00511	.99710	1.
.599	.94430	.91813	.99711	1.00539	.99726	1.
.598	.94590	.91787	.99728	1.00567	.99742	1.
.597	.94751	.91762	.99744	1.00595	.99758	1.
.596	.94911	.91736	.99759	1.00623	.99773	1.
.595	.95072	.91711	.99774	1.00651	.99787	1.
.594	.95233	.91685	.99789	1.00679	.99801	1.
.593	.95393	.91659	.99803	1.00708	.99815	1.
.592	.95554	.91633	.99817	1.00736	.99828	1.
.591	.95715	.91608	.99830	1.00764	.99840	1.
.590	.95876	.91582	.99843	1.00793	.99852	1.
.589	.96036	.91556	.99855	1.00821	.99864	1.
.588	.96197	.91530	.99866	1.00850	.99875	1.
.587	.96358	.91504	.99877	1.00878	.99886	1.
.586	.96519	.91478	.99888	1.00907	.99896	1.
.585	.96681	.91452	.99898	1.00936	.99905	1.
.584	.96842	.91426	.99908	1.00965	.99914	1.
.583	.97003	.91400	.99917	1.00993	.99923	1.
.582	.97164	.91374	.99926	1.01022	.99931	1.
.581	.97325	.91347	.99934	1.01051	.99939	1.
.580	.97487	.91321	.99942	1.01080	.99946	1.
.579	.97648	.91295	.99949	1.01109	.99953	1.
.578	.97810	.91269	.99956	1.01139	.99959	1.
.577	.97971	.91242	.99962	1.01168	.99965	1.
.576	.98133	.91216	.99968	1.01197	.99971	1.
.575	.98294	.91189	.99973	1.01226	.99975	1.
.574	.98456	.91163	.99978	1.01256	.99980	1.
.573	.98618	.91137	.99983	1.01285	.99984	1.
.572	.98780	.91110	.99986	1.01315	.99988	1.
.571	.98942	.91083	.99990	1.01344	.99991	1.
.570	.99104	.91057	.99993	1.01374	.99993	1.
.569	.99266	.91030	.99995	1.01403	.99996	1.
.568	.99428	.91003	.99997	1.01433	.99997	1.
.567	.99590	.90977	.99998	1.01463	.99999	1.
.566	.99752	.90950	.99999	1.01493	.99999	1.
.565	.99915	.90923	.00000	1.01523	.00000	1.
.564	1.00077	.90896	.00000	1.01553	.00000	1.00000
.563	1.00239	.90869	.99999	1.01583	.00000	1.00000
.562	1.00402	.90843	.99999	1.01613	.99999	1.00000
.561	1.00565	.90816	.99997	1.01643	.99997	1.00000
.560	1.00727	.90789	.99995	1.01673	.99996	1.00000
.559	1.00890	.90762	.99993	1.01704	.99994	1.00000
.558	1.01053	.90734	.99990	1.01734	.99991	1.00000
.557	1.01216	.90707	.99987	1.01764	.99988	1.00000
.556	1.01379	.90680	.99983	1.01795	.99985	1.00000
.555	1.01542	.90653	.99979	1.01825	.99981	1.00000
.554	1.01705	.90626	.99974	1.01856	.99976	1.00001
.553	1.01868	.90598	.99969	1.01887	.99972	1.00001
.552	1.02032	.90571	.99963	1.01917	.99967	1.00001
.551	1.02195	.90544	.99957	1.01948	.99961	1.00001

TABLE III. Isentropic Exponent = 1.20

p/p_t	M	T/T_t	GAMMA	ISOTHERMAL $T_t/T_t{}^*$	RAYLEIGH $T_t/T_t{}^*$	SHOCK p_{t1}/p_{t2}
.550	1.02358	.90516	.99950	1.01979	.99955	1.00002
.549	1.02522	.90489	.99943	1.02010	.99949	1.00002
.548	1.02686	.90461	.99935	1.02041	.99942	1.00002
.547	1.02849	.90434	.99927	1.02072	.99935	1.00003
.546	1.03013	.90406	.99919	1.02103	.99928	1.00003
.545	1.03177	.90379	.99910	1.02134	.99920	1.00004
.544	1.03341	.90351	.99900	1.02166	.99911	1.00005
.543	1.03505	.90323	.99890	1.02197	.99903	1.00005
.542	1.03670	.90296	.99880	1.02228	.99893	1.00006
.541	1.03834	.90268	.99869	1.02260	.99884	1.00007
.540	1.03998	.90240	.99857	1.02291	.99874	1.00008
.539	1.04163	.90212	.99846	1.02323	.99864	1.00009
.538	1.04327	.90184	.99833	1.02355	.99853	1.00010
.537	1.04492	.90156	.99820	1.02386	.99842	1.00011
.536	1.04657	.90128	.99807	1.02418	.99830	1.00012
.535	1.04822	.90100	.99793	1.02450	.99819	1.00014
.534	1.04987	.90072	.99779	1.02482	.99806	1.00015
.533	1.05152	.90044	.99764	1.02514	.99794	1.00017
.532	1.05317	.90016	.99749	1.02546	.99781	1.00018
.531	1.05482	.89988	.99734	1.02578	.99767	1.00020
.530	1.05648	.89959	.99718	1.02611	.99754	1.00022
.529	1.05813	.89931	.99701	1.02643	.99739	1.00023
.528	1.05979	.89903	.99684	1.02675	.99725	1.00025
.527	1.06144	.89874	.99667	1.02708	.99710	1.00028
.526	1.06310	.89846	.99649	1.02740	.99695	1.00030
.525	1.06476	.89817	.99630	1.02773	.99679	1.00032
.524	1.06642	.89789	.99611	1.02805	.99663	1.00035
.523	1.06808	.89760	.99592	1.02838	.99647	1.00037
.522	1.06975	.89732	.99572	1.02871	.99630	1.00040
.521	1.07141	.89703	.99552	1.02904	.99613	1.00043
.520	1.07307	.89674	.99531	1.02937	.99595	1.00046
.519	1.07474	.89645	.99510	1.02970	.99578	1.00049
.518	1.07641	.89617	.99488	1.03003	.99559	1.00052
.517	1.07808	.89588	.99466	1.03036	.99541	1.00055
.516	1.07974	.89559	.99443	1.03069	.99522	1.00059
.515	1.08142	.89530	.99420	1.03103	.99503	1.00062
.514	1.08309	.89501	.99397	1.03136	.99483	1.00066
.513	1.08476	.89472	.99373	1.03170	.99463	1.00070
.512	1.08643	.89443	.99348	1.03203	.99443	1.00074
.511	1.08811	.89414	.99323	1.03237	.99422	1.00078
.510	1.08979	.89384	.99298	1.03270	.99401	1.00082
.509	1.09147	.89355	.99272	1.03304	.99380	1.00087
.508	1.09314	.89326	.99246	1.03338	.99358	1.00091
.507	1.09483	.89297	.99219	1.03372	.99336	1.00096
.506	1.09651	.89267	.99192	1.03406	.99314	1.00101
.505	1.09819	.89238	.99164	1.03440	.99291	1.00106
.504	1.09987	.89208	.99136	1.03474	.99268	1.00111
.503	1.10156	.89179	.99107	1.03509	.99245	1.00117
.502	1.10325	.89149	.99078	1.03543	.99221	1.00122
.501	1.10494	.89120	.99049	1.03577	.99197	1.00128

TABLE III.   Isentropic Exponent = 1.20

p/p_t	M	T/T_t	GAMMA	ISOTHERMAL T_t/T_t^*	RAYLEIGH T_t/T_t^*	SHOCK p_{t1}/p_{t2}
.500	1.10663	.89090	.99019	1.03612	.99173	1.00134
.499	1.10832	.89060	.98988	1.03646	.99148	1.00140
.498	1.11001	.89030	.98957	1.03681	.99123	1.00147
.497	1.11170	.89001	.98926	1.03716	.99098	1.00153
.496	1.11340	.88971	.98894	1.03751	.99072	1.00160
.495	1.11510	.88941	.98861	1.03786	.99046	1.00166
.494	1.11679	.88911	.98829	1.03821	.99019	1.00174
.493	1.11849	.88881	.98795	1.03856	.98993	1.00181
.492	1.12019	.88851	.98762	1.03891	.98966	1.00188
.491	1.12190	.88821	.98727	1.03926	.98938	1.00196
.490	1.12360	.88790	.98693	1.03961	.98911	1.00204
.489	1.12531	.88760	.98658	1.03997	.98883	1.00211
.488	1.12701	.88730	.98622	1.04032	.98855	1.00220
.487	1.12872	.88700	.98586	1.04068	.98826	1.00228
.486	1.13043	.88669	.98549	1.04103	.98797	1.00237
.485	1.13214	.88639	.98512	1.04139	.98768	1.00245
.484	1.13386	.88608	.98475	1.04175	.98738	1.00254
.483	1.13557	.88578	.98437	1.04211	.98708	1.00264
.482	1.13729	.88547	.98399	1.04247	.98678	1.00273
.481	1.13900	.88516	.98360	1.04283	.98648	1.00283
.480	1.14072	.88486	.98321	1.04319	.98617	1.00293
.479	1.14244	.88455	.98281	1.04355	.98586	1.00303
.478	1.14417	.88424	.98241	1.04392	.98554	1.00313
.477	1.14589	.88393	.98200	1.04428	.98523	1.00324
.476	1.14762	.88362	.98159	1.04465	.98491	1.00334
.475	1.14934	.88331	.98117	1.04501	.98458	1.00345
.474	1.15107	.88300	.98075	1.04538	.98426	1.00356
.473	1.15280	.88269	.98033	1.04575	.98393	1.00368
.472	1.15453	.88238	.97990	1.04612	.98360	1.00380
.471	1.15627	.88207	.97946	1.04649	.98326	1.00391
.470	1.15800	.88176	.97902	1.04686	.98292	1.00404
.469	1.15974	.88145	.97858	1.04723	.98258	1.00416
.468	1.16148	.88113	.97813	1.04760	.98224	1.00429
.467	1.16322	.88082	.97768	1.04798	.98189	1.00442
.466	1.16496	.88050	.97722	1.04835	.98154	1.00455
.465	1.16671	.88019	.97676	1.04873	.98119	1.00468
.464	1.16845	.87987	.97629	1.04910	.98083	1.00482
.463	1.17020	.87956	.97582	1.04948	.98047	1.00496
.462	1.17195	.87924	.97534	1.04986	.98011	1.00510
.461	1.17370	.87892	.97486	1.05024	.97975	1.00524
.460	1.17546	.87860	.97438	1.05062	.97938	1.00539
.459	1.17721	.87828	.97389	1.05100	.97901	1.00554
.458	1.17897	.87797	.97339	1.05138	.97864	1.00569
.457	1.18073	.87765	.97289	1.05176	.97826	1.00584
.456	1.18249	.87733	.97239	1.05215	.97788	1.00600
.455	1.18425	.87700	.97188	1.05253	.97750	1.00616
.454	1.18601	.87668	.97137	1.05292	.97711	1.00632
.453	1.18778	.87636	.97085	1.05331	.97673	1.00649
.452	1.18955	.87604	.97033	1.05369	.97634	1.00666
.451	1.19132	.87571	.96980	1.05408	.97594	1.00683

TABLE III. Isentropic Exponent = 1.20

p/p_t	M	T/T_t	GAMMA	ISOTHERMAL T_t/T_t^*	RAYLEIGH T_t/T_t^*	SHOCK p_{t1}/p_{t2}
.450	1.19309	.87539	.96927	1.05447	.97555	1.00700
.449	1.19487	.87507	.96873	1.05486	.97515	1.00718
.448	1.19664	.87474	.96819	1.05526	.97475	1.00736
.447	1.19842	.87442	.96765	1.05565	.97435	1.00754
.446	1.20020	.87409	.96710	1.05604	.97394	1.00773
.445	1.20198	.87376	.96654	1.05644	.97353	1.00791
.444	1.20377	.87343	.96598	1.05684	.97312	1.00811
.443	1.20555	.87311	.96542	1.05723	.97270	1.00830
.442	1.20734	.87278	.96485	1.05763	.97229	1.00850
.441	1.20913	.87245	.96428	1.05803	.97187	1.00870
.440	1.21092	.87212	.96370	1.05843	.97144	1.00890
.439	1.21272	.87179	.96312	1.05883	.97102	1.00911
.438	1.21451	.87146	.96253	1.05923	.97059	1.00932
.437	1.21631	.87112	.96194	1.05964	.97016	1.00953
.436	1.21811	.87079	.96134	1.06004	.96973	1.00975
.435	1.21992	.87046	.96074	1.06045	.96929	1.00996
.434	1.22172	.87012	.96013	1.06086	.96885	1.01019
.433	1.22353	.86979	.95952	1.06126	.96841	1.01041
.432	1.22534	.86946	.95891	1.06167	.96797	1.01064
.431	1.22715	.86912	.95829	1.06208	.96752	1.01087
.430	1.22896	.86878	.95766	1.06249	.96707	1.01111
.429	1.23078	.86845	.95703	1.06291	.96662	1.01135
.428	1.23260	.86811	.95640	1.06332	.96617	1.01159
.427	1.23442	.86777	.95576	1.06373	.96571	1.01183
.426	1.23624	.86743	.95511	1.06415	.96525	1.01208
.425	1.23807	.86709	.95447	1.06457	.96479	1.01233
.424	1.23990	.86675	.95381	1.06499	.96432	1.01259
.423	1.24173	.86641	.95316	1.06540	.96386	1.01285
.422	1.24356	.86607	.95249	1.06582	.96339	1.01311
.421	1.24539	.86573	.95183	1.06625	.96291	1.01338
.420	1.24723	.86538	.95115	1.06667	.96244	1.01365
.419	1.24907	.86504	.95048	1.06709	.96196	1.01392
.418	1.25091	.86469	.94980	1.06752	.96148	1.01419
.417	1.25276	.86435	.94911	1.06794	.96100	1.01447
.416	1.25460	.86400	.94842	1.06837	.96052	1.01476
.415	1.25645	.86366	.94773	1.06880	.96003	1.01505
.414	1.25830	.86331	.94703	1.06923	.95954	1.01534
.413	1.26016	.86296	.94632	1.06966	.95905	1.01563
.412	1.26201	.86261	.94561	1.07009	.95855	1.01593
.411	1.26387	.86226	.94490	1.07053	.95806	1.01623
.410	1.26574	.86191	.94418	1.07096	.95756	1.01654
.409	1.26760	.86156	.94346	1.07140	.95706	1.01685
.408	1.26947	.86121	.94273	1.07183	.95655	1.01716
.407	1.27134	.86086	.94199	1.07227	.95604	1.01748
.406	1.27321	.86051	.94126	1.07271	.95554	1.01780
.405	1.27508	.86015	.94051	1.07315	.95502	1.01812
.404	1.27696	.85980	.93977	1.07360	.95451	1.01845
.403	1.27884	.85944	.93901	1.07404	.95399	1.01879
.402	1.28072	.85909	.93826	1.07448	.95348	1.01912
.401	1.28261	.85873	.93750	1.07493	.95295	1.01947

TABLE III. Isentropic Exponent = 1.20

p/p_t	M	T/T_t	GAMMA	ISOTHERMAL T_t/T_t^*	RAYLEIGH T_t/T_t^*	SHOCK p_{t1}/p_{t2}
.400	1.28450	.85837	.93673	1.07538	.95243	1.01981
.399	1.28639	.85802	.93596	1.07583	.95191	1.02016
.398	1.28828	.85766	.93518	1.07628	.95138	1.02051
.397	1.29018	.85730	.93440	1.07673	.95085	1.02087
.396	1.29208	.85694	.93362	1.07718	.95031	1.02123
.395	1.29398	.85658	.93283	1.07764	.94978	1.02160
.394	1.29588	.85621	.93203	1.07809	.94924	1.02197
.393	1.29779	.85585	.93123	1.07855	.94870	1.02234
.392	1.29970	.85549	.93043	1.07901	.94816	1.02272
.391	1.30161	.85512	.92962	1.07946	.94761	1.02310
.390	1.30353	.85476	.92880	1.07993	.94707	1.02349
.389	1.30545	.85439	.92798	1.08039	.94652	1.02388
.388	1.30737	.85403	.92716	1.08085	.94597	1.02428
.387	1.30930	.85366	.92633	1.08132	.94541	1.02468
.386	1.31123	.85329	.92550	1.08178	.94486	1.02508
.385	1.31316	.85292	.92466	1.08225	.94430	1.02549
.384	1.31509	.85255	.92382	1.08272	.94374	1.02591
.383	1.31703	.85218	.92297	1.08319	.94317	1.02633
.382	1.31897	.85181	.92212	1.08366	.94261	1.02675
.381	1.32091	.85144	.92126	1.08414	.94204	1.02718
.380	1.32286	.85107	.92040	1.08461	.94147	1.02761
.379	1.32481	.85069	.91953	1.08509	.94090	1.02805
.378	1.32676	.85032	.91866	1.08557	.94033	1.02849
.377	1.32871	.84994	.91778	1.08604	.93975	1.02893
.376	1.33067	.84957	.91690	1.08653	.93917	1.02938
.375	1.33264	.84919	.91601	1.08701	.93859	1.02984
.374	1.33460	.84881	.91512	1.08749	.93801	1.03030
.373	1.33657	.84843	.91422	1.08798	.93742	1.03077
.372	1.33854	.84805	.91332	1.08846	.93683	1.03124
.371	1.34052	.84767	.91241	1.08895	.93624	1.03171
.370	1.34249	.84729	.91150	1.08944	.93565	1.03219
.369	1.34448	.84691	.91059	1.08993	.93506	1.03268
.368	1.34646	.84653	.90966	1.09043	.93446	1.03317
.367	1.34845	.84614	.90874	1.09092	.93386	1.03367
.366	1.35044	.84576	.90781	1.09142	.93326	1.03417
.365	1.35244	.84537	.90687	1.09192	.93266	1.03467
.364	1.35444	.84499	.90593	1.09242	.93205	1.03518
.363	1.35644	.84460	.90498	1.09292	.93145	1.03570
.362	1.35844	.84421	.90403	1.09342	.93084	1.03622
.361	1.36045	.84382	.90308	1.09392	.93022	1.03675
.360	1.36247	.84343	.90212	1.09443	.92961	1.03728
.359	1.36448	.84304	.90115	1.09494	.92899	1.03782
.358	1.36650	.84265	.90018	1.09545	.92838	1.03836
.357	1.36852	.84226	.89920	1.09596	.92775	1.03891
.356	1.37055	.84186	.89822	1.09647	.92713	1.03946
.355	1.37258	.84147	.89724	1.09698	.92651	1.04002
.354	1.37462	.84107	.89625	1.09750	.92588	1.04059
.353	1.37665	.84068	.89525	1.09802	.92525	1.04116
.352	1.37870	.84028	.89425	1.09854	.92462	1.04174
.351	1.38074	.83988	.89324	1.09906	.92399	1.04232

TABLE III.   Isentropic Exponent = 1.20

p/p_t	M	T/T_t	GAMMA	ISOTHERMAL $T_t/T_t{}^*$	RAYLEIGH $T_t/T_t{}^*$	SHOCK p_{t1}/p_{t2}
.350	1.38279	.83948	.89223	1.09958	.92335	1.04291
.349	1.38484	.83908	.89122	1.10010	.92271	1.04350
.348	1.38690	.83868	.89020	1.10063	.92207	1.04410
.347	1.38896	.83828	.88917	1.10116	.92143	1.04470
.346	1.39103	.83788	.88814	1.10169	.92079	1.04532
.345	1.39309	.83747	.88710	1.10222	.92014	1.04593
.344	1.39517	.83707	.88606	1.10275	.91949	1.04656
.343	1.39724	.83666	.88502	1.10329	.91884	1.04719
.342	1.39932	.83625	.88396	1.10382	.91819	1.04782
.341	1.40141	.83584	.88291	1.10436	.91754	1.04846
.340	1.40350	.83544	.88185	1.10490	.91688	1.04911
.339	1.40559	.83503	.88078	1.10545	.91622	1.04976
.338	1.40768	.83461	.87971	1.10599	.91556	1.05042
.337	1.40978	.83420	.87863	1.10654	.91490	1.05109
.336	1.41189	.83379	.87755	1.10709	.91423	1.05176
.335	1.41400	.83338	.87646	1.10764	.91356	1.05244
.334	1.41611	.83296	.87537	1.10819	.91289	1.05313
.333	1.41823	.83254	.87427	1.10874	.91222	1.05382
.332	1.42035	.83213	.87317	1.10930	.91155	1.05452
.331	1.42248	.83171	.87206	1.10986	.91087	1.05523
.330	1.42461	.83129	.87095	1.11042	.91020	1.05594
.329	1.42674	.83087	.86983	1.11098	.90952	1.05666
.328	1.42888	.83045	.86871	1.11154	.90883	1.05738
.327	1.43102	.83003	.86758	1.11211	.90815	1.05811
.326	1.43317	.82960	.86644	1.11267	.90746	1.05885
.325	1.43532	.82918	.86530	1.11324	.90678	1.05960
.324	1.43748	.82875	.86416	1.11382	.90608	1.06035
.323	1.43964	.82832	.86301	1.11439	.90539	1.06111
.322	1.44181	.82790	.86186	1.11497	.90470	1.06188
.321	1.44398	.82747	.86070	1.11554	.90400	1.06266
.320	1.44615	.82704	.85953	1.11613	.90330	1.06344
.319	1.44833	.82661	.85836	1.11671	.90260	1.06423
.318	1.45052	.82617	.85718	1.11729	.90190	1.06502
.317	1.45271	.82574	.85600	1.11788	.90120	1.06583
.316	1.45490	.82531	.85482	1.11847	.90049	1.06664
.315	1.45710	.82487	.85362	1.11906	.89978	1.06746
.314	1.45930	.82443	.85243	1.11965	.89907	1.06828
.313	1.46151	.82399	.85122	1.12025	.89836	1.06912
.312	1.46372	.82355	.85002	1.12084	.89764	1.06996
.311	1.46594	.82311	.84880	1.12144	.89692	1.07081
.310	1.46816	.82267	.84758	1.12205	.89620	1.07167
.309	1.47039	.82223	.84636	1.12265	.89548	1.07253
.308	1.47263	.82179	.84513	1.12326	.89476	1.07340
.307	1.47486	.82134	.84390	1.12387	.89403	1.07428
.306	1.47711	.82089	.84265	1.12448	.89331	1.07517
.305	1.47936	.82045	.84141	1.12509	.89258	1.07607
.304	1.48161	.82000	.84016	1.12571	.89185	1.07698
.303	1.48387	.81955	.83890	1.12633	.89111	1.07789
.302	1.48613	.81910	.83764	1.12695	.89038	1.07881
.301	1.48840	.81864	.83637	1.12757	.88964	1.07974

TABLE III.   Isentropic Exponent = 1.20

p/p_t	M	T/T_t	GAMMA	ISOTHERMAL T_t/T_t^*	RAYLEIGH T_t/T_t^*	SHOCK p_{t1}/p_{t2}
.300	1.49068	.81819	.83510	1.12820	.88890	1.08068
.299	1.49296	.81773	.83382	1.12882	.88816	1.08163
.298	1.49524	.81728	.83254	1.12945	.88741	1.08259
.297	1.49753	.81682	.83125	1.13009	.88667	1.08355
.296	1.49983	.81636	.82995	1.13072	.88592	1.08453
.295	1.50213	.81590	.82865	1.13136	.88517	1.08551
.294	1.50444	.81544	.82734	1.13200	.88442	1.08650
.293	1.50675	.81498	.82603	1.13264	.88366	1.08750
.292	1.50907	.81451	.82472	1.13329	.88291	1.08851
.291	1.51140	.81405	.82339	1.13394	.88215	1.08953
.290	1.51373	.81358	.82206	1.13459	.88139	1.09056
.289	1.51606	.81311	.82073	1.13524	.88063	1.09160
.288	1.51841	.81264	.81939	1.13590	.87986	1.09264
.287	1.52075	.81217	.81804	1.13656	.87910	1.09370
.286	1.52311	.81170	.81669	1.13722	.87833	1.09476
.285	1.52547	.81122	.81534	1.13788	.87756	1.09584
.284	1.52783	.81075	.81397	1.13855	.87678	1.09693
.283	1.53021	.81027	.81261	1.13922	.87601	1.09802
.282	1.53258	.80979	.81123	1.13989	.87523	1.09913
.281	1.53497	.80932	.80985	1.14057	.87445	1.10024
.280	1.53736	.80883	.80847	1.14124	.87367	1.10137
.279	1.53975	.80835	.80708	1.14192	.87289	1.10250
.278	1.54216	.80787	.80568	1.14261	.87210	1.10365
.277	1.54457	.80738	.80428	1.14329	.87132	1.10481
.276	1.54698	.80690	.80287	1.14398	.87053	1.10597
.275	1.54940	.80641	.80146	1.14468	.86974	1.10715
.274	1.55183	.80592	.80004	1.14537	.86894	1.10834
.273	1.55427	.80543	.79861	1.14607	.86815	1.10954
.272	1.55671	.80494	.79718	1.14677	.86735	1.11075
.271	1.55916	.80444	.79574	1.14747	.86655	1.11197
.270	1.56161	.80395	.79430	1.14818	.86575	1.11320
.269	1.56407	.80345	.79285	1.14889	.86494	1.11445
.268	1.56654	.80295	.79139	1.14961	.86414	1.11570
.267	1.56902	.80245	.78993	1.15032	.86333	1.11697
.266	1.57150	.80195	.78847	1.15104	.86252	1.11825
.265	1.57399	.80145	.78699	1.15176	.86171	1.11954
.264	1.57649	.80094	.78551	1.15249	.86089	1.12084
.263	1.57899	.80043	.78403	1.15322	.86008	1.12215
.262	1.58150	.79993	.78254	1.15395	.85926	1.12348
.261	1.58402	.79942	.78104	1.15469	.85844	1.12482
.260	1.58654	.79891	.77954	1.15543	.85761	1.12617
.259	1.58908	.79839	.77803	1.15617	.85679	1.12753
.258	1.59162	.79788	.77651	1.15691	.85596	1.12890
.257	1.59416	.79736	.77499	1.15766	.85513	1.13029
.256	1.59672	.79684	.77347	1.15842	.85430	1.13169
.255	1.59928	.79632	.77193	1.15917	.85347	1.13310
.254	1.60185	.79580	.77039	1.15993	.85263	1.13453
.253	1.60443	.79528	.76885	1.16069	.85179	1.13597
.252	1.60701	.79476	.76730	1.16146	.85095	1.13742
.251	1.60961	.79423	.76574	1.16223	.85011	1.13889

TABLE III. Isentropic Exponent = 1.20

p/p_t	M	T/T_t	GAMMA	ISOTHERMAL T_t/T_t^*	RAYLEIGH T_t/T_t^*	SHOCK p_{t1}/p_{t2}
.250	1.61221	.79370	.76417	1.16300	.84926	1.14037
.249	1.61482	.79317	.76260	1.16378	.84842	1.14186
.248	1.61743	.79264	.76103	1.16456	.84757	1.14337
.247	1.62006	.79211	.75944	1.16535	.84672	1.14489
.246	1.62269	.79157	.75785	1.16613	.84586	1.14643
.245	1.62533	.79103	.75626	1.16693	.84501	1.14798
.244	1.62798	.79049	.75466	1.16772	.84415	1.14954
.243	1.63064	.78995	.75305	1.16852	.84329	1.15112
.242	1.63331	.78941	.75144	1.16933	.84243	1.15271
.241	1.63598	.78887	.74981	1.17013	.84156	..15432
.240	1.63867	.78832	.74819	1.17094	.84070	1.15595
.239	1.64136	.78777	.74655	1.17176	.83983	1.15759
.238	1.64406	.78722	.74491	1.17258	.83896	1.15924
.237	1.64677	.78667	.74327	1.17340	.83808	1.16091
.236	1.64949	.78611	.74161	1.17423	.83721	1.16260
.235	1.65221	.78556	.73995	1.17506	.83633	1.16430
.234	1.65495	.78500	.73829	1.17590	.83545	1.16602
.233	1.65770	.78444	.73662	1.17673	.83457	1.16775
.232	1.66045	.78388	.73494	1.17758	.83368	1.16950
.231	1.66322	.78331	.73325	1.17843	.83280	1.17127
.230	1.66599	.78275	.73156	1.17928	.83191	1.17305
.229	1.66877	.78218	.72986	1.18014	.83101	1.17486
.228	1.67157	.78161	.72815	1.18100	.83012	1.17667
.227	1.67437	.78104	.72644	1.18186	.82922	1.17851
.226	1.67718	.78046	.72472	1.18273	.82833	1.18037
.225	1.68000	.77988	.72300	1.18361	.82742	1.18224
.224	1.68283	.77931	.72127	1.18449	.82652	1.18413
.223	1.68568	.77873	.71953	1.18537	.82562	1.18604
.222	1.68853	.77814	.71778	1.18626	.82471	1.18796
.221	1.69139	.77756	.71603	1.18715	.82380	1.18991
.220	1.69426	.77697	.71427	1.18805	.82288	1.19187
.219	1.69714	.77638	.71250	1.18895	.82197	1.19386
.218	1.70004	.77579	.71073	1.18986	.82105	1.19586
.217	1.70294	.77519	.70895	1.19077	.82013	1.19788
.216	1.70586	.77460	.70716	1.19169	.81921	1.19993
.215	1.70878	.77400	.70537	1.19261	.81828	1.20199
.214	1.71172	.77340	.70357	1.19354	.81736	1.20407
.213	1.71466	.77279	.70176	1.19447	.81643	1.20618
.212	1.71762	.77219	.69994	1.19541	.81550	1.20830
.211	1.72059	.77158	.69812	1.19635	.81456	1.21045
.210	1.72357	.77097	.69629	1.19730	.81362	1.21262
.209	1.72656	.77036	.69445	1.19825	.81268	1.21481
.208	1.72957	.76974	.69261	1.19921	.81174	1.21702
.207	1.73258	.76912	.69076	1.20017	.81080	1.21925
.206	1.73561	.76850	.68890	1.20114	.80985	1.22151
.205	1.73865	.76788	.68704	1.20211	.80890	1.22379
.204	1.74170	.76725	.68516	1.20309	.80795	1.22609
.203	1.74476	.76662	.68328	1.20408	.80699	1.22842
.202	1.74784	.76599	.68140	1.20507	.80604	1.23077
.201	1.75093	.76536	.67950	1.20607	.80508	1.23314

TABLE III.   Isentropic Exponent = 1.20

p/p_t	M	T/T_t	GAMMA	ISOTHERMAL T_t/T_t^*	RAYLEIGH T_t/T_t^*	SHOCK p_{t1}/p_{t2}
.200	1.75403	.76472	.67760	1.20707	.80411	1.23554
.199	1.75714	.76409	.67569	1.20808	.80315	1.23796
.198	1.76026	.76344	.67377	1.20909	.80218	1.24041
.197	1.76340	.76280	.67185	1.21012	.80121	1.24288
.196	1.76655	.76215	.66992	1.21114	.80024	1.24538
.195	1.76972	.76150	.66798	1.21218	.79926	1.24791
.194	1.77290	.76085	.66603	1.21321	.79828	1.25046
.193	1.77609	.76020	.66408	1.21426	.79730	1.25304
.192	1.77929	.75954	.66211	1.21531	.79632	1.25565
.191	1.78251	.75888	.66014	1.21637	.79533	1.25828
.190	1.78574	.75821	.65817	1.21743	.79434	1.26094
.189	1.78899	.75755	.65618	1.21851	.79335	1.26363
.188	1.79225	.75688	.65419	1.21958	.79235	1.26635
.187	1.79552	.75621	.65219	1.22067	.79135	1.26910
.186	1.79881	.75553	.65018	1.22176	.79035	1.27188
.185	1.80212	.75485	.64816	1.22286	.78935	1.27468
.184	1.80544	.75417	.64614	1.22396	.78834	1.27752
.183	1.80877	.75349	.64410	1.22508	.78733	1.28039
.182	1.81212	.75280	.64206	1.22619	.78632	1.28329
.181	1.81548	.75211	.64002	1.22732	.78531	1.28622
.180	1.81886	.75141	.63796	1.22845	.78429	1.28919
.179	1.82226	.75072	.63589	1.22960	.78327	1.29218
.178	1.82567	.75002	.63382	1.23074	.78224	1.29521
.177	1.82910	.74931	.63174	1.23190	.78122	1.29828
.176	1.83254	.74860	.62965	1.23306	.78019	1.30138
.175	1.83600	.74789	.62755	1.23424	.77915	1.30451
.174	1.83947	.74718	.62545	1.23542	.77812	1.30768
.173	1.84297	.74646	.62333	1.23660	.77708	1.31088
.172	1.84648	.74574	.62121	1.23780	.77604	1.31412
.171	1.85000	.74502	.61908	1.23900	.77499	1.31740
.170	1.85355	.74429	.61694	1.24021	.77394	1.32072
.169	1.85711	.74356	.61479	1.24143	.77289	1.32407
.168	1.86069	.74282	.61263	1.24266	.77184	1.32746
.167	1.86429	.74208	.61047	1.24390	.77078	1.33090
.166	1.86790	.74134	.60829	1.24514	.76972	1.33437
.165	1.87154	.74059	.60611	1.24640	.76865	1.33788
.164	1.87519	.73984	.60392	1.24766	.76759	1.34144
.163	1.87886	.73909	.60172	1.24894	.76651	1.34503
.162	1.88256	.73833	.59951	1.25022	.76544	1.34867
.161	1.88627	.73757	.59729	1.25151	.76436	1.35235
.160	1.89000	.73681	.59506	1.25281	.76328	1.35608
.159	1.89375	.73604	.59283	1.25412	.76220	1.35985
.158	1.89752	.73526	.59058	1.25544	.76111	1.36367
.157	1.90131	.73449	.58833	1.25677	.76002	1.36754
.156	1.90512	.73370	.58607	1.25811	.75892	1.37145
.155	1.90895	.73292	.58379	1.25945	.75782	1.37541
.154	1.91280	.73213	.58151	1.26081	.75672	1.37942
.153	1.91668	.73133	.57922	1.26218	.75561	1.38347
.152	1.92058	.73053	.57692	1.26356	.75450	1.38758
.151	1.92449	.72973	.57461	1.26495	.75339	1.39175

TABLE III.   Isentropic Exponent = 1.20

p/p_t	M	T/T_t	GAMMA	ISOTHERMAL T_t/T^*	RAYLEIGH T_t/T_t^*	SHOCK p_{t1}/p_{t2}
.150	1.92844	.72892	.57229	1.26636	.75227	1.39596
.149	1.93240	.72811	.56996	1.26777	.75115	1.40023
.148	1.93639	.72729	.56762	1.26919	.75003	1.40455
.147	1.94040	.72647	.56527	1.27063	.74890	1.40893
.146	1.94443	.72565	.56291	1.27207	.74777	1.41336
.145	1.94849	.72482	.56055	1.27353	.74663	1.41785
.144	1.95257	.72398	.55817	1.27500	.74549	1.42240
.143	1.95667	.72314	.55578	1.27648	.74435	1.42701
.142	1.96080	.72230	.55338	1.27798	.74320	1.43168
.141	1.96496	.72144	.55097	1.27948	.74205	1.43641
.140	1.96914	.72059	.54856	1.28100	.74090	1.44121
.139	1.97335	.71973	.54613	1.28253	.73974	1.44607
.138	1.97759	.71886	.54369	1.28408	.73857	1.45100
.137	1.98185	.71799	.54124	1.28564	.73740	1.45599
.136	1.98613	.71712	.53878	1.28721	.73623	1.46105
.135	1.99045	.71624	.53631	1.28879	.73505	1.46618
.134	1.99479	.71535	.53383	1.29039	.73387	1.47139
.133	1.99917	.71446	.53134	1.29200	.73269	1.47666
.132	2.00357	.71356	.52884	1.29363	.73150	1.48201
.131	2.00800	.71265	.52633	1.29527	.73030	1.48744
.130	2.01246	.71174	.52380	1.29692	.72910	1.49294
.129	2.01695	.71083	.52127	1.29859	.72790	1.49852
.128	2.02147	.70991	.51872	1.30028	.72669	1.50418
.127	2.02602	.70898	.51617	1.30198	.72548	1.50993
.126	2.03061	.70805	.51360	1.30370	.72426	1.51576
.125	2.03522	.70711	.51102	1.30543	.72304	1.52167
.124	2.03987	.70616	.50843	1.30718	.72181	1.52767
.123	2.04455	.70521	.50583	1.30894	.72058	1.53376
.122	2.04927	.70425	.50322	1.31072	.71934	1.53994
.121	2.05402	.70328	.50059	1.31252	.71810	1.54622
.120	2.05881	.70231	.49796	1.31434	.71685	1.55259
.119	2.06363	.70133	.49531	1.31617	.71560	1.55906
.118	2.06848	.70035	.49265	1.31803	.71434	1.56563
.117	2.07338	.69935	.48998	1.31990	.71308	1.57231
.116	2.07831	.69836	.48729	1.32179	.71181	1.57908
.115	2.08328	.69735	.48460	1.32370	.71054	1.58597
.114	2.08828	.69633	.48189	1.32562	.70926	1.59297
.113	2.09333	.69531	.47917	1.32757	.70797	1.60007
.112	2.09842	.69428	.47643	1.32954	.70668	1.60730
.111	2.10354	.69325	.47369	1.33153	.70539	1.61464
.110	2.10871	.69220	.47093	1.33354	.70408	1.62210
.109	2.11393	.69115	.46816	1.33557	.70278	1.62969
.108	2.11918	.69009	.46537	1.33762	.70146	1.63740
.107	2.12448	.68902	.46257	1.33970	.70014	1.64524
.106	2.12982	.68794	.45976	1.34180	.69882	1.65322
.105	2.13521	.68685	.45694	1.34392	.69748	1.66134
.104	2.14064	.68576	.45410	1.34606	.69614	1.66959
.103	2.14613	.68466	.45125	1.34823	.69480	1.67800
.102	2.15166	.68354	.44838	1.35043	.69345	1.68654
.101	2.15724	.68242	.44550	1.35265	.69209	1.69525

TABLE III. Isentropic Exponent = 1.20

p/p_t	M	T/T_t	GAMMA	ISOTHERMAL T_t/T_t^*	RAYLEIGH T_t/T_t^*	SHOCK p_{t1}/p_{t2}
.100	2.16287	.68129	.44261	1.35489	.69072	1.70410
.099	2.16855	.68015	.43970	1.35716	.68935	1.71312
.098	2.17428	.67900	.43678	1.35946	.68797	1.72231
.097	2.18007	.67784	.43385	1.36179	.68659	1.73166
.096	2.18591	.67667	.43090	1.36414	.68519	1.74119
.095	2.19181	.67549	.42793	1.36652	.68379	1.75090
.094	2.19776	.67430	.42495	1.36894	.68238	1.76079
.093	2.20377	.67310	.42196	1.37138	.68097	1.77088
.092	2.20984	.67189	.41895	1.37385	.67955	1.78115
.091	2.21597	.67067	.41592	1.37636	.67812	1.79164
.090	2.22216	.66943	.41288	1.37889	.67668	1.80233
.089	2.22842	.66819	.40982	1.38146	.67523	1.81323
.088	2.23474	.66693	.40675	1.38407	.67377	1.82436
.087	2.24113	.66566	.40366	1.38671	.67231	1.83571
.086	2.24758	.66438	.40056	1.38938	.67084	1.84730
.085	2.25411	.66309	.39744	1.39209	.66936	1.85913
.084	2.26070	.66178	.39430	1.39484	.66787	1.87121
.083	2.26737	.66046	.39114	1.39763	.66637	1.88355
.082	2.27411	.65913	.38797	1.40045	.66486	1.89615
.081	2.28093	.65778	.38478	1.40332	.66334	1.90904
.080	2.28783	.65642	.38158	1.40623	.66182	1.92221
.079	2.29480	.65505	.37835	1.40918	.66028	1.93567
.078	2.30186	.65366	.37511	1.41218	.65873	1.94944
.077	2.30901	.65225	.37185	1.41522	.65718	1.96353
.076	2.31624	.65083	.36857	1.41830	.65561	1.97794
.075	2.32355	.64940	.36527	1.42144	.65403	1.99270
.074	2.33096	.64795	.36196	1.42462	.65244	2.00780
.073	2.33847	.64648	.35862	1.42785	.65084	2.02328
.072	2.34607	.64499	.35527	1.43114	.64923	2.03913
.071	2.35377	.64349	.35189	1.43448	.64761	2.05538
.070	2.36157	.64197	.34850	1.43788	.64597	2.07203
.069	2.36947	.64043	.34508	1.44133	.64432	2.08911
.068	2.37748	.63888	.34165	1.44484	.64267	2.10664
.067	2.38561	.63730	.33819	1.44841	.64099	2.12462
.066	2.39385	.63571	.33471	1.45205	.63931	2.14309
.065	2.40220	.63409	.33121	1.45575	.63761	2.16205
.064	2.41068	.63246	.32769	1.45951	.63590	2.18154
.063	2.41929	.63080	.32415	1.46335	.63417	2.20157
.062	2.42802	.62912	.32058	1.46726	.63243	2.22217
.061	2.43689	.62742	.31699	1.47124	.63067	2.24336
.060	2.44589	.62569	.31338	1.47530	.62890	2.26517
.059	2.45504	.62394	.30974	1.47943	.62712	2.28763
.058	2.46433	.62216	.30608	1.48366	.62531	2.31077
.057	2.47378	.62036	.30239	1.48796	.62349	2.33462
.056	2.48339	.61854	.29868	1.49236	.62166	2.35921
.055	2.49316	.61668	.29495	1.49685	.61980	2.38460
.054	2.50310	.61480	.29118	1.50143	.61793	2.41080
.053	2.51322	.61289	.28739	1.50612	.61604	2.43787
.052	2.52352	.61094	.28358	1.51091	.61413	2.46585
.051	2.53401	.60897	.27973	1.51580	.61220	2.49479

TABLE III. Isentropic Exponent = 1.20

p/p_t	M	T/T_t	GAMMA	ISOTHERMAL T_t/T_t^*	RAYLEIGH T_t/T_t^*	SHOCK p_{t1}/p_{t2}
.050	2.54470	.60696	.27586	1.52081	.61025	2.52475
.049	2.55559	.60492	.27196	1.52594	.60828	2.55577
.048	2.56670	.60285	.26802	1.53120	.60628	2.58792
.047	2.57804	.60074	.26406	1.53658	.60427	2.62127
.046	2.58960	.59859	.26007	1.54210	.60223	2.65588
.045	2.60141	.59640	.25604	1.54776	.60016	2.69184
.044	2.61348	.59417	.25199	1.55356	.59807	2.72923
.043	2.62581	.59190	.24790	1.55953	.59596	2.76813
.042	2.63842	.58958	.24377	1.56566	.59381	2.80865
.041	2.65133	.58722	.23961	1.57196	.59164	2.85089
.040	2.66454	.58480	.23542	1.57844	.58944	2.89497
.039	2.67807	.58234	.23118	1.58511	.58721	2.94103
.038	2.69195	.57983	.22691	1.59199	.58494	2.98920
.037	2.70618	.57725	.22260	1.59908	.58264	3.03963
.036	2.72079	.57462	.21826	1.60640	.58031	3.09251
.035	2.73580	.57193	.21386	1.61396	.57794	3.14802
.034	2.75123	.56918	.20943	1.62178	.57553	3.20638
.033	2.76711	.56635	.20495	1.62987	.57308	3.26781
.032	2.78347	.56345	.20043	1.63825	.57059	3.33259
.031	2.80033	.56048	.19586	1.64694	.56806	3.40101
.030	2.81773	.55743	.19124	1.65596	.56547	3.47339
.029	2.83571	.55428	.18657	1.66535	.56284	3.55012
.028	2.85431	.55105	.18185	1.67512	.56015	3.63162
.027	2.87357	.54772	.17708	1.68530	.55741	3.71837
.026	2.89355	.54429	.17225	1.69593	.55461	3.81093
.025	2.91429	.54074	.16735	1.70706	.55174	3.90993
.024	2.93588	.53708	.16240	1.71871	.54881	4.01610
.023	2.95836	.53328	.15738	1.73094	.54580	4.13029
.022	2.98184	.52934	.15230	1.74382	.54271	4.25352
.021	3.00639	.52525	.14714	1.75739	.53954	4.38695
.020	3.03213	.52100	.14191	1.77174	.53628	4.53198
.019	3.05919	.51657	.13660	1.78695	.53292	4.69028
.018	3.08769	.51193	.13121	1.80312	.52945	4.86387
.017	3.11782	.50708	.12573	1.82038	.52586	5.05521
.016	3.14978	.50198	.12015	1.83887	.52214	5.26733
.015	3.18379	.49661	.11447	1.85876	.51828	5.50402
.014	3.22016	.49093	.10868	1.88025	.51425	5.77009
.013	3.25923	.48491	.10278	1.90362	.51004	6.07172
.012	3.30144	.47848	.09674	1.92919	.50562	6.41700
.011	3.34736	.47159	.09057	1.95737	.50097	6.81678
.010	3.39770	.46416	.08424	1.98871	.49604	7.28594
.009	3.45341	.45608	.07774	2.02394	.49078	7.84554
.008	3.51578	.44721	.07104	2.06406	.48514	8.52640
.007	3.58663	.43737	.06412	2.11051	.47901	9.37570
.006	3.66864	.42628	.05695	2.16544	.47229	10.46981
.005	3.76600	.41352	.04946	2.23225	.46478	11.94157
.004	3.88575	.39842	.04159	2.31683	.45618	14.04628
.003	4.04126	.37977	.03323	2.43063	.44598	17.35104
.002	4.26295	.35495	.02417	2.60056	.43307	23.45088
.001	4.65004	.31623	.01397	2.91903	.41432	39.56351

TABLE IV. Isentropic Exponent = 1.30

p/p_t	M	T/T_t	GAMMA	ISOTHERMAL T_t/T_t^*	RAYLEIGH T_t/T_t^*	SHOCK p_{t1}/p_{t2}
1.0000	.00000	1.00000	.00000	.89655	.00000	1.
.9995	.02774	.99988	.04738	.89666	.00353	1.
.9990	.03924	.99977	.06698	.89676	.00705	1.
.9985	.04806	.99965	.08201	.89686	.01057	1.
.9980	.05550	.99954	.09467	.89697	.01406	1.
.9975	.06206	.99942	.10582	.89707	.01755	1.
.9970	.06800	.99931	.11588	.89717	.02103	1.
.9965	.07346	.99919	.12513	.89728	.02450	1.
.9960	.07854	.99908	.13373	.89738	.02795	1.
.9955	.08332	.99896	.14181	.89749	.03140	1.
.9950	.08784	.99884	.14943	.89759	.03483	1.
.9945	.09214	.99873	.15668	.89769	.03826	1.
.9940	.09625	.99861	.16360	.89780	.04167	1.
.9935	.10020	.99850	.17023	.89790	.04507	1.
.9930	.10400	.99838	.17661	.89801	.04846	1.
.9925	.10767	.99826	.18275	.89811	.05184	1.
.9920	.11121	.99815	.18869	.89822	.05521	1.
.9915	.11465	.99803	.19444	.89832	.05857	1.
.9910	.11800	.99792	.20002	.89842	.06192	1.
.9905	.12125	.99780	.20544	.89853	.06526	1.
.9900	.12442	.99768	.21072	.89863	.06859	1.
.9895	.12751	.99757	.21586	.89874	.07190	1.
.9890	.13053	.99745	.22087	.89884	.07521	1.
.9885	.13349	.99733	.22577	.89895	.07851	1.
.9880	.13638	.99722	.23056	.89905	.08179	1.
.9875	.13921	.99710	.23525	.89916	.08507	1.
.9870	.14199	.99698	.23983	.89926	.08833	1.
.9865	.14472	.99687	.24433	.89937	.09159	1.
.9860	.14740	.99675	.24874	.89947	.09483	1.
.9855	.15003	.99664	.25307	.89958	.09807	1.
.9850	.15262	.99652	.25732	.89968	.10129	1.
.9845	.15517	.99640	.26150	.89979	.10451	1.
.9840	.15767	.99628	.26560	.89990	.10771	1.
.9835	.16014	.99617	.26964	.90000	.11091	1.
.9830	.16258	.99605	.27362	.90011	.11409	1.
.9825	.15498	.99593	.27753	.90021	.11726	1.
.9820	.16734	.99582	.28139	.90032	.12043	1.
.9815	.16968	.99570	.28518	.90042	.12358	1.
.9810	.17198	.99558	.28893	.90053	.12673	1.
.9805	.17426	.99547	.29262	.90064	.12986	1.
.9800	.17650	.99535	.29626	.90074	.13299	1.
.9795	.17872	.99523	.29985	.90085	.13610	1.
.9790	.18092	.99511	.30339	.90095	.13921	1.
.9785	.18309	.99500	.30689	.90106	.14230	1.
.9780	.18524	.99488	.31035	.90117	.14539	1.
.9775	.18736	.99476	.31376	.90127	.14846	1.
.9770	.18946	.99464	.31714	.90138	.15153	1.
.9765	.19154	.99453	.32047	.90149	.15459	1.
.9760	.19359	.99441	.32377	.90159	.15764	1.
.9755	.19563	.99429	.32703	.90170	.16067	1.

TABLE IV. Isentropic Exponent = 1.30

p/p_t	M	T/T_t	GAMMA	ISOTHERMAL $T_t/T_t{}^*$	RAYLEIGH $T_t/T_t{}^*$	SHOCK p_{t1}/p_{t2}
.9750	.19765	.99417	.33025	.90181	.16370	1.
.9745	.19965	.99406	.33343	.90191	.16672	1.
.9740	.20163	.99394	.33659	.90202	.16973	1.
.9735	.20359	.99382	.33971	.90213	.17273	1.
.9730	.20553	.99370	.34280	.90223	.17572	1.
.9725	.20746	.99359	.34585	.90234	.17870	1.
.9720	.20937	.99347	.34888	.90245	.18167	1.
.9715	.21126	.99335	.35188	.90255	.18463	1.
.9710	.21314	.99323	.35484	.90266	.18759	1.
.9705	.21500	.99311	.35778	.90277	.19053	1.
.9700	.21685	.99300	.36069	.90288	.19347	1.
.9695	.21869	.99288	.36358	.90298	.19639	1.
.9690	.22051	.99276	.36644	.90309	.19931	1.
.9685	.22231	.99264	.36927	.90320	.20221	1.
.9680	.22411	.99252	.37208	.90331	.20511	1.
.9675	.22589	.99240	.37486	.90341	.20800	1.
.9670	.22765	.99229	.37762	.90352	.21088	1.
.9665	.22941	.99217	.38036	.90363	.21375	1.
.9660	.23115	.99205	.38307	.90374	.21661	1.
.9655	.23288	.99193	.38576	.90385	.21947	1.
.9650	.23460	.99181	.38843	.90395	.22231	1.
.9645	.23631	.99169	.39108	.90406	.22515	1.
.9640	.23800	.99157	.39371	.90417	.22797	1.
.9635	.23969	.99146	.39631	.90428	.23079	1.
.9630	.24136	.99134	.39890	.90439	.23360	1.
.9625	.24303	.99122	.40146	.90449	.23640	1.
.9620	.24468	.99110	.40401	.90460	.23919	1.
.9615	.24632	.99098	.40654	.90471	.24197	1.
.9610	.24796	.99086	.40904	.90482	.24474	1.
.9605	.24958	.99074	.41153	.90493	.24751	1.
.9600	.25120	.99062	.41401	.90504	.25026	1.
.9595	.25280	.99050	.41646	.90515	.25301	1.
.9590	.25440	.99039	.41890	.90526	.25575	1.
.9585	.25599	.99027	.42132	.90536	.25848	1.
.9580	.25756	.99015	.42372	.90547	.26120	1.
.9575	.25913	.99003	.42610	.90558	.26392	1.
.9570	.26070	.98991	.42847	.90569	.26662	1.
.9565	.26225	.98979	.43083	.90580	.26932	1.
.9560	.26379	.98967	.43316	.90591	.27200	1.
.9555	.26533	.98955	.43549	.90602	.27468	1.
.9550	.26686	.98943	.43779	.90613	.27735	1.
.9545	.26838	.98931	.44009	.90624	.28002	1.
.9540	.26990	.98919	.44236	.90635	.28267	1.
.9535	.27140	.98907	.44463	.90646	.28532	1.
.9530	.27290	.98895	.44687	.90657	.28796	1.
.9525	.27439	.98883	.44911	.90668	.29058	1.
.9520	.27588	.98871	.45133	.90679	.29321	1.
.9515	.27736	.98859	.45354	.90690	.29582	1.
.9510	.27883	.98847	.45573	.90701	.29842	1.
.9505	.28029	.98835	.45791	.90712	.30102	1.

TABLE IV.   Isentropic Exponent = 1.30

p/p_t	M	T/T_t	GAMMA	ISOTHERMAL T_t/T_t^*	RAYLEIGH T_t/T_t^*	SHOCK p_{t1}/p_{t2}
.950	.28175	.98823	.46008	.90723	.30361	1.
.949	.28464	.98799	.46437	.90745	.30876	1.
.948	.28751	.98775	.46861	.90767	.31388	1.
.947	.29036	.98751	.47281	.90789	.31897	1.
.946	.29318	.98727	.47696	.90811	.32403	1.
.945	.29598	.98703	.48106	.90833	.32905	1.
.944	.29875	.98679	.48511	.90855	.33404	1.
.943	.30150	.98655	.48913	.90878	.33901	1.
.942	.30424	.98631	.49310	.90900	.34394	1.
.941	.30695	.98606	.49702	.90922	.34884	1.
.940	.30964	.98582	.50091	.90945	.35371	1.
.939	.31231	.98558	.50475	.90967	.35854	1.
.938	.31496	.98534	.50856	.90989	.36335	1.
.937	.31759	.98510	.51233	.91012	.36813	1.
.936	.32021	.98485	.51606	.91034	.37288	1.
.935	.32281	.98461	.51975	.91057	.37760	1.
.934	.32539	.98437	.52341	.91079	.38229	1.
.933	.32795	.98412	.52703	.91102	.38695	1.
.932	.33050	.98388	.53062	.91124	.39158	1.
.931	.33303	.98364	.53418	.91147	.39618	1.
.930	.33554	.98339	.53770	.91169	.40075	1.
.929	.33804	.98315	.54119	.91192	.40530	1.
.928	.34052	.98290	.54465	.91215	.40981	1.
.927	.34299	.98266	.54807	.91237	.41430	1.
.926	.34545	.98241	.55147	.91260	.41876	1.
.925	.34789	.98217	.55483	.91283	.42319	1.
.924	.35032	.98192	.55817	.91306	.42759	1.
.923	.35273	.98168	.56148	.91328	.43197	1.
.922	.35513	.98143	.56476	.91351	.43631	1.
.921	.35752	.98119	.56801	.91374	.44063	1.
.920	.35989	.98094	.57123	.91397	.44493	1.
.919	.36225	.98070	.57443	.91420	.44919	1.
.918	.36460	.98045	.57760	.91443	.45343	1.
.917	.36694	.98020	.58074	.91466	.45765	1.
.916	.36927	.97996	.58386	.91489	.46183	1.
.915	.37158	.97971	.58695	.91512	.46599	1.
.914	.37389	.97946	.59002	.91535	.47013	1.
.913	.37618	.97921	.59306	.91558	.47423	1.
.912	.37846	.97897	.59608	.91581	.47832	1.
.911	.38073	.97872	.59908	.91605	.48237	1.
.910	.38299	.97847	.60205	.91628	.48640	1.
.909	.38524	.97822	.60500	.91651	.49041	1.
.908	.38748	.97797	.60793	.91674	.49439	1.
.907	.38972	.97773	.61083	.91698	.49834	1.
.906	.39194	.97748	.61371	.91721	.50227	1.
.905	.39415	.97723	.61657	.91744	.50618	1.
.904	.39635	.97698	.61941	.91768	.51006	1.
.903	.39854	.97673	.62223	.91791	.51391	1.
.902	.40073	.97648	.62503	.91815	.51774	1.
.901	.40290	.97623	.62780	.91838	.52155	1.

TABLE IV. Isentropic Exponent = 1.30

p/p_t	M	T/T_t	GAMMA	ISOTHERMAL T_t/T_t^*	RAYLEIGH T_t/T_t^*	SHOCK p_{t1}/p_{t2}
.900	.40507	.97598	.63056	.91862	.52533	1.
.899	.40723	.97573	.63329	.91885	.52909	1.
.898	.40937	.97548	.63601	.91909	.53282	1.
.897	.41152	.97523	.63871	.91933	.53654	1.
.896	.41365	.97498	.64138	.91956	.54022	1.
.895	.41577	.97473	.64404	.91980	.54389	1.
.894	.41789	.97447	.64668	.92004	.54753	1.
.893	.42000	.97422	.64930	.92027	.55115	1.
.892	.42210	.97397	.65190	.92051	.55474	1.
.891	.42419	.97372	.65449	.92075	.55831	1.
.890	.42628	.97347	.65705	.92099	.56186	1.
.889	.42836	.97321	.65960	.92123	.56539	1.
.888	.43043	.97296	.66213	.92147	.56889	1.
.887	.43250	.97271	.66465	.92171	.57237	1.
.886	.43455	.97245	.66714	.92195	.57583	1.
.885	.43661	.97220	.66962	.92219	.57927	1.
.884	.43865	.97195	.67209	.92243	.58269	1.
.883	.44069	.97169	.67453	.92267	.58608	1.
.882	.44272	.97144	.67696	.92291	.58945	1.
.881	.44474	.97119	.67938	.92315	.59280	1.
.880	.44676	.97093	.68177	.92339	.59613	1.
.879	.44877	.97068	.68416	.92364	.59944	1.
.878	.45078	.97042	.68652	.92388	.60273	1.
.877	.45278	.97017	.68887	.92412	.60599	1.
.876	.45477	.96991	.69121	.92437	.60924	1.
.875	.45676	.96965	.69353	.92461	.61246	1.
.874	.45874	.96940	.69583	.92485	.61566	1.
.873	.46072	.96914	.69812	.92510	.61885	1.
.872	.46269	.96889	.70040	.92534	.62201	1.
.871	.46466	.96863	.70266	.92559	.62515	1.
.870	.46661	.96837	.70491	.92583	.62827	1.
.869	.46857	.96812	.70714	.92608	.63137	1.
.868	.47052	.96786	.70936	.92632	.63445	1.
.867	.47246	.96760	.71156	.92657	.63751	1.
.866	.47440	.96734	.71375	.92682	.64055	1.
.865	.47633	.96709	.71593	.92706	.64358	1.
.864	.47826	.96683	.71809	.92731	.64658	1.
.863	.48018	.96657	.72024	.92756	.64956	1.
.862	.48210	.96631	.72237	.92781	.65252	1.
.861	.48401	.96605	.72450	.92806	.65547	1.
.860	.48592	.96579	.72660	.92831	.65839	1.
.859	.48783	.96553	.72870	.92856	.66130	1.
.858	.48972	.96527	.73078	.92880	.66419	1.
.857	.49162	.96501	.73285	.92905	.66705	1.
.856	.49351	.96475	.73491	.92931	.66990	1.
.855	.49539	.96449	.73696	.92956	.67273	1.
.854	.49727	.96423	.73899	.92981	.67555	1.
.853	.49915	.96397	.74101	.93006	.67834	1.
.852	.50102	.96371	.74302	.93031	.68112	1.
.851	.50289	.96345	.74501	.93056	.68387	1.

TABLE IV. Isentropic Exponent = 1.30

p/p_t	M	T/T_t	GAMMA	ISOTHERMAL T_t/T_t^*	RAYLEIGH T_t/T_t^*	SHOCK p_{t1}/p_{t2}
.850	.50475	.96319	.74700	.93081	.68661	1.
.849	.50661	.96293	.74897	.93107	.68934	1.
.848	.50847	.96267	.75093	.93132	.69204	1.
.847	.51032	.96240	.75287	.93157	.69473	1.
.846	.51217	.96214	.75481	.93183	.69739	1.
.845	.51401	.96188	.75673	.93208	.70004	1.
.844	.51585	.96162	.75865	.93234	.70268	1.
.843	.51769	.96135	.76055	.93259	.70529	1.
.842	.51952	.96109	.76244	.93285	.70789	1.
.841	.52134	.96083	.76432	.93310	.71047	1.
.840	.52317	.96056	.76618	.93336	.71304	1.
.839	.52499	.96030	.76804	.93362	.71559	1.
.838	.52681	.96004	.76989	.93387	.71812	1.
.837	.52862	.95977	.77172	.93413	.72063	1.
.836	.53043	.95951	.77354	.93439	.72313	1.
.835	.53224	.95924	.77536	.93465	.72561	1.
.834	.53404	.95898	.77716	.93491	.72807	1.
.833	.53584	.95871	.77895	.93516	.73052	1.
.832	.53763	.95844	.78073	.93542	.73295	1.
.831	.53943	.95818	.78250	.93568	.73537	1.
.830	.54121	.95791	.78426	.93594	.73776	1.
.829	.54300	.95765	.78601	.93620	.74015	1.
.828	.54478	.95738	.78775	.93646	.74251	1.
.827	.54656	.95711	.78948	.93673	.74486	1.
.826	.54834	.95684	.79120	.93699	.74720	1.
.825	.55011	.95658	.79290	.93725	.74952	1.
.824	.55188	.95631	.79460	.93751	.75182	1.
.823	.55365	.95604	.79629	.93777	.75411	1.
.822	.55542	.95577	.79797	.93804	.75638	1.
.821	.55718	.95551	.79964	.93830	.75864	1.
.820	.55894	.95524	.80130	.93857	.76088	1.
.819	.56069	.95497	.80295	.93883	.76311	1.
.818	.56244	.95470	.80459	.93909	.76532	1.
.817	.56419	.95443	.80622	.93936	.76752	1.
.816	.56594	.95416	.80784	.93962	.76970	1.
.815	.56768	.95389	.80945	.93989	.77187	1.
.814	.56943	.95362	.81105	.94016	.77402	1.
.813	.57117	.95335	.81265	.94042	.77616	1.
.812	.57290	.95308	.81423	.94069	.77828	1.
.811	.57464	.95281	.81580	.94096	.78039	1.
.810	.57637	.95254	.81737	.94123	.78248	1.
.809	.57809	.95226	.81892	.94150	.78456	1.
.808	.57982	.95199	.82047	.94176	.78663	1.
.807	.58154	.95172	.82201	.94203	.78868	1.
.806	.58327	.95145	.82354	.94230	.79071	1.
.805	.58498	.95118	.82506	.94257	.79274	1.
.804	.58670	.95090	.82657	.94284	.79474	1.
.803	.58841	.95063	.82807	.94311	.79674	1.
.802	.59013	.95036	.82956	.94339	.79872	1.
.801	.59183	.95008	.83105	.94366	.80069	1.

TABLE IV. Isentropic Exponent = 1.30

p/p_t	M	T/T_t	GAMMA	ISOTHERMAL T_t/T^*	RAYLEIGH T_t/T^*	SHOCK p_{t1}/p_{t2}
.800	.59354	.94981	.83253	.94393	.80264	1.
.799	.59525	.94953	.83399	.94420	.80458	1.
.798	.59695	.94926	.83545	.94447	.80651	1.
.797	.59865	.94899	.83690	.94475	.80842	1.
.796	.60035	.94871	.83834	.94502	.81032	1.
.795	.60204	.94844	.83978	.94530	.81220	1.
.794	.60373	.94816	.84120	.94557	.81408	1.
.793	.60543	.94788	.84262	.94585	.81594	1.
.792	.60711	.94761	.84403	.94612	.81778	1.
.791	.60880	.94733	.84543	.94640	.81962	1.
.790	.61049	.94706	.84682	.94667	.82144	1.
.789	.61217	.94678	.84820	.94695	.82324	1.
.788	.61385	.94650	.84958	.94723	.82504	1.
.787	.61553	.94622	.85095	.94750	.82682	1.
.786	.61721	.94595	.85231	.94778	.82859	1.
.785	.61888	.94567	.85366	.94806	.83034	1.
.784	.62056	.94539	.85500	.94834	.83209	1.
.783	.62223	.94511	.85634	.94862	.83382	1.
.782	.62390	.94483	.85766	.94890	.83554	1.
.781	.62557	.94455	.85898	.94918	.83724	1.
.780	.62723	.94428	.86030	.94946	.83894	1.
.779	.62890	.94400	.86160	.94974	.84062	1.
.778	.63056	.94372	.86290	.95002	.84229	1.
.777	.63222	.94344	.86419	.95030	.84395	1.
.776	.63388	.94316	.86547	.95059	.84559	1.
.775	.63553	.94288	.86674	.95087	.84723	1.
.774	.63719	.94259	.86801	.95115	.84885	1.
.773	.63884	.94231	.86926	.95144	.85046	1.
.772	.64050	.94203	.87052	.95172	.85206	1.
.771	.64215	.94175	.87176	.95201	.85364	1.
.770	.64380	.94147	.87299	.95229	.85522	1.
.769	.64544	.94119	.87422	.95258	.85678	1.
.768	.64709	.94090	.87544	.95286	.85833	1.
.767	.64873	.94062	.87666	.95315	.85987	1.
.766	.65038	.94034	.87786	.95344	.86140	1.
.765	.65202	.94005	.87906	.95372	.86292	1.
.764	.65366	.93977	.88025	.95401	.86442	1.
.763	.65530	.93949	.88144	.95430	.86592	1.
.762	.65693	.93920	.88262	.95459	.86740	1.
.761	.65857	.93892	.88379	.95488	.86887	1.
.760	.66020	.93863	.88495	.95517	.87033	1.
.759	.66183	.93835	.88610	.95546	.87178	1.
.758	.66347	.93806	.88725	.95575	.87322	1.
.757	.66510	.93778	.88839	.95604	.87465	1.
.756	.66672	.93749	.88953	.95633	.87607	1.
.755	.66835	.93720	.89066	.95662	.87747	1.
.754	.66998	.93692	.89178	.95692	.87887	1.
.753	.67160	.93663	.89289	.95721	.88025	1.
.752	.67323	.93634	.89400	.95750	.88163	1.
.751	.67485	.93606	.89510	.95780	.88299	1.

TABLE IV. Isentropic Exponent $= 1.30$

p/p_t	M	T/T_t	GAMMA	ISOTHERMAL T_t/T_t^*	RAYLEIGH T_t/T_t^*	SHOCK p_{t1}/p_{t2}
.750	.67647	.93577	.89619	.95809	.88435	1.
.749	.67809	.93548	.89728	.95839	.88569	1.
.748	.67971	.93519	.89836	.95868	.88702	1.
.747	.68132	.93490	.89943	.95898	.88834	1.
.746	.68294	.93461	.90049	.95928	.88966	1.
.745	.68455	.93432	.90155	.95957	.89096	1.
.744	.68617	.93403	.90260	.95987	.89225	1.
.743	.68778	.93374	.90365	.96017	.89353	1.
.742	.68939	.93345	.90469	.96047	.89480	1.
.741	.69100	.93316	.90572	.96077	.89606	1.
.740	.69261	.93287	.90675	.96106	.89731	1.
.739	.69422	.93258	.90777	.96136	.89855	1.
.738	.69583	.93229	.90878	.96167	.89979	1.
.737	.69743	.93200	.90978	.96197	.90101	1.
.736	.69904	.93171	.91078	.96227	.90222	1.
.735	.70064	.93141	.91178	.96257	.90342	1.
.734	.70225	.93112	.91276	.96287	.90461	1.
.733	.70385	.93083	.91374	.96318	.90580	1.
.732	.70545	.93054	.91472	.96348	.90697	1.
.731	.70705	.93024	.91568	.96378	.90813	1.
.730	.70865	.92995	.91664	.96409	.90929	1.
.729	.71025	.92965	.91760	.96439	.91043	1.
.728	.71185	.92936	.91855	.96470	.91157	1.
.727	.71344	.92907	.91949	.96500	.91269	1.
.726	.71504	.92877	.92042	.96531	.91381	1.
.725	.71663	.92848	.92135	.96562	.91492	1.
.724	.71823	.92818	.92227	.96592	.91601	1.
.723	.71982	.92788	.92319	.96623	.91710	1.
.722	.72141	.92759	.92410	.96654	.91818	1.
.721	.72301	.92729	.92501	.96685	.91925	1.
.720	.72460	.92699	.92590	.96716	.92032	1.
.719	.72619	.92670	.92680	.96747	.92137	1.
.718	.72778	.92640	.92768	.96778	.92241	1.
.717	.72937	.92610	.92856	.96809	.92345	1.
.716	.73095	.92580	.92943	.96840	.92447	1.
.715	.73254	.92550	.93030	.96872	.92549	1.
.714	.73413	.92521	.93116	.96903	.92650	1.
.713	.73571	.92491	.93202	.96934	.92750	1.
.712	.73730	.92461	.93287	.96966	.92849	1.
.711	.73888	.92431	.93371	.96997	.92947	1.
.710	.74047	.92401	.93455	.97029	.93045	1.
.709	.74205	.92371	.93538	.97060	.93141	1.
.708	.74363	.92341	.93620	.97092	.93237	1.
.707	.74521	.92310	.93702	.97124	.93332	1.
.706	.74680	.92280	.93784	.97155	.93426	1.
.705	.74838	.92250	.93864	.97187	.93519	1.
.704	.74996	.92220	.93945	.97219	.93611	1.
.703	.75154	.92190	.94024	.97251	.93703	1.
.702	.75311	.92159	.94103	.97283	.93794	1.
.701	.75469	.92129	.94181	.97315	.93883	1.

TABLE IV. Isentropic Exponent = 1.30

p/p$_t$	M	T/T$_t$	GAMMA	ISOTHERMAL T$_t$/T*	RAYLEIGH T$_t$/T*	SHOCK p$_{t1}$/p$_{t2}$
.700	.75627	.92099	.94259	.97347	.93972	1.
.699	.75785	.92068	.94336	.97379	.94061	1.
.698	.75943	.92038	.94413	.97411	.94148	1.
.697	.76100	.92007	.94489	.97443	.94235	1.
.696	.76258	.91977	.94565	.97476	.94321	1.
.695	.76415	.91946	.94640	.97508	.94406	1.
.694	.76573	.91916	.94714	.97540	.94490	1.
.693	.76730	.91885	.94788	.97573	.94573	1.
.692	.76888	.91855	.94861	.97605	.94656	1.
.691	.77045	.91824	.94934	.97638	.94738	1.
.690	.77202	.91793	.95006	.97671	.94819	1.
.689	.77360	.91763	.95077	.97703	.94899	1.
.688	.77517	.91732	.95148	.97736	.94979	1.
.687	.77674	.91701	.95218	.97769	.95058	1.
.686	.77831	.91670	.95288	.97802	.95136	1.
.685	.77989	.91639	.95358	.97835	.95213	1.
.684	.78146	.91609	.95426	.97868	.95290	1.
.683	.78303	.91578	.95494	.97901	.95366	1.
.682	.78460	.91547	.95562	.97934	.95441	1.
.681	.78617	.91516	.95629	.97967	.95515	1.
.680	.78774	.91485	.95695	.98000	.95589	1.
.679	.78931	.91454	.95761	.98034	.95661	1.
.678	.79088	.91422	.95827	.98067	.95733	1.
.677	.79245	.91391	.95892	.98100	.95805	1.
.676	.79401	.91360	.95956	.98134	.95875	1.
.675	.79558	.91329	.96020	.98167	.95945	1.
.674	.79715	.91298	.96083	.98201	.96015	1.
.673	.79872	.91266	.96145	.98235	.96083	1.
.672	.80029	.91235	.96208	.98268	.96151	1.
.671	.80185	.91204	.96269	.98302	.96218	1.
.670	.80342	.91172	.96330	.98336	.96284	1.
.669	.80499	.91141	.96391	.98370	.96350	1.
.668	.80656	.91110	.96451	.98404	.96415	1.
.667	.80812	.91078	.96510	.98438	.96479	1.
.666	.80969	.91047	.96569	.98472	.96543	1.
.665	.81126	.91015	.96627	.98506	.96606	1.
.664	.81282	.90983	.96685	.98540	.96668	1.
.663	.81439	.90952	.96742	.98574	.96730	1.
.662	.81596	.90920	.96799	.98609	.96790	1.
.661	.81752	.90888	.96855	.98643	.96851	1.
.660	.81909	.90857	.96911	.98678	.96910	1.
.659	.82066	.90825	.96966	.98712	.96969	1.
.658	.82222	.90793	.97020	.98747	.97027	1.
.657	.82379	.90761	.97075	.98782	.97085	1.
.656	.82535	.90729	.97128	.98816	.97142	1.
.655	.82692	.90697	.97181	.98851	.97198	1.
.654	.82848	.90665	.97234	.98886	.97253	1.
.653	.83005	.90633	.97286	.98921	.97308	1.
.652	.83162	.90601	.97337	.98956	.97363	1.
.651	.83318	.90569	.97388	.98991	.97416	1.

TABLE IV. Isentropic Exponent = 1.30

p/p_t	M	T/T_t	GAMMA	ISOTHERMAL T_t/T_t^*	RAYLEIGH T_t/T_t^*	SHOCK p_{t1}/p_{t2}
.650	.83475	.90537	.97439	.99026	.97469	1.
.649	.83631	.90505	.97488	.99061	.97521	1.
.648	.83788	.90473	.97538	.99096	.97573	1.
.647	.83945	.90440	.97587	.99132	.97624	1.
.646	.84101	.90408	.97635	.99167	.97675	1.
.645	.84258	.90376	.97683	.99203	.97725	1.
.644	.84414	.90343	.97730	.99238	.97774	1.
.643	.84571	.90311	.97777	.99274	.97822	1.
.642	.84728	.90279	.97823	.99309	.97870	1.
.641	.84884	.90246	.97869	.99345	.97918	1.
.640	.85041	.90214	.97915	.99381	.97965	1.
.639	.85198	.90181	.97959	.99417	.98011	1.
.638	.85354	.90149	.98004	.99453	.98056	1.
.637	.85511	.90116	.98047	.99489	.98101	1.
.636	.85668	.90083	.98091	.99525	.98146	1.
.635	.85825	.90051	.98133	.99561	.98189	1.
.634	.85981	.90018	.98176	.99597	.98232	1.
.633	.86138	.89985	.98217	.99633	.98275	1.
.632	.86295	.89952	.98259	.99670	.98317	1.
.631	.86452	.89919	.98299	.99706	.98358	1.
.630	.86608	.89886	.98340	.99743	.98399	1.
.629	.86765	.89853	.98379	.99779	.98439	1.
.628	.86922	.89820	.98419	.99816	.98479	1.
.627	.87079	.89787	.98457	.99853	.98518	1.
.626	.87236	.89754	.98496	.99889	.98557	1.
.625	.87393	.89721	.98533	.99926	.98595	1.
.624	.87550	.89688	.98571	.99963	.98632	1.
.623	.87707	.89655	.98608	1.00000	.98669	1.
.622	.87864	.89622	.98644	1.00037	.98705	1.
.621	.88021	.89588	.98680	1.00075	.98741	1.
.620	.88178	.89555	.98715	1.00112	.98776	1.
.619	.88335	.89522	.98750	1.00149	.98811	1.
.618	.88492	.89488	.98784	1.00186	.98845	1.
.617	.88650	.89455	.98818	1.00224	.98878	1.
.616	.88807	.89421	.98851	1.00261	.98911	1.
.615	.88964	.89388	.98884	1.00299	.98944	1.
.614	.89121	.89354	.98916	1.00337	.98975	1.
.613	.89279	.89321	.98948	1.00374	.99007	1.
.612	.89436	.89287	.98980	1.00412	.99038	1.
.611	.89594	.89253	.99010	1.00450	.99068	1.
.610	.89751	.89220	.99041	1.00488	.99098	1.
.609	.89909	.89186	.99071	1.00526	.99127	1.
.608	.90066	.89152	.99100	1.00564	.99155	1.
.607	.90224	.89118	.99129	1.00602	.99184	1.
.606	.90381	.89084	.99158	1.00641	.99211	1.
.605	.90539	.89050	.99185	1.00679	.99238	1.
.604	.90697	.89016	.99213	1.00718	.99265	1.
.603	.90855	.88982	.99240	1.00756	.99291	1.
.602	.91012	.88948	.99267	1.00795	.99317	1.
.601	.91170	.88914	.99293	1.00833	.99342	1.

TABLE IV. Isentropic Exponent = 1.30

p/p_t	M	T/T_t	GAMMA	ISOTHERMAL T_t/T^*	RAYLEIGH T_t/T^*	SHOCK p_{t1}/p_{t2}
.600	.91328	.88880	.99318	1.00872	.99366	1.
.599	.91486	.88846	.99343	1.00911	.99390	1.
.598	.91644	.88812	.99368	1.00950	.99414	1.
.597	.91802	.88777	.99392	1.00989	.99437	1.
.596	.91960	.88743	.99416	1.01028	.99460	1.
.595	.92118	.88709	.99439	1.01067	.99482	1.
.594	.92277	.88674	.99462	1.01106	.99503	1.
.593	.92435	.88640	.99484	1.01146	.99524	1.
.592	.92593	.88605	.99505	1.01185	.99545	1.
.591	.92752	.88571	.99527	1.01225	.99565	1.
.590	.92910	.88536	.99548	1.01264	.99585	1.
.589	.93069	.88501	.99568	1.01304	.99604	1.
.588	.93227	.88467	.99588	1.01344	.99622	1.
.587	.93386	.88432	.99607	1.01383	.99641	1.
.586	.93545	.88397	.99626	1.01423	.99658	1.
.585	.93703	.88362	.99644	1.01463	.99676	1.
.584	.93862	.88327	.99662	1.01503	.99692	1.
.583	.94021	.88292	.99680	1.01543	.99709	1.
.582	.94180	.88257	.99697	1.01584	.99725	1.
.581	.94339	.88222	.99713	1.01624	.99740	1.
.580	.94498	.88187	.99730	1.01664	.99755	1.
.579	.94658	.88152	.99745	1.01705	.99769	1.
.578	.94817	.88117	.99760	1.01745	.99783	1.
.577	.94976	.88082	.99775	1.01786	.99797	1.
.576	.95136	.88047	.99789	1.01827	.99810	1.
.575	.95295	.88011	.99803	1.01868	.99822	1.
.574	.95455	.87976	.99816	1.01909	.99835	1.
.573	.95614	.87941	.99829	1.01950	.99846	1.
.572	.95774	.87905	.99841	1.01991	.99858	1.
.571	.95934	.87870	.99853	1.02032	.99868	1.
.570	.96093	.87834	.99865	1.02073	.99879	1.
.569	.96253	.87799	.99876	1.02115	.99889	1.
.568	.96413	.87763	.99886	1.02156	.99898	1.
.567	.96573	.87727	.99896	1.02198	.99907	1.
.566	.96734	.87692	.99906	1.02239	.99916	1.
.565	.96894	.87656	.99915	1.02281	.99924	1.
.564	.97054	.87620	.99923	1.02323	.99932	1.
.563	.97215	.87584	.99932	1.02365	.99939	1.
.562	.97375	.87548	.99939	1.02407	.99946	1.
.561	.97536	.87512	.99947	1.02449	.99953	1.
.560	.97696	.87476	.99953	1.02491	.99959	1.
.559	.97857	.87440	.99960	1.02533	.99964	1.
.558	.98018	.87404	.99966	1.02576	.99970	1.
.557	.98179	.87368	.99971	1.02618	.99974	1.
.556	.98340	.87332	.99976	1.02661	.99979	1.
.555	.98501	.87295	.99980	1.02703	.99983	1.
.554	.98663	.87259	.99984	1.02746	.99986	1.
.553	.98824	.87223	.99988	1.02789	.99989	1.
.552	.98985	.87186	.99991	1.02832	.99992	1.
.551	.99147	.87150	.99994	1.02875	.99994	1.

TABLE IV.   Isentropic Exponent = 1.30

p/p_t	M	T/T_t	GAMMA	ISOTHERMAL T_t/T_t^*	RAYLEIGH T_t/T_t^*	SHOCK p_{t1}/p_{t2}
.550	.99308	.87113	.99996	1.02918	.99996	1.
.549	.99470	.87077	.99998	1.02961	.99998	1.
.548	.99632	.87040	.99999	1.03005	.99999	1.
.547	.99794	.87003	.00000	1.03048	.00000	1.
.546	.99956	.86967	.00000	1.03092	.00000	1.
.545	1.00118	.86930	.00000	1.03135	.00000	1.00000
.544	1.00280	.86893	.99999	1.03179	.99999	1.00000
.543	1.00443	.86856	.99998	1.03223	.99999	1.00000
.542	1.00605	.86819	.99997	1.03267	.99997	1.00000
.541	1.00768	.86782	.99995	1.03311	.99996	1.00000
.540	1.00930	.86745	.99993	1.03355	.99994	1.00000
.539	1.01093	.86708	.99990	1.03399	.99991	1.00000
.538	1.01256	.86671	.99986	1.03443	.99988	1.00000
.537	1.01419	.86634	.99983	1.03488	.99985	1.00000
.536	1.01582	.86596	.99978	1.03532	.99981	1.00000
.535	1.01745	.86559	.99974	1.03577	.99977	1.00001
.534	1.01908	.86522	.99969	1.03622	.99973	1.00001
.533	1.02072	.86484	.99963	1.03666	.99968	1.00001
.532	1.02235	.86447	.99957	1.03711	.99963	1.00001
.531	1.02399	.86409	.99951	1.03756	.99958	1.00002
.530	1.02563	.86372	.99944	1.03802	.99952	1.00002
.529	1.02727	.86334	.99936	1.03847	.99946	1.00003
.528	1.02891	.86296	.99928	1.03892	.99939	1.00003
.527	1.03055	.86259	.99920	1.03938	.99932	1.00004
.526	1.03219	.86221	.99911	1.03983	.99925	1.00004
.525	1.03384	.86183	.99902	1.04029	.99917	1.00005
.524	1.03548	.86145	.99893	1.04075	.99909	1.00005
.523	1.03713	.86107	.99882	1.04121	.99901	1.00006
.522	1.03877	.86069	.99872	1.04167	.99892	1.00007
.521	1.04042	.86031	.99861	1.04213	.99883	1.00008
.520	1.04207	.85993	.99849	1.04259	.99873	1.00009
.519	1.04372	.85955	.99837	1.04305	.99863	1.00010
.518	1.04538	.85916	.99825	1.04352	.99853	1.00011
.517	1.04703	.85878	.99812	1.04398	.99842	1.00012
.516	1.04869	.85840	.99799	1.04445	.99831	1.00014
.515	1.05034	.85801	.99785	1.04492	.99820	1.00015
.514	1.05200	.85763	.99771	1.04538	.99809	1.00017
.513	1.05366	.85724	.99757	1.04585	.99797	1.00018
.512	1.05532	.85686	.99741	1.04633	.99784	1.00020
.511	1.05698	.85647	.99726	1.04680	.99772	1.00022
.510	1.05865	.85608	.99710	1.04727	.99759	1.00024
.509	1.06031	.85570	.99694	1.04775	.99745	1.00026
.508	1.06198	.85531	.99677	1.04822	.99731	1.00028
.507	1.06365	.85492	.99659	1.04870	.99717	1.00030
.506	1.06532	.85453	.99642	1.04918	.99703	1.00032
.505	1.06699	.85414	.99623	1.04966	.99688	1.00035
.504	1.06866	.85375	.99605	1.05014	.99673	1.00037
.503	1.07033	.85336	.99585	1.05062	.99658	1.00040
.502	1.07201	.85297	.99566	1.05110	.99642	1.00043
.501	1.07368	.85257	.99546	1.05158	.99626	1.00046

TABLE IV. Isentropic Exponent = 1.30

p/p_t	M	T/T_t	GAMMA	ISOTHERMAL T_t/T_t^*	RAYLEIGH T_t/T_t^*	SHOCK p_{t1}/p_{t2}
.500	1.07536	.85218	.99525	1.05207	.99610	1.00049
.499	1.07704	.85179	.99504	1.05255	.99593	1.00052
.498	1.07872	.85139	.99483	1.05304	.99576	1.00055
.497	1.08041	.85100	.99461	1.05353	.99558	1.00059
.496	1.08209	.85060	.99439	1.05402	.99541	1.00062
.495	1.08377	.85021	.99416	1.05451	.99523	1.00066
.494	1.08546	.84981	.99393	1.05500	.99504	1.00070
.493	1.08715	.84941	.99369	1.05550	.99486	1.00074
.492	1.08884	.84901	.99345	1.05599	.99467	1.00078
.491	1.09053	.84862	.99321	1.05649	.99447	1.00083
.490	1.09223	.84822	.99296	1.05698	.99428	1.00087
.489	1.09392	.84782	.99270	1.05748	.99408	1.00092
.488	1.09562	.84742	.99244	1.05798	.99388	1.00097
.487	1.09732	.84702	.99218	1.05848	.99367	1.00101
.486	1.09902	.84661	.99191	1.05899	.99346	1.00107
.485	1.10072	.84621	.99164	1.05949	.99325	1.00112
.484	1.10242	.84581	.99136	1.05999	.99303	1.00117
.483	1.10413	.84540	.99108	1.06050	.99281	1.00123
.482	1.10584	.84500	.99080	1.06101	.99259	1.00129
.481	1.10755	.84460	.99051	1.06152	.99237	1.00135
.480	1.10926	.84419	.99021	1.06203	.99214	1.00141
.479	1.11097	.84378	.98991	1.06254	.99191	1.00147
.478	1.11268	.84338	.98961	1.06305	.99168	1.00154
.477	1.11440	.84297	.98930	1.06356	.99144	1.00160
.476	1.11612	.84256	.98899	1.06408	.99120	1.00167
.475	1.11783	.84215	.98867	1.06460	.99096	1.00174
.474	1.11956	.84174	.98835	1.06511	.99071	1.00181
.473	1.12128	.84133	.98802	1.06563	.99046	1.00189
.472	1.12300	.84092	.98769	1.06615	.99021	1.00196
.471	1.12473	.84051	.98735	1.06667	.98996	1.00204
.470	1.12646	.84010	.98701	1.06720	.98970	1.00212
.469	1.12819	.83969	.98667	1.06772	.98944	1.00220
.468	1.12992	.83927	.98632	1.06825	.98918	1.00229
.467	1.13166	.83886	.98597	1.06878	.98891	1.00237
.466	1.13339	.83844	.98561	1.06931	.98864	1.00246
.465	1.13513	.83803	.98525	1.06984	.98837	1.00255
.464	1.13687	.83761	.98488	1.07037	.98809	1.00264
.463	1.13861	.83719	.98451	1.07090	.98781	1.00274
.462	1.14036	.83678	.98413	1.07143	.98753	1.00284
.461	1.14210	.83636	.98375	1.07197	.98725	1.00293
.460	1.14385	.83594	.98336	1.07251	.98696	1.00304
.459	1.14560	.83552	.98297	1.07305	.98667	1.00314
.458	1.14735	.83510	.98258	1.07359	.98638	1.00324
.457	1.14911	.83468	.98218	1.07413	.98609	1.00335
.456	1.15086	.83426	.98178	1.07467	.98579	1.00346
.455	1.15262	.83383	.98137	1.07522	.98549	1.00357
.454	1.15438	.83341	.98096	1.07576	.98518	1.00369
.453	1.15614	.83299	.98054	1.07631	.98488	1.00381
.452	1.15791	.83256	.98012	1.07686	.98457	1.00393
.451	1.15967	.83214	.97969	1.07741	.98426	1.00405

TABLE IV. Isentropic Exponent = 1.30

p/p_t	M	T/T_t	GAMMA	ISOTHERMAL T_t/T_t^*	RAYLEIGH T_t/T_t^*	SHOCK p_{t1}/p_{t2}
.450	1.16144	.83171	.97926	1.07796	.98394	1.00417
.449	1.16321	.83128	.97883	1.07852	.98362	1.00430
.448	1.15498	.83086	.97839	1.07907	.98330	1.00443
.447	1.16676	.83043	.97794	1.07963	.98298	1.00456
.446	1.16854	.83000	.97749	1.08018	.98266	1.00469
.445	1.17032	.82957	.97704	1.08074	.98233	1.00483
.444	1.17210	.82914	.97658	1.08131	.98200	1.00497
.443	1.17388	.82871	.97612	1.08187	.98166	1.00511
.442	1.17567	.82827	.97565	1.08243	.98133	1.00526
.441	1.17746	.82784	.97518	1.08300	.98099	1.00541
.440	1.17925	.82741	.97470	1.08357	.98065	1.00556
.439	1.18104	.82697	.97422	1.08414	.98030	1.00571
.438	1.18283	.82654	.97374	1.08471	.97995	1.00586
.437	1.18463	.82610	.97325	1.08528	.97960	1.00602
.436	1.18643	.82567	.97275	1.08585	.97925	1.00618
.435	1.18823	.82523	.97225	1.08643	.97890	1.00635
.434	1.19004	.82479	.97175	1.08701	.97854	1.00651
.433	1.19185	.82435	.97124	1.08758	.97818	1.00668
.432	1.19365	.82391	.97073	1.08816	.97782	1.00685
.431	1.19547	.82347	.97021	1.08875	.97745	1.00703
.430	1.19728	.82303	.96969	1.08933	.97708	1.00721
.429	1.19910	.82259	.96916	1.08992	.97671	1.00739
.428	1.20092	.82215	.96863	1.09050	.97634	1.00757
.427	1.20274	.82170	.96809	1.09109	.97596	1.00776
.426	1.20456	.82126	.96755	1.09168	.97559	1.00795
.425	1.20639	.82081	.96701	1.09227	.97521	1.00814
.424	1.20822	.82037	.96646	1.09287	.97482	1.00833
.423	1.21005	.81992	.96590	1.09346	.97444	1.00853
.422	1.21188	.81947	.96534	1.09406	.97405	1.00873
.421	1.21372	.81902	.96478	1.09466	.97366	1.00894
.420	1.21556	.81857	.96421	1.09526	.97326	1.00915
.419	1.21740	.81812	.96364	1.09586	.97287	1.00936
.418	1.21925	.81767	.96306	1.09647	.97247	1.00957
.417	1.22109	.81722	.96248	1.09707	.97207	1.00979
.416	1.22294	.81677	.96189	1.09768	.97167	1.01001
.415	1.22480	.81631	.96130	1.09829	.97126	1.01023
.414	1.22665	.81586	.96070	1.09890	.97085	1.01046
.413	1.22851	.81540	.96010	1.09952	.97044	1.01069
.412	1.23037	.81495	.95950	1.10013	.97003	1.01092
.411	1.23223	.81449	.95888	1.10075	.96961	1.01116
.410	1.23410	.81403	.95827	1.10137	.96920	1.01140
.409	1.23597	.81358	.95765	1.10199	.96878	1.01165
.408	1.23784	.81312	.95703	1.10261	.96835	1.01189
.407	1.23972	.81266	.95640	1.10324	.96793	1.01214
.406	1.24159	.81219	.95576	1.10386	.96750	1.01240
.405	1.24347	.81173	.95512	1.10449	.96707	1.01266
.404	1.24536	.81127	.95448	1.10512	.96664	1.01292
.403	1.24724	.81081	.95383	1.10575	.96620	1.01318
.402	1.24913	.81034	.95318	1.10639	.96577	1.01345
.401	1.25102	.80987	.95252	1.10703	.96533	1.01372

TABLE IV.   Isentropic Exponent = 1.30

p/p_t	M	T/T_t	GAMMA	ISOTHERMAL T_t/T_t^*	RAYLEIGH T_t/T_t^*	SHOCK p_{t1}/p_{t2}
.400	1.25292	.80941	.95186	1.10766	.96489	1.01400
.399	1.25482	.80894	.95119	1.10830	.96444	1.01428
.398	1.25672	.80847	.95052	1.10895	.96399	1.01456
.397	1.25862	.80800	.94985	1.10959	.96355	1.01485
.396	1.26053	.80753	.94917	1.11024	.96309	1.01514
.395	1.26244	.80706	.94848	1.11088	.96264	1.01543
.394	1.26435	.80659	.94779	1.11153	.96219	1.01573
.393	1.26627	.80612	.94709	1.11219	.96173	1.01603
.392	1.26818	.80564	.94639	1.11284	.96127	1.01633
.391	1.27011	.80517	.94569	1.11350	.96080	1.01664
.390	1.27203	.80469	.94498	1.11415	.96034	1.01696
.389	1.27396	.80422	.94427	1.11481	.95987	1.01727
.388	1.27589	.80374	.94355	1.11548	.95940	1.01760
.387	1.27783	.80326	.94282	1.11614	.95893	1.01792
.386	1.27977	.80278	.94209	1.11681	.95846	1.01825
.385	1.28171	.80230	.94136	1.11748	.95798	1.01858
.384	1.28365	.80182	.94062	1.11815	.95750	1.01892
.383	1.28560	.80134	.93988	1.11882	.95702	1.01926
.382	1.28755	.80085	.93913	1.11950	.95654	1.01961
.381	1.28950	.80037	.93838	1.12017	.95605	1.01996
.380	1.29146	.79988	.93762	1.12085	.95556	1.02031
.379	1.29342	.79940	.93686	1.12153	.95507	1.02067
.378	1.29539	.79891	.93609	1.12222	.95458	1.02103
.377	1.29736	.79842	.93532	1.12290	.95408	1.02140
.376	1.29933	.79793	.93454	1.12359	.95359	1.02177
.375	1.30130	.79744	.93376	1.12428	.95309	1.02214
.374	1.30328	.79695	.93297	1.12498	.95259	1.02252
.373	1.30526	.79646	.93218	1.12567	.95208	1.02290
.372	1.30725	.79597	.93138	1.12637	.95158	1.02329
.371	1.30924	.79547	.93058	1.12707	.95107	1.02369
.370	1.31123	.79498	.92978	1.12777	.95056	1.02408
.369	1.31323	.79448	.92897	1.12848	.95005	1.02448
.368	1.31523	.79398	.92815	1.12918	.94953	1.02489
.367	1.31723	.79348	.92733	1.12989	.94902	1.02530
.366	1.31924	.79298	.92650	1.13060	.94850	1.02572
.365	1.32125	.79248	.92567	1.13132	.94798	1.02614
.364	1.32326	.79198	.92484	1.13203	.94745	1.02656
.363	1.32528	.79148	.92400	1.13275	.94693	1.02699
.362	1.32730	.79098	.92315	1.13347	.94640	1.02743
.361	1.32933	.79047	.92230	1.13420	.94587	1.02787
.360	1.33136	.78997	.92144	1.13492	.94534	1.02831
.359	1.33339	.78946	.92058	1.13565	.94481	1.02876
.358	1.33543	.78895	.91972	1.13638	.94427	1.02921
.357	1.33747	.78844	.91885	1.13712	.94373	1.02967
.356	1.33952	.78793	.91797	1.13785	.94319	1.03013
.355	1.34157	.78742	.91709	1.13859	.94265	1.03060
.354	1.34362	.78691	.91620	1.13934	.94210	1.03108
.353	1.34568	.78639	.91531	1.14008	.94156	1.03156
.352	1.34774	.78588	.91442	1.14083	.94101	1.03204
.351	1.34980	.78536	.91352	1.14158	.94046	1.03253

TABLE IV. Isentropic Exponent = 1.30

p/p_t	M	T/T_t	GAMMA	ISOTHERMAL T_t/T_t^*	RAYLEIGH T_t/T_t^*	SHOCK p_{t1}/p_{t2}
.350	1.35187	.78485	.91261	1.14233	.93991	1.03302
.349	1.35395	.78433	.91170	1.14308	.93935	1.03352
.348	1.35602	.78381	.91079	1.14384	.93879	1.03403
.347	1.35811	.78329	.90986	1.14460	.93823	1.03454
.346	1.36019	.78277	.90894	1.14536	.93767	1.03505
.345	1.36228	.78225	.90801	1.14613	.93711	1.03557
.344	1.36438	.78172	.90707	1.14689	.93654	1.03610
.343	1.36647	.78120	.90613	1.14767	.93597	1.03663
.342	1.36858	.78067	.90518	1.14844	.93540	1.03717
.341	1.37069	.78014	.90423	1.14922	.93483	1.03771
.340	1.37280	.77961	.90328	1.14999	.93426	1.03826
.339	1.37491	.77908	.90231	1.15078	.93368	1.03882
.338	1.37703	.77855	.90135	1.15156	.93310	1.03937
.337	1.37916	.77802	.90037	1.15235	.93252	1.03994
.336	1.38129	.77749	.89940	1.15314	.93194	1.04051
.335	1.38342	.77695	.89842	1.15393	.93136	1.04109
.334	1.38556	.77642	.89743	1.15473	.93077	1.04167
.333	1.38770	.77588	.89643	1.15553	.93018	1.04226
.332	1.38985	.77534	.89544	1.15633	.92959	1.04286
.331	1.39200	.77480	.89443	1.15714	.92900	1.04346
.330	1.39416	.77426	.89342	1.15794	.92840	1.04406
.329	1.39632	.77372	.89241	1.15876	.92781	1.04468
.328	1.39849	.77318	.89139	1.15957	.92721	1.04530
.327	1.40066	.77263	.89037	1.16039	.92661	1.04592
.326	1.40284	.77209	.88934	1.16121	.92600	1.04655
.325	1.40502	.77154	.88830	1.16203	.92540	1.04719
.324	1.40720	.77099	.88726	1.16286	.92479	1.04783
.323	1.40939	.77044	.88622	1.16369	.92418	1.04848
.322	1.41159	.76989	.88517	1.16452	.92357	1.04914
.321	1.41379	.76934	.88411	1.16536	.92296	1.04980
.320	1.41600	.76878	.88305	1.16620	.92234	1.05047
.319	1.41821	.76823	.88198	1.16704	.92172	1.05115
.318	1.42042	.76767	.88091	1.16788	.92111	1.05183
.317	1.42264	.76711	.87983	1.16873	.92048	1.05252
.316	1.42487	.76655	.87875	1.16959	.91986	1.05322
.315	1.42710	.76599	.87766	1.17044	.91923	1.05392
.314	1.42934	.76543	.87657	1.17130	.91861	1.05463
.313	1.43158	.76487	.87547	1.17216	.91798	1.05535
.312	1.43383	.76430	.87437	1.17303	.91735	1.05608
.311	1.43608	.76374	.87326	1.17390	.91671	1.05681
.310	1.43834	.76317	.87214	1.17477	.91608	1.05754
.309	1.44060	.76260	.87102	1.17565	.91544	1.05829
.308	1.44287	.76203	.86989	1.17653	.91480	1.05904
.307	1.44514	.76146	.86876	1.17741	.91416	1.05980
.306	1.44742	.76089	.86762	1.17830	.91351	1.06057
.305	1.44971	.76031	.86648	1.17919	.91287	1.06134
.304	1.45200	.75974	.86533	1.18008	.91222	1.06213
.303	1.45430	.75916	.86418	1.18098	.91157	1.06292
.302	1.45660	.75858	.86302	1.18188	.91092	1.06371
.301	1.45891	.75800	.86186	1.18279	.91027	1.06452

TABLE IV. Isentropic Exponent = 1.30

p/p_t	M	T/T_t	GAMMA	ISOTHERMAL T_t/T_t^*	RAYLEIGH T_t/T_t^*	SHOCK p_{t1}/p_{t2}
.300	1.46122	.75742	.86068	1.18369	.90961	1.06533
.299	1.46354	.75683	.85951	1.18461	.90895	1.06615
.298	1.46587	.75625	.85833	1.18552	.90829	1.06698
.297	1.46820	.75566	.85714	1.18644	.90763	1.06781
.296	1.47053	.75508	.85595	1.18737	.90697	1.06866
.295	1.47288	.75449	.85475	1.18829	.90630	1.06951
.294	1.47523	.75390	.85354	1.18923	.90563	1.07037
.293	1.47758	.75330	.85233	1.19016	.90496	1.07124
.292	1.47994	.75271	.85112	1.19110	.90429	1.07211
.291	1.48231	.75211	.84990	1.19204	.90362	1.07300
.290	1.48469	.75152	.84867	1.19299	.90294	1.07389
.289	1.48707	.75092	.84744	1.19394	.90226	1.07479
.288	1.48945	.75032	.84620	1.19490	.90158	1.07570
.287	1.49185	.74971	.84495	1.19586	.90090	1.07662
.286	1.49425	.74911	.84370	1.19682	.90022	1.07755
.285	1.49665	.74851	.84245	1.19779	.89953	1.07848
.284	1.49907	.74790	.84119	1.19876	.89884	1.07943
.283	1.50148	.74729	.83992	1.19974	.89815	1.08038
.282	1.50391	.74668	.83865	1.20072	.89746	1.08135
.281	1.50634	.74607	.83737	1.20170	.89677	1.08232
.280	1.50878	.74545	.83608	1.20269	.89607	1.08330
.279	1.51123	.74484	.83479	1.20369	.89537	1.08429
.278	1.51368	.74422	.83349	1.20468	.89467	1.08529
.277	1.51614	.74360	.83219	1.20569	.89397	1.08630
.276	1.51861	.74298	.83088	1.20669	.89326	1.08732
.275	1.52108	.74236	.82957	1.20770	.89256	1.08834
.274	1.52356	.74174	.82825	1.20872	.89185	1.08938
.273	1.52605	.74111	.82692	1.20974	.89114	1.09043
.272	1.52854	.74048	.82559	1.21076	.89043	1.09149
.271	1.53105	.73986	.82425	1.21179	.88971	1.09255
.270	1.53356	.73922	.82291	1.21283	.88900	1.09363
.269	1.53607	.73859	.82156	1.21387	.88828	1.09472
.268	1.53860	.73796	.82020	1.21491	.88756	1.09582
.267	1.54113	.73732	.81884	1.21596	.88683	1.09693
.266	1.54367	.73668	.81747	1.21701	.88611	1.09804
.265	1.54622	.73604	.81609	1.21807	.88538	1.09917
.264	1.54877	.73540	.81471	1.21913	.88465	1.10031
.263	1.55133	.73476	.81332	1.22020	.88392	1.10146
.262	1.55390	.73411	.81193	1.22128	.88319	1.10263
.261	1.55648	.73346	.81053	1.22235	.88245	1.10380
.260	1.55906	.73281	.80912	1.22344	.88172	1.10498
.259	1.56166	.73216	.80771	1.22453	.88098	1.10618
.258	1.56426	.73151	.80629	1.22562	.88024	1.10738
.257	1.56687	.73085	.80487	1.22672	.87949	1.10860
.256	1.56949	.73020	.80344	1.22782	.87875	1.10983
.255	1.57211	.72954	.80200	1.22893	.87800	1.11107
.254	1.57475	.72888	.80056	1.23005	.87725	1.11232
.253	1.57739	.72821	.79911	1.23117	.87650	1.11358
.252	1.58004	.72755	.79765	1.23229	.87574	1.11486
.251	1.58270	.72688	.79619	1.23342	.87499	1.11615

TABLE IV. Isentropic Exponent = 1.30

p/p_t	M	T/T_t	GAMMA	ISOTHERMAL T_t/T_t^*	RAYLEIGH T_t/T_t^*	SHOCK p_{t1}/p_{t2}
.250	1.58537	.72621	.79472	1.23456	.87423	1.11745
.249	1.58805	.72554	.79324	1.23570	.87347	1.11876
.248	1.59073	.72487	.79176	1.23685	.87271	1.12009
.247	1.59343	.72419	.79027	1.23800	.87194	1.12143
.246	1.59613	.72351	.78878	1.23916	.87118	1.12278
.245	1.59884	.72283	.78728	1.24033	.87041	1.12414
.244	1.60156	.72215	.78577	1.24150	.86964	1.12552
.243	1.60429	.72147	.78425	1.24268	.86886	1.12691
.242	1.60703	.72078	.78273	1.24386	.86809	1.12831
.241	1.60978	.72009	.78120	1.24505	.86731	1.12973
.240	1.61254	.71940	.77967	1.24625	.86653	1.13116
.239	1.61531	.71871	.77813	1.24745	.86575	1.13260
.238	1.61809	.71801	.77658	1.24865	.86497	1.13406
.237	1.62087	.71732	.77503	1.24987	.86418	1.13553
.236	1.62367	.71662	.77346	1.25109	.86339	1.13701
.235	1.62647	.71592	.77190	1.25231	.86260	1.13851
.234	1.62929	.71521	.77032	1.25355	.86181	1.14003
.233	1.63212	.71450	.76874	1.25479	.86102	1.14156
.232	1.63495	.71380	.76715	1.25603	.86022	1.14310
.231	1.63780	.71308	.76556	1.25729	.85942	1.14466
.230	1.64066	.71237	.76395	1.25855	.85862	1.14624
.229	1.64352	.71166	.76234	1.25981	.85782	1.14783
.228	1.64640	.71094	.76073	1.26108	.85701	1.14943
.227	1.64929	.71022	.75910	1.26236	.85620	1.15105
.226	1.65219	.70949	.75747	1.26365	.85539	1.15269
.225	1.65509	.70877	.75584	1.26495	.85458	1.15434
.224	1.65801	.70804	.75419	1.26625	.85377	1.15601
.223	1.66094	.70731	.75254	1.26755	.85295	1.15770
.222	1.66389	.70658	.75088	1.26887	.85213	1.15940
.221	1.66684	.70584	.74922	1.27019	.85131	1.16112
.220	1.66980	.70510	.74754	1.27152	.85049	1.16286
.219	1.67278	.70436	.74586	1.27286	.84966	1.16461
.218	1.67577	.70362	.74418	1.27420	.84883	1.16638
.217	1.67876	.70287	.74248	1.27556	.84800	1.16817
.216	1.68177	.70212	.74078	1.27692	.84717	1.16997
.215	1.68480	.70137	.73907	1.27829	.84633	1.17180
.214	1.68783	.70062	.73736	1.27966	.84550	1.17364
.213	1.69088	.69986	.73563	1.28105	.84466	1.17550
.212	1.69393	.69910	.73390	1.28244	.84381	1.17738
.211	1.69700	.69834	.73216	1.28384	.84297	1.17928
.210	1.70009	.69757	.73042	1.28525	.84212	1.18120
.209	1.70318	.69680	.72866	1.28666	.84127	1.18314
.208	1.70629	.69603	.72690	1.28809	.84042	1.18509
.207	1.70941	.69526	.72513	1.28952	.83957	1.18707
.206	1.71254	.69448	.72336	1.29096	.83871	1.18907
.205	1.71569	.69370	.72157	1.29241	.83785	1.19108
.204	1.71885	.69292	.71978	1.29387	.83699	1.19312
.203	1.72202	.69214	.71798	1.29534	.83613	1.19518
.202	1.72521	.69135	.71617	1.29682	.83526	1.19726
.201	1.72841	.69056	.71436	1.29830	.83439	1.19937

TABLE IV.   Isentropic Exponent = 1.30

p/p_t	M	T/T_t	GAMMA	ISOTHERMAL T_t/T_t^*	RAYLEIGH T_t/T_t^*	SHOCK p_{t1}/p_{t2}
.200	1.73162	.68976	.71254	1.29980	.83352	1.20149
.199	1.73485	.68896	.71071	1.30130	.83265	1.20363
.198	1.73809	.68816	.70887	1.30282	.83177	1.20580
.197	1.74134	.68736	.70702	1.30434	.83089	1.20799
.196	1.74461	.68655	.70517	1.30587	.83001	1.21021
.195	1.74790	.68574	.70331	1.30742	.82913	1.21245
.194	1.75119	.68493	.70144	1.30897	.82824	1.21471
.193	1.75451	.68411	.69956	1.31053	.82735	1.21699
.192	1.75784	.68329	.69767	1.31210	.82646	1.21930
.191	1.76118	.68247	.69578	1.31368	.82557	1.22164
.190	1.76454	.68165	.69387	1.31528	.82467	1.22399
.189	1.76791	.68082	.69196	1.31688	.82377	1.22638
.188	1.77130	.67998	.69004	1.31849	.82287	1.22879
.187	1.77470	.67915	.68811	1.32012	.82196	1.23122
.186	1.77813	.67831	.68618	1.32175	.82106	1.23369
.185	1.78156	.67746	.68423	1.32340	.82015	1.23617
.184	1.78502	.67662	.68228	1.32505	.81923	1.23869
.183	1.78849	.67577	.68032	1.32672	.81832	1.24123
.182	1.79197	.67491	.67835	1.32840	.81740	1.24381
.181	1.79547	.67405	.67637	1.33009	.81648	1.24641
.180	1.79899	.67319	.67438	1.33179	.81555	1.24903
.179	1.80253	.67233	.67239	1.33350	.81463	1.25169
.178	1.80609	.67146	.67038	1.33523	.81370	1.25438
.177	1.80966	.67059	.66837	1.33696	.81277	1.25710
.176	1.81325	.66971	.66635	1.33871	.81183	1.25984
.175	1.81686	.66883	.66432	1.34048	.81089	1.26262
.174	1.82048	.66795	.66228	1.34225	.80995	1.26543
.173	1.82413	.66706	.66023	1.34404	.80901	1.26827
.172	1.82779	.66617	.65817	1.34584	.80806	1.27115
.171	1.83147	.66527	.65610	1.34765	.80711	1.27406
.170	1.83518	.66437	.65403	1.34947	.80616	1.27700
.169	1.83890	.66347	.65194	1.35131	.80521	1.27997
.168	1.84264	.66256	.64985	1.35316	.80425	1.28298
.167	1.84640	.66165	.64774	1.35503	.80329	1.28602
.166	1.85018	.66073	.64563	1.35691	.80232	1.28910
.165	1.85398	.65981	.64351	1.35880	.80136	1.29222
.164	1.85780	.65889	.64138	1.36071	.80039	1.29537
.163	1.86164	.65796	.63924	1.36263	.79941	1.29856
.162	1.86551	.65702	.63708	1.36457	.79844	1.30179
.161	1.86939	.65608	.63492	1.36652	.79746	1.30506
.160	1.87330	.65514	.63275	1.36848	.79647	1.30836
.159	1.87723	.65419	.63057	1.37047	.79549	1.31171
.158	1.88118	.65324	.62838	1.37246	.79450	1.31509
.157	1.88515	.65229	.62618	1.37448	.79351	1.31852
.156	1.88915	.65133	.62397	1.37650	.79251	1.32199
.155	1.89317	.65036	.62175	1.37855	.79151	1.32550
.154	1.89721	.64939	.61952	1.38061	.79051	1.32906
.153	1.90127	.64841	.61728	1.38269	.78950	1.33266
.152	1.90536	.64743	.61503	1.38478	.78849	1.33630
.151	1.90948	.64645	.61277	1.38689	.78748	1.33999

TABLE IV. Isentropic Exponent = 1.30

p/p_t	M	T/T_t	GAMMA	ISOTHERMAL T_t/T_t^*	RAYLEIGH T_t/T_t^*	SHOCK p_{t1}/p_{t2}
.150	1.91362	.64546	.61050	1.38902	.78646	1.34373
.149	1.91778	.64446	.60822	1.39116	.78544	1.34751
.148	1.92197	.64346	.60593	1.39333	.78442	1.35134
.147	1.92619	.64245	.60363	1.39551	.78339	1.35522
.146	1.93043	.64144	.60132	1.39771	.78236	1.35915
.145	1.93470	.64043	.59899	1.39993	.78133	1.36313
.144	1.93899	.63940	.59666	1.40217	.78029	1.36717
.143	1.94332	.63838	.59431	1.40442	.77925	1.37125
.142	1.94767	.63734	.59196	1.40670	.77821	1.37539
.141	1.95204	.63631	.58959	1.40899	.77716	1.37959
.140	1.95645	.63526	.58721	1.41131	.77611	1.38384
.139	1.96089	.63421	.58482	1.41365	.77505	1.38815
.138	1.96535	.63316	.58242	1.41601	.77399	1.39251
.137	1.96984	.63209	.58001	1.41838	.77293	1.39693
.136	1.97437	.63103	.57759	1.42078	.77186	1.40142
.135	1.97892	.62995	.57515	1.42321	.77079	1.40596
.134	1.98351	.62887	.57271	1.42565	.76971	1.41057
.133	1.98813	.62779	.57025	1.42812	.76863	1.41525
.132	1.99278	.62669	.56778	1.43061	.76755	1.41998
.131	1.99746	.62560	.56530	1.43312	.76646	1.42479
.130	2.00218	.62449	.56281	1.43565	.76536	1.42966
.129	2.00693	.62338	.56030	1.43822	.76427	1.43460
.128	2.01171	.62226	.55778	1.44080	.76317	1.43961
.127	2.01653	.62113	.55525	1.44341	.76206	1.44470
.126	2.02138	.62000	.55271	1.44605	.76095	1.44986
.125	2.02627	.61886	.55016	1.44871	.75984	1.45509
.124	2.03120	.61772	.54759	1.45140	.75872	1.46040
.123	2.03616	.61656	.54501	1.45411	.75760	1.46579
.122	2.04116	.61540	.54242	1.45685	.75647	1.47126
.121	2.04620	.61424	.53981	1.45962	.75534	1.47681
.120	2.05128	.61306	.53719	1.46242	.75420	1.48245
.119	2.05639	.61188	.53456	1.46525	.75306	1.48817
.118	2.06155	.61069	.53191	1.46810	.75191	1.49398
.117	2.06675	.60949	.52926	1.47099	.75076	1.49988
.116	2.07199	.60828	.52658	1.47391	.74960	1.50587
.115	2.07727	.60707	.52390	1.47685	.74844	1.51196
.114	2.08260	.60585	.52120	1.47983	.74728	1.51814
.113	2.08797	.60462	.51849	1.48284	.74611	1.52442
.112	2.09338	.60338	.51576	1.48589	.74493	1.53080
.111	2.09884	.60213	.51302	1.48897	.74375	1.53728
.110	2.10435	.60087	.51026	1.49208	.74256	1.54387
.109	2.10991	.59961	.50749	1.49523	.74137	1.55057
.108	2.11551	.59833	.50471	1.49841	.74017	1.55738
.107	2.12116	.59705	.50191	1.50163	.73897	1.56430
.106	2.12686	.59576	.49910	1.50489	.73776	1.57134
.105	2.13261	.59446	.49627	1.50819	.73654	1.57850
.104	2.13842	.59315	.49342	1.51152	.73532	1.58579
.103	2.14428	.59182	.49057	1.51489	.73409	1.59320
.102	2.15019	.59049	.48769	1.51831	.73286	1.60074
.101	2.15616	.58915	.48480	1.52176	.73162	1.60841

TABLE IV.   Isentropic Exponent = 1.30

p/p_t	M	T/T_t	GAMMA	ISOTHERMAL T_t/T_t^*	RAYLEIGH T_t/T_t^*	SHOCK p_{t1}/p_{t2}
.100	2.16218	.58780	.48189	1.52526	.73038	1.61621
.099	2.16826	.58644	.47897	1.52880	.72913	1.62416
.098	2.17440	.58507	.47603	1.53239	.72787	1.63225
.097	2.18060	.58368	.47308	1.53602	.72660	1.64049
.096	2.18686	.58229	.47011	1.53970	.72533	1.64888
.095	2.19319	.58088	.46712	1.54342	.72406	1.65743
.094	2.19958	.57947	.46412	1.54720	.72277	1.66614
.093	2.20603	.57804	.46110	1.55102	.72148	1.67501
.092	2.21255	.57660	.45806	1.55490	.72018	1.68406
.091	2.21914	.57515	.45500	1.55882	.71888	1.69328
.090	2.22580	.57368	.45193	1.56280	.71757	1.70268
.089	2.23253	.57220	.44883	1.56684	.71625	1.71226
.088	2.23933	.57071	.44572	1.57093	.71492	1.72204
.087	2.24621	.56921	.44260	1.57508	.71359	1.73202
.086	2.25316	.56769	.43945	1.57928	.71224	1.74219
.085	2.26019	.56616	.43628	1.58355	.71089	1.75258
.084	2.26730	.56462	.43310	1.58788	.70953	1.76319
.083	2.27450	.56306	.42989	1.59228	.70817	1.77402
.082	2.28178	.56149	.42667	1.59674	.70679	1.77402
.081	2.28914	.55990	.42343	1.60127	.70541	1.79638
.080	2.29660	.55830	.42016	1.60586	.70402	1.80792
.079	2.30414	.55668	.41688	1.61053	.70262	1.81972
.078	2.31178	.55505	.41357	1.61527	.70121	1.83179
.077	2.31952	.55340	.41025	1.62009	.69979	1.84413
.076	2.32735	.55173	.40690	1.62498	.69836	1.85675
.075	2.33528	.55005	.40353	1.62996	.69692	1.86966
.074	2.34332	.54834	.40014	1.63502	.69547	1.88288
.073	2.35146	.54663	.39673	1.64016	.69401	1.89642
.072	2.35972	.54489	.39329	1.64539	.69255	1.91028
.071	2.36808	.54313	.38983	1.65071	.69107	1.92448
.070	2.37656	.54136	.38635	1.65612	.68958	1.93903
.069	2.38517	.53956	.38284	1.66163	.68808	1.95394
.068	2.39389	.53775	.37931	1.66723	.68657	1.96924
.067	2.40274	.53591	.37576	1.67294	.68504	1.98493
.066	2.41172	.53406	.37218	1.67876	.68351	2.00104
.065	2.42084	.53218	.36858	1.68468	.68196	2.01757
.064	2.43010	.53028	.36494	1.69072	.68040	2.03455
.063	2.43950	.52835	.36129	1.69688	.67883	2.05200
.062	2.44904	.52641	.35760	1.70316	.67724	2.06993
.061	2.45875	.52443	.35389	1.70956	.67564	2.08837
.060	2.46861	.52244	.35016	1.71609	.67403	2.10734
.059	2.47863	.52042	.34639	1.72276	.67241	2.12686
.058	2.48882	.51837	.34259	1.72957	.67077	2.14696
.057	2.49919	.51629	.33877	1.73653	.66911	2.16767
.056	2.50974	.51419	.33491	1.74363	.66744	2.18902
.055	2.52048	.51205	.33103	1.75090	.66575	2.21104
.054	2.53142	.50989	.32711	1.75833	.66405	2.23375
.053	2.54256	.50769	.32316	1.76593	.66233	2.25721
.052	2.55391	.50547	.31918	1.77371	.66059	2.28144
.051	2.56548	.50321	.31517	1.78167	.65884	2.30648

TABLE IV. Isentropic Exponent = 1.30

p/p_t	M	T/T_t	GAMMA	ISOTHERMAL T_t/T_t^*	RAYLEIGH T_t/T_t^*	SHOCK p_{t1}/p_{t2}
.050	2.57728	.50091	.31112	1.78984	.65706	2.33238
.049	2.58932	.49858	.30703	1.79820	.65527	2.35920
.048	2.60160	.49622	.30292	1.80678	.65346	2.38697
.047	2.61415	.49381	.29876	1.81558	.65163	2.41575
.046	2.62696	.49137	.29457	1.82461	.64978	2.44560
.045	2.64006	.48888	.29034	1.83389	.64790	2.47659
.044	2.65346	.48635	.28606	1.84342	.64600	2.50879
.043	2.66716	.48378	.28175	1.85323	.64408	2.54227
.042	2.68119	.48116	.27740	1.86332	.64214	2.57711
.041	2.69556	.47849	.27301	1.87371	.64017	2.61340
.040	2.71029	.47577	.26857	1.88442	.63817	2.65125
.039	2.72539	.47300	.26408	1.89546	.63615	2.69075
.038	2.74090	.47017	.25955	1.90685	.63410	2.73203
.037	2.75682	.46729	.25497	1.91863	.63201	2.77521
.036	2.77318	.46434	.25034	1.93080	.62990	2.82045
.035	2.79001	.46133	.24566	1.94339	.62775	2.86788
.034	2.80734	.45826	.24093	1.95643	.62557	2.91770
.033	2.82520	.45511	.23614	1.96996	.62336	2.97009
.032	2.84361	.45189	.23130	1.98400	.62111	3.02527
.031	2.86262	.44859	.22640	1.99859	.61881	3.08349
.030	2.88227	.44521	.22144	2.01377	.61648	3.14501
.029	2.90260	.44174	.21641	2.02958	.61410	3.21015
.028	2.92366	.43818	.21132	2.04608	.61167	3.27925
.027	2.94551	.43452	.20616	2.06333	.60920	3.35270
.026	2.96820	.43075	.20092	2.08138	.60667	3.43096
.025	2.99181	.42687	.19562	2.10030	.60408	3.51454
.024	3.01642	.42287	.19023	2.12018	.60144	3.60405
.023	3.04210	.41873	.18476	2.14111	.59873	3.70018
.022	3.06896	.41446	.17920	2.16318	.59596	3.80374
.021	3.09712	.41003	.17356	2.18653	.59311	3.91567
.020	3.12670	.40544	.16781	2.21129	.59017	4.03712
.019	3.15785	.40067	.16197	2.23762	.58716	4.16942
.018	3.19076	.39570	.15601	2.26571	.58405	4.31420
.017	3.22562	.39052	.14994	2.29580	.58083	4.47344
.016	3.26269	.38509	.14374	2.32814	.57750	4.64957
.015	3.30226	.37940	.13741	2.36307	.57404	4.84561
.014	3.34469	.37341	.13094	2.40100	.57045	5.06540
.013	3.39041	.36708	.12431	2.44241	.56669	5.31385
.012	3.43998	.36036	.11750	2.48795	.56276	5.59736
.011	3.49409	.35319	.11051	2.53841	.55862	5.92452
.010	3.55365	.34551	.10330	2.59486	.55424	6.30701
.009	3.61985	.33721	.09586	2.65872	.54959	6.76134
.008	3.69433	.32817	.08816	2.73198	.54460	7.31155
.007	3.77939	.31821	.08014	2.81748	.53920	7.99426
.006	3.87847	.30709	.07176	2.91951	.53329	8.86836
.005	3.99692	.29444	.06293	3.04497	.52672	10.03568
.004	4.14390	.27966	.05356	3.20587	.51923	11.69027
.003	4.33684	.26170	.04346	3.42593	.51040	14.25890
.002	4.61595	.23832	.03231	3.76197	.49933	18.92569
.001	5.11462	.20309	.01939	4.41453	.48348	30.93437

TABLE V. Isentropic Exponent = 1.40

p/p_t	M	T/T_t	GAMMA	ISOTHERMAL T_t/T_t^*	RAYLEIGH T_t/T_t^*	SHOCK p_{t1}/p_{t2}
1.0000	.00000	1.00000	.00000	.87500	.00000	1.
.9995	.02673	.99986	.04617	.87513	.00342	1.
.9990	.03781	.99971	.06528	.87525	.00684	1.
.9985	.04631	.99957	.07993	.87538	.01024	1.
.9980	.05349	.99943	.09227	.87550	.01363	1.
.9975	.05981	.99929	.10313	.87563	.01701	1.
.9970	.06553	.99914	.11294	.87575	.02038	1.
.9965	.07079	.99900	.12196	.87588	.02374	1.
.9960	.07569	.99886	.13034	.87600	.02709	1.
.9955	.08029	.99871	.13821	.87613	.03043	1.
.9950	.08465	.99857	.14565	.87625	.03376	1.
.9945	.08880	.99843	.15272	.87638	.03708	1.
.9940	.09276	.99828	.15947	.87651	.04039	1.
.9935	.09656	.99814	.16593	.87663	.04369	1.
.9930	.10023	.99799	.17215	.87676	.04698	1.
.9925	.10376	.99785	.17814	.87688	.05026	1.
.9920	.10718	.99771	.18394	.87701	.05353	1.
.9915	.11050	.99756	.18955	.87714	.05679	1.
.9910	.11372	.99742	.19499	.87726	.06004	1.
.9905	.11685	.99728	.20028	.87739	.06328	1.
.9900	.11991	.99713	.20543	.87752	.06651	1.
.9895	.12289	.99699	.21044	.87764	.06973	1.
.9890	.12580	.99684	.21534	.87777	.07294	1.
.9885	.12865	.99670	.22012	.87790	.07614	1.
.9880	.13144	.99656	.22479	.87802	.07933	1.
.9875	.13417	.99641	.22936	.87815	.08251	1.
.9870	.13685	.99627	.23384	.87828	.08568	1.
.9865	.13948	.99612	.23823	.87840	.08884	1.
.9860	.14206	.99598	.24254	.87853	.09199	1.
.9855	.14460	.99584	.24676	.87866	.09513	1.
.9850	.14710	.99569	.25091	.87879	.09827	1.
.9845	.14955	.99555	.25499	.87891	.10139	1.
.9840	.15197	.99540	.25900	.87904	.10450	1.
.9835	.15435	.99526	.26294	.87917	.10761	1.
.9830	.15670	.99511	.26683	.87930	.11070	1.
.9825	.15901	.99497	.27065	.87942	.11378	1.
.9820	.16129	.99482	.27441	.87955	.11686	1.
.9815	.16355	.99468	.27812	.87968	.11992	1.
.9810	.16577	.99453	.28178	.87981	.12298	1.
.9805	.16796	.99439	.28538	.87994	.12603	1.
.9800	.17013	.99424	.28894	.88007	.12907	1.
.9795	.17227	.99410	.29245	.88019	.13209	1.
.9790	.17439	.99395	.29591	.88032	.13511	1.
.9785	.17648	.99381	.29933	.88045	.13812	1.
.9780	.17855	.99366	.30271	.88058	.14112	1.
.9775	.18060	.99352	.30605	.88071	.14412	1.
.9770	.18262	.99337	.30934	.88084	.14710	1.
.9765	.18463	.99323	.31260	.88097	.15007	1.
.9760	.18661	.99308	.31582	.88109	.15304	1.
.9755	.18858	.99294	.31901	.88122	.15599	1.

TABLE V. Isentropic Exponent = 1.40

p/p_t	M	T/T_t	GAMMA	ISOTHERMAL T_t/T_t^*	RAYLEIGH T_t/T_t^*	SHOCK p_{t1}/p_{t2}
.9750	.19052	.99279	.32216	.88135	.15894	1.
.9745	.19245	.99265	.32527	.88148	.16188	1.
.9740	.19436	.99250	.32836	.88161	.16480	1.
.9735	.19625	.99236	.33141	.88174	.16772	1.
.9730	.19813	.99221	.33443	.88187	.17063	1.
.9725	.19999	.99206	.33742	.88200	.17353	1.
.9720	.20183	.99192	.34038	.88213	.17643	1.
.9715	.20366	.99177	.34331	.88226	.17931	1.
.9710	.20547	.99163	.34621	.88239	.18219	1.
.9705	.20727	.99148	.34909	.88252	.18505	1.
.9700	.20905	.99134	.35193	.88265	.18791	1.
.9695	.21082	.99119	.35476	.88278	.19076	1.
.9690	.21258	.99104	.35755	.88291	.19360	1.
.9685	.21432	.99090	.36033	.88304	.19643	1.
.9680	.21605	.99075	.36307	.88317	.19925	1.
.9675	.21777	.99060	.36580	.88330	.20207	1.
.9670	.21947	.99046	.36850	.88343	.20487	1.
.9665	.22117	.99031	.37118	.88356	.20767	1.
.9660	.22285	.99017	.37383	.88369	.21046	1.
.9655	.22452	.99002	.37647	.88382	.21324	1.
.9650	.22618	.98987	.37908	.88395	.21601	1.
.9645	.22782	.98973	.38167	.88408	.21878	1.
.9640	.22946	.98958	.38424	.88421	.22153	1.
.9635	.23109	.98943	.38679	.88435	.22428	1.
.9630	.23270	.98929	.38933	.88448	.22701	1.
.9625	.23431	.98914	.39184	.88461	.22974	1.
.9620	.23591	.98899	.39433	.88474	.23247	1.
.9615	.23749	.98885	.39681	.88487	.23518	1.
.9610	.23907	.98870	.39926	.88500	.23788	1.
.9605	.24064	.98855	.40170	.88513	.24058	1.
.9600	.24220	.98840	.40412	.88527	.24327	1.
.9595	.24375	.98826	.40653	.88540	.24595	1.
.9590	.24529	.98811	.40891	.88553	.24862	1.
.9585	.24682	.98796	.41128	.88566	.25128	1.
.9580	.24834	.98782	.41364	.88579	.25394	1.
.9575	.24986	.98767	.41598	.88593	.25659	1.
.9570	.25137	.98752	.41830	.88606	.25923	1.
.9565	.25286	.98737	.42061	.88619	.26186	1.
.9560	.25436	.98723	.42290	.88632	.26448	1.
.9555	.25584	.98708	.42517	.88645	.26710	1.
.9550	.25732	.98693	.42744	.88659	.26971	1.
.9545	.25879	.98678	.42968	.88672	.27231	1.
.9540	.26025	.98664	.43192	.88685	.27490	1.
.9535	.26170	.98649	.43413	.88699	.27748	1.
.9530	.26315	.98634	.43634	.88712	.28006	1.
.9525	.26459	.98619	.43853	.88725	.28263	1.
.9520	.26602	.98604	.44071	.88738	.28519	1.
.9515	.26745	.98590	.44287	.88752	.28774	1.
.9510	.26887	.98575	.44502	.88765	.29028	1.
.9505	.27028	.98560	.44716	.88778	.29282	1.

TABLE V. Isentropic Exponent = 1.40

p/p_t	M	T/T_t	GAMMA	ISOTHERMAL T_t/T_t^*	RAYLEIGH T_t/T_t^*	SHOCK p_{t1}/p_{t2}
.950	.27169	.98545	.44929	.88792	.29535	1.
.949	.27449	.98516	.45350	.88818	.30038	1.
.948	.27726	.98486	.45767	.88845	.30539	1.
.947	.28001	.98456	.46178	.88872	.31036	1.
.946	.28273	.98426	.46585	.88899	.31531	1.
.945	.28543	.98397	.46988	.88926	.32022	1.
.944	.28811	.98367	.47386	.88953	.32510	1.
.943	.29077	.98337	.47780	.88980	.32996	1.
.942	.29341	.98307	.48170	.89007	.33478	1.
.941	.29603	.98278	.48556	.89034	.33958	1.
.940	.29863	.98248	.48938	.89061	.34434	1.
.939	.30121	.98218	.49315	.89088	.34908	1.
.938	.30377	.98188	.49690	.89115	.35379	1.
.937	.30632	.98158	.50060	.89142	.35847	1.
.936	.30884	.98128	.50427	.89169	.36312	1.
.935	.31135	.98098	.50790	.89196	.36774	1.
.934	.31385	.98068	.51150	.89224	.37234	1.
.933	.31632	.98038	.51506	.89251	.37691	1.
.932	.31878	.98008	.51859	.89278	.38144	1.
.931	.32123	.97978	.52209	.89306	.38596	1.
.930	.32366	.97948	.52555	.89333	.39044	1.
.929	.32607	.97918	.52899	.89361	.39490	1.
.928	.32847	.97888	.53239	.89388	.39932	1.
.927	.33086	.97858	.53576	.89416	.40373	1.
.926	.33323	.97827	.53911	.89443	.40810	1.
.925	.33559	.97797	.54242	.89471	.41245	1.
.924	.33794	.97767	.54571	.89499	.41677	1.
.923	.34027	.97737	.54897	.89526	.42107	1.
.922	.34259	.97706	.55220	.89554	.42534	1.
.921	.34490	.97676	.55540	.89582	.42958	1.
.920	.34720	.97646	.55858	.89610	.43380	1.
.919	.34948	.97615	.56173	.89637	.43799	1.
.918	.35176	.97585	.56485	.89665	.44216	1.
.917	.35402	.97555	.56795	.89693	.44630	1.
.916	.35627	.97524	.57103	.89721	.45041	1.
.915	.35851	.97494	.57408	.89749	.45450	1.
.914	.36073	.97463	.57711	.89777	.45857	1.
.913	.36295	.97433	.58011	.89805	.46261	1.
.912	.36516	.97402	.58309	.89833	.46663	1.
.911	.36735	.97372	.58605	.89862	.47062	1.
.910	.36954	.97341	.58898	.89890	.47458	1.
.909	.37172	.97311	.59189	.89918	.47853	1.
.908	.37389	.97280	.59478	.89946	.48244	1.
.907	.37604	.97250	.59765	.89975	.48634	1.
.906	.37819	.97219	.60050	.90003	.49021	1.
.905	.38033	.97188	.60332	.90031	.49406	1.
.904	.38246	.97158	.60613	.90060	.49788	1.
.903	.38459	.97127	.60891	.90088	.50168	1.
.902	.38670	.97096	.61168	.90117	.50545	1.
.901	.38880	.97065	.61442	.90145	.50921	1.

TABLE V. Isentropic Exponent = 1.40

p/p_t	M	T/T_t	GAMMA	ISOTHERMAL T_t/T_t^*	RAYLEIGH T_t/T_t^*	SHOCK p_{t1}/p_{t2}
.900	.39090	.97035	.61715	.90174	.51294	1.
.899	.39299	.97004	.61985	.90203	.51664	1.
.898	.39507	.96973	.62254	.90231	.52033	1.
.897	.39714	.96942	.62521	.90260	.52399	1.
.896	.39921	.96911	.62786	.90289	.52763	1.
.895	.40126	.96880	.63049	.90318	.53125	1.
.894	.40331	.96849	.63310	.90347	.53484	1.
.893	.40535	.96818	.63570	.90375	.53841	1.
.892	.40739	.96787	.63827	.90404	.54196	1.
.891	.40942	.96756	.64083	.90433	.54549	1.
.890	.41144	.96725	.64338	.90462	.54900	1.
.889	.41345	.96694	.64590	.90491	.55248	1.
.888	.41546	.96663	.64841	.90521	.55595	1.
.887	.41745	.96632	.65090	.90550	.55939	1.
.886	.41945	.96601	.65338	.90579	.56281	1.
.885	.42143	.96570	.65584	.90608	.56621	1.
.884	.42341	.96539	.65828	.90637	.56959	1.
.883	.42539	.96507	.66071	.90667	.57295	1.
.882	.42736	.96476	.66312	.90696	.57629	1.
.881	.42932	.96445	.66551	.90725	.57960	1.
.880	.43127	.96414	.66789	.90755	.58290	1.
.879	.43322	.96382	.67026	.90784	.58617	1.
.878	.43516	.96351	.67261	.90814	.58943	1.
.877	.43710	.96319	.67494	.90844	.59266	1.
.876	.43903	.96288	.67726	.90873	.59588	1.
.875	.44096	.96257	.67957	.90903	.59907	1.
.874	.44288	.96225	.68186	.90932	.60225	1.
.873	.44479	.96194	.68414	.90962	.60540	1.
.872	.44670	.96162	.68640	.90992	.60854	1.
.871	.44861	.96131	.68865	.91022	.61166	1.
.870	.45051	.96099	.69088	.91052	.61475	1.
.869	.45240	.96068	.69310	.91082	.61783	1.
.868	.45429	.96036	.69531	.91112	.62089	1.
.867	.45617	.96004	.69751	.91142	.62393	1.
.866	.45805	.95973	.69969	.91172	.62695	1.
.865	.45993	.95941	.70185	.91202	.62995	1.
.864	.46180	.95909	.70401	.91232	.63293	1.
.863	.46366	.95878	.70615	.91262	.63590	1.
.862	.46552	.95846	.70827	.91292	.63884	1.
.861	.46737	.95814	.71039	.91323	.64177	1.
.860	.46922	.95782	.71249	.91353	.64468	1.
.859	.47107	.95750	.71458	.91383	.64757	1.
.858	.47291	.95719	.71666	.91414	.65044	1.
.857	.47475	.95687	.71872	.91444	.65329	1.
.856	.47658	.95655	.72078	.91475	.65613	1.
.855	.47841	.95623	.72282	.91505	.65895	1.
.854	.48023	.95591	.72485	.91536	.66175	1.
.853	.48205	.95559	.72686	.91567	.66453	1.
.852	.48387	.95527	.72887	.91597	.66730	1.
.851	.48568	.95495	.73086	.91628	.67004	1.

TABLE V. Isentropic Exponent = 1.40

p/p_t	M	T/T_t	GAMMA	ISOTHERMAL T_t/T_t^*	RAYLEIGH T_t/T_t^*	SHOCK p_{t1}/p_{t2}
.850	.48749	.95463	.73284	.91659	.67277	1.
.849	.48929	.95431	.73481	.91690	.67549	1.
.848	.49109	.95399	.73677	.91720	.67818	1.
.847	.49289	.95366	.73872	.91751	.68086	1.
.846	.49468	.95334	.74065	.91782	.68352	1.
.845	.49647	.95302	.74258	.91813	.68617	1.
.844	.49825	.95270	.74449	.91844	.68880	1.
.843	.50003	.95237	.74639	.91876	.69141	1.
.842	.50181	.95205	.74828	.91907	.69400	1.
.841	.50359	.95173	.75016	.91938	.69658	1.
.840	.50536	.95141	.75203	.91969	.69914	1.
.839	.50712	.95108	.75389	.92001	.70169	1.
.838	.50889	.95076	.75574	.92032	.70422	1.
.837	.51065	.95043	.75758	.92063	.70673	1.
.836	.51240	.95011	.75941	.92095	.70923	1.
.835	.51416	.94978	.76122	.92126	.71171	1.
.834	.51591	.94946	.76303	.92158	.71417	1.
.833	.51765	.94913	.76483	.92189	.71662	1.
.832	.51940	.94881	.76662	.92221	.71906	1.
.831	.52114	.94848	.76839	.92253	.72148	1.
.830	.52287	.94816	.77016	.92284	.72388	1.
.829	.52461	.94783	.77191	.92316	.72627	1.
.828	.52634	.94750	.77366	.92348	.72864	1.
.827	.52807	.94717	.77540	.92380	.73099	1.
.826	.52979	.94685	.77713	.92412	.73333	1.
.825	.53152	.94652	.77884	.92444	.73566	1.
.824	.53324	.94619	.78055	.92476	.73797	1.
.823	.53495	.94586	.78225	.92508	.74027	1.
.822	.53667	.94554	.78394	.92540	.74255	1.
.821	.53838	.94521	.78562	.92572	.74481	1.
.820	.54009	.94488	.78729	.92605	.74706	1.
.819	.54179	.94455	.78895	.92637	.74930	1.
.818	.54349	.94422	.79060	.92669	.75152	1.
.817	.54519	.94389	.79224	.92702	.75373	1.
.816	.54689	.94356	.79387	.92734	.75592	1.
.815	.54859	.94323	.79550	.92767	.75810	1.
.814	.55028	.94290	.79711	.92799	.76027	1.
.813	.55197	.94257	.79872	.92832	.76242	1.
.812	.55366	.94223	.80032	.92864	.76455	1.
.811	.55534	.94190	.80190	.92897	.76668	1.
.810	.55703	.94157	.80348	.92930	.76878	1.
.809	.55871	.94124	.80505	.92963	.77088	1.
.808	.56038	.94091	.80662	.92996	.77296	1.
.807	.56206	.94057	.80817	.93028	.77503	1.
.806	.56373	.94024	.80971	.93061	.77708	1.
.805	.56540	.93991	.81125	.93094	.77912	1.
.804	.56707	.93957	.81278	.93127	.78114	1.
.803	.56874	.93924	.81430	.93161	.78316	1.
.802	.57040	.93890	.81581	.93194	.78516	1.
.801	.57206	.93857	.81731	.93227	.78714	1.

TABLE V. Isentropic Exponent = 1.40

p/p_t	M	T/T_t	GAMMA	ISOTHERMAL T_t/T_t^*	RAYLEIGH T_t/T_t^*	SHOCK p_{t1}/p_{t2}
.800	.57372	.93823	.81880	.93260	.78911	1.
.799	.57538	.93790	.82029	.93294	.79107	1.
.798	.57704	.93756	.82177	.93327	.79302	1.
.797	.57869	.93723	.82324	.93360	.79495	1.
.796	.58034	.93689	.82470	.93394	.79687	1.
.795	.58199	.93656	.82615	.93427	.79878	1.
.794	.58364	.93622	.82760	.93461	.80068	1.
.793	.58528	.93588	.82903	.93495	.80256	1.
.792	.58693	.93554	.83046	.93528	.80443	1.
.791	.58857	.93521	.83188	.93562	.80628	1.
.790	.59021	.93487	.83330	.93596	.80813	1.
.789	.59185	.93453	.83470	.93630	.80996	1.
.788	.59348	.93419	.83610	.93664	.81178	1.
.787	.59512	.93385	.83749	.93698	.81358	1.
.786	.59675	.93351	.83887	.93732	.81538	1.
.785	.59838	.93317	.84025	.93766	.81716	1.
.784	.60001	.93283	.84161	.93800	.81893	1.
.783	.60163	.93249	.84297	.93834	.82069	1.
.782	.60326	.93215	.84433	.93869	.82243	1.
.781	.60488	.93181	.84567	.93903	.82417	1.
.780	.60650	.93147	.84701	.93937	.82589	1.
.779	.60812	.93113	.84834	.93972	.82760	1.
.778	.60974	.93079	.84966	.94006	.82930	1.
.777	.61136	.93045	.85097	.94041	.83098	1.
.776	.61297	.93010	.85228	.94075	.83266	1.
.775	.61459	.92976	.85358	.94110	.83432	1.
.774	.61620	.92942	.85487	.94145	.83597	1.
.773	.61781	.92908	.85616	.94180	.83761	1.
.772	.61942	.92873	.85743	.94214	.83924	1.
.771	.62103	.92839	.85871	.94249	.84086	1.
.770	.62263	.92804	.85997	.94284	.84246	1.
.769	.62424	.92770	.86123	.94319	.84406	1.
.768	.62584	.92736	.86247	.94354	.84564	1.
.767	.62744	.92701	.86372	.94389	.84721	1.
.766	.62904	.92666	.86495	.94425	.84878	1.
.765	.63064	.92632	.86618	.94460	.85033	1.
.764	.63224	.92597	.86740	.94495	.85186	1.
.763	.63384	.92563	.86862	.94531	.85339	1.
.762	.63543	.92528	.86982	.94566	.85491	1.
.761	.63703	.92493	.87102	.94602	.85642	1.
.760	.63862	.92458	.87222	.94637	.85791	1.
.759	.64021	.92424	.87340	.94673	.85940	1.
.758	.64180	.92389	.87459	.94708	.86087	1.
.757	.64339	.92354	.87576	.94744	.86234	1.
.756	.64497	.92319	.87692	.94780	.86379	1.
.755	.64656	.92284	.87808	.94816	.86523	1.
.754	.64815	.92249	.87924	.94852	.86667	1.
.753	.64973	.92214	.88038	.94888	.86809	1.
.752	.65131	.92179	.88152	.94924	.86950	1.
.751	.65289	.92144	.88266	.94960	.87090	1.

TABLE V. Isentropic Exponent = 1.40

p/p_t	M	T/T_t	GAMMA	ISOTHERMAL T_t/T_t^*	RAYLEIGH T_t/T_t^*	SHOCK p_{t1}/p_{t2}
.750	.65447	.92109	.88378	.94996	.87229	1.
.749	.65605	.92074	.88490	.95032	.87367	1.
.748	.65763	.92039	.88602	.95068	.87504	1.
.747	.65921	.92004	.88712	.95105	.87641	1.
.746	.66079	.91969	.88823	.95141	.87776	1.
.745	.66236	.91933	.88932	.95178	.87910	1.
.744	.66393	.91898	.89041	.95214	.88043	1.
.743	.66551	.91863	.89149	.95251	.88175	1.
.742	.66708	.91827	.89256	.95287	.88306	1.
.741	.66865	.91792	.89363	.95324	.88436	1.
.740	.67022	.91757	.89469	.95361	.88565	1.
.739	.67179	.91721	.89575	.95398	.88694	1.
.738	.67336	.91686	.89680	.95435	.88821	1.
.737	.67493	.91650	.89784	.95472	.88947	1.
.736	.67649	.91615	.89888	.95509	.89073	1.
.735	.67806	.91579	.89991	.95546	.89197	1.
.734	.67962	.91543	.90093	.95583	.89320	1.
.733	.68119	.91508	.90195	.95620	.89443	1.
.732	.68275	.91472	.90296	.95658	.89565	1.
.731	.68431	.91436	.90397	.95695	.89685	1.
.730	.68587	.91401	.90497	.95732	.89805	1.
.729	.68743	.91365	.90596	.95770	.89924	1.
.728	.68899	.91329	.90695	.95807	.90042	1.
.727	.69055	.91293	.90793	.95845	.90159	1.
.726	.69211	.91257	.90891	.95883	.90275	1.
.725	.69367	.91221	.90988	.95921	.90390	1.
.724	.69522	.91185	.91084	.95958	.90504	1.
.723	.69678	.91149	.91180	.95996	.90618	1.
.722	.69833	.91113	.91275	.96034	.90730	1.
.721	.69989	.91077	.91370	.96072	.90842	1.
.720	.70144	.91041	.91464	.96110	.90953	1.
.719	.70300	.91005	.91557	.96149	.91063	1.
.718	.70455	.90969	.91650	.96187	.91172	1.
.717	.70610	.90933	.91742	.96225	.91280	1.
.716	.70765	.90896	.91834	.96263	.91387	1.
.715	.70920	.90860	.91925	.96302	.91493	1.
.714	.71075	.90824	.92015	.96340	.91599	1.
.713	.71230	.90787	.92105	.96379	.91704	1.
.712	.71385	.90751	.92194	.96418	.91808	1.
.711	.71540	.90715	.92283	.96456	.91911	1.
.710	.71695	.90678	.92371	.96495	.92013	1.
.709	.71849	.90642	.92459	.96534	.92114	1.
.708	.72004	.90605	.92546	.96573	.92215	1.
.707	.72159	.90568	.92632	.96612	.92315	1.
.706	.72313	.90532	.92718	.96651	.92413	1.
.705	.72468	.90495	.92803	.96690	.92512	1.
.704	.72622	.90458	.92888	.96729	.92609	1.
.703	.72777	.90422	.92972	.96769	.92705	1.
.702	.72931	.90385	.93056	.96808	.92801	1.
.701	.73085	.90348	.93139	.96848	.92896	1.

TABLE V. Isentropic Exponent = 1.40

p/p_t	M	T/T_t	GAMMA	ISOTHERMAL T_t/T_t^*	RAYLEIGH T_t/T_t^*	SHOCK p_{t1}/p_{t2}
.700	.73240	.90311	.93222	.96887	.92990	1.
.699	.73394	.90274	.93303	.96927	.93083	1.
.698	.73548	.90238	.93385	.96966	.93176	1.
.697	.73702	.90201	.93466	.97006	.93267	1.
.696	.73856	.90164	.93546	.97046	.93358	1.
.695	.74010	.90127	.93626	.97086	.93448	1.
.694	.74164	.90089	.93705	.97126	.93538	1.
.693	.74318	.90052	.93783	.97166	.93626	1.
.692	.74472	.90015	.93862	.97206	.93714	1.
.691	.74626	.89978	.93939	.97246	.93801	1.
.690	.74780	.89941	.94016	.97286	.93888	1.
.689	.74934	.89904	.94093	.97327	.93973	1.
.688	.75088	.89866	.94168	.97367	.94058	1.
.687	.75242	.89829	.94244	.97407	.94142	1.
.686	.75396	.89792	.94319	.97448	.94225	1.
.685	.75550	.89754	.94393	.97489	.94308	1.
.684	.75703	.89717	.94467	.97529	.94390	1.
.683	.75857	.89679	.94540	.97570	.94471	1.
.682	.76011	.89642	.94613	.97611	.94551	1.
.681	.76165	.89604	.94685	.97652	.94631	1.
.680	.76318	.89566	.94756	.97693	.94710	1.
.679	.76472	.89529	.94827	.97734	.94788	1.
.678	.76626	.89491	.94898	.97775	.94866	1.
.677	.76779	.89453	.94968	.97816	.94942	1.
.676	.76933	.89416	.95038	.97858	.95018	1.
.675	.77086	.89378	.95106	.97899	.95094	1.
.674	.77240	.89340	.95175	.97941	.95168	1.
.673	.77394	.89302	.95243	.97982	.95242	1.
.672	.77547	.89264	.95310	.98024	.95316	1.
.671	.77701	.89226	.95377	.98065	.95388	1.
.670	.77854	.89188	.95444	.98107	.95460	1.
.669	.78008	.89150	.95509	.98149	.95531	1.
.668	.78161	.89112	.95575	.98191	.95602	1.
.667	.78315	.89074	.95640	.98233	.95672	1.
.666	.78468	.89036	.95704	.98275	.95741	1.
.665	.78622	.88997	.95768	.98317	.95809	1.
.664	.78775	.88959	.95831	.98360	.95877	1.
.663	.78929	.88921	.95894	.98402	.95944	1.
.662	.79082	.88883	.95956	.98445	.96011	1.
.661	.79236	.88844	.96018	.98487	.96077	1.
.660	.79389	.88806	.96079	.98530	.96142	1.
.659	.79543	.88767	.96140	.98572	.96206	1.
.658	.79696	.88729	.96200	.98615	.96270	1.
.657	.79850	.88690	.96260	.98658	.96334	1.
.656	.80003	.88652	.96319	.98701	.96396	1.
.655	.80157	.88613	.96378	.98744	.96458	1.
.654	.80310	.88574	.96436	.98787	.96519	1.
.653	.80464	.88536	.96494	.98830	.96580	1.
.652	.80617	.88497	.96551	.98874	.96640	1.
.651	.80771	.88458	.96608	.98917	.96700	1.

TABLE V.   Isentropic Exponent = 1.40

p/p_t	M	T/T_t	GAMMA	ISOTHERMAL T_t/T_t^*	RAYLEIGH T_t/T_t^*	SHOCK p_{t1}/p_{t2}
.650	.80925	.88419	.96664	.98960	.96758	1.
.649	.81078	.88380	.96720	.99004	.96816	1.
.648	.81232	.88341	.96775	.99048	.96874	1.
.647	.81385	.88302	.96830	.99091	.96931	1.
.646	.81539	.88263	.96884	.99135	.96987	1.
.645	.81693	.88224	.96938	.99179	.97043	1.
.644	.81846	.88185	.96991	.99223	.97098	1.
.643	.82000	.88146	.97044	.99267	.97152	1.
.642	.82154	.88107	.97096	.99311	.97206	1.
.641	.82307	.88068	.97148	.99355	.97260	1.
.640	.82461	.88028	.97199	.99400	.97312	1.
.639	.82615	.87989	.97250	.99444	.97364	1.
.638	.82769	.87950	.97300	.99489	.97416	1.
.637	.82923	.87910	.97350	.99533	.97467	1.
.636	.83076	.87871	.97399	.99578	.97517	1.
.635	.83230	.87831	.97448	.99623	.97567	1.
.634	.83384	.87792	.97497	.99668	.97616	1.
.633	.83538	.87752	.97544	.99713	.97664	1.
.632	.83692	.87713	.97592	.99758	.97712	1.
.631	.83846	.87673	.97639	.99803	.97760	1.
.630	.84000	.87633	.97685	.99848	.97807	1.
.629	.84154	.87593	.97731	.99893	.97853	1.
.628	.84308	.87554	.97777	.99939	.97899	1.
.627	.84462	.87514	.97822	.99984	.97944	1.
.626	.84616	.87474	.97866	1.00030	.97988	1.
.625	.84771	.87434	.97910	1.00076	.98032	1.
.624	.84925	.87394	.97954	1.00121	.98076	1.
.623	.85079	.87354	.97997	1.00167	.98119	1.
.622	.85233	.87314	.98040	1.00213	.98161	1.
.621	.85388	.87274	.98082	1.00259	.98203	1.
.620	.85542	.87234	.98123	1.00305	.98244	1.
.619	.85696	.87193	.98165	1.00352	.98285	1.
.618	.85851	.87153	.98205	1.00398	.98325	1.
.617	.86005	.87113	.98246	1.00445	.98365	1.
.616	.86160	.87072	.98285	1.00491	.98404	1.
.615	.86314	.87032	.98325	1.00538	.98442	1.
.614	.86469	.86991	.98364	1.00585	.98480	1.
.613	.86624	.86951	.98402	1.00631	.98518	1.
.612	.86778	.86910	.98440	1.00678	.98555	1.
.611	.86933	.86870	.98477	1.00725	.98591	1.
.610	.87088	.86829	.98514	1.00773	.98627	1.
.609	.87243	.86788	.98551	1.00820	.98663	1.
.608	.87398	.86748	.98587	1.00867	.98698	1.
.607	.87553	.86707	.98622	1.00915	.98732	1.
.606	.87708	.86666	.98658	1.00962	.98766	1.
.605	.87863	.86625	.98692	1.01010	.98799	1.
.604	.88018	.86584	.98726	1.01058	.98832	1.
.603	.88173	.86543	.98760	1.01105	.98864	1.
.602	.88329	.86502	.98793	1.01153	.98896	1.
.601	.88484	.86461	.98826	1.01201	.98928	1.

TABLE V. Isentropic Exponent = 1.40

p/p$_t$	M	T/T$_t$	GAMMA	ISOTHERMAL T$_t$/T$_t$*	RAYLEIGH T$_t$/T$_t$*	SHOCK p$_{t1}$/p$_{t2}$
.600	.88639	.86420	.98858	1.01250	.98959	1.
.599	.88795	.86379	.98890	1.01298	.98989	1.
.598	.88950	.86338	.98922	1.01346	.99019	1.
.597	.89106	.86296	.98953	1.01395	.99048	1.
.596	.89261	.86255	.98983	1.01443	.99077	1.
.595	.89417	.86214	.99013	1.01492	.99105	1.
.594	.89573	.86172	.99043	1.01541	.99133	1.
.593	.89729	.86131	.99072	1.01590	.99161	1.
.592	.89885	.86089	.99101	1.01639	.99188	1.
.591	.90041	.86048	.99129	1.01688	.99214	1.
.590	.90197	.86006	.99157	1.01737	.99240	1.
.589	.90353	.85964	.99184	1.01786	.99266	1.
.588	.90509	.85923	.99211	1.01836	.99291	1.
.587	.90665	.85881	.99237	1.01885	.99315	1.
.586	.90821	.85839	.99263	1.01935	.99339	1.
.585	.90978	.85797	.99288	1.01985	.99363	1.
.584	.91134	.85755	.99313	1.02035	.99386	1.
.583	.91291	.85713	.99338	1.02085	.99409	1.
.582	.91447	.85671	.99362	1.02135	.99431	1.
.581	.91604	.85629	.99386	1.02185	.99453	1.
.580	.91761	.85587	.99409	1.02235	.99474	1.
.579	.91918	.85545	.99432	1.02286	.99495	1.
.578	.92075	.85503	.99454	1.02336	.99515	1.
.577	.92232	.85460	.99476	1.02387	.99535	1.
.576	.92389	.85418	.99497	1.02437	.99555	1.
.575	.92546	.85376	.99518	1.02488	.99574	1.
.574	.92703	.85333	.99539	1.02539	.99593	1.
.573	.92861	.85291	.99559	1.02590	.99611	1.
.572	.93018	.85248	.99578	1.02642	.99629	1.
.571	.93176	.85205	.99597	1.02693	.99646	1.
.570	.93333	.85163	.99616	1.02744	.99663	1.
.569	.93491	.85120	.99634	1.02796	.99679	1.
.568	.93649	.85077	.99652	1.02848	.99695	1.
.567	.93807	.85034	.99669	1.02899	.99711	1.
.566	.93965	.84992	.99686	1.02951	.99726	1.
.565	.94123	.84949	.99703	1.03003	.99741	1.
.564	.94281	.84906	.99719	1.03055	.99755	1.
.563	.94439	.84863	.99734	1.03108	.99769	1.
.562	.94597	.84820	.99750	1.03160	.99782	1.
.561	.94756	.84776	.99764	1.03213	.99795	1.
.560	.94914	.84733	.99778	1.03265	.99808	1.
.559	.95073	.84690	.99792	1.03318	.99820	1.
.558	.95232	.84647	.99806	1.03371	.99832	1.
.557	.95391	.84603	.99818	1.03424	.99843	1.
.556	.95550	.84560	.99831	1.03477	.99854	1.
.555	.95709	.84516	.99843	1.03530	.99865	1.
.554	.95868	.84473	.99854	1.03584	.99875	1.
.553	.96027	.84429	.99866	1.03637	.99884	1.
.552	.96186	.84386	.99876	1.03691	.99894	1.
.551	.96346	.84342	.99886	1.03744	.99903	1.

TABLE V. Isentropic Exponent = 1.40

p/p_t	M	T/T_t	GAMMA	ISOTHERMAL T_t/T_t^*	RAYLEIGH T_t/T_t^*	SHOCK p_{t1}/p_{t2}
.550	.96505	.84298	.99896	1.03798	.99911	1.
.549	.96665	.84254	.99906	1.03852	.99919	1.
.548	.96825	.84210	.99915	1.03906	.99927	1.
.547	.96985	.84166	.99923	1.03961	.99934	1.
.546	.97145	.84122	.99931	1.04015	.99941	1.
.545	.97305	.84078	.99939	1.04069	.99948	1.
.544	.97465	.84034	.99946	1.04124	.99954	1.
.543	.97626	.83990	.99952	1.04179	.99960	1.
.542	.97786	.83946	.99959	1.04234	.99965	1.
.541	.97947	.83902	.99964	1.04289	.99970	1.
.540	.98107	.83857	.99970	1.04344	.99974	1.
.539	.98268	.83813	.99975	1.04399	.99979	1.
.538	.98429	.83768	.99979	1.04455	.99983	1.
.537	.98590	.83724	.99983	1.04510	.99986	1.
.536	.98752	.83679	.99987	1.04566	.99989	1.
.535	.98913	.83635	.99990	1.04622	.99992	1.
.534	.99074	.83590	.99993	1.04678	.99994	1.
.533	.99236	.83545	.99995	1.04734	.99996	1.
.532	.99398	.83500	.99997	1.04790	.99997	1.
.531	.99559	.83456	.99998	1.04846	.99999	1.
.530	.99721	.83411	.99999	1.04903	.99999	1.
.529	.99884	.83366	.00000	1.04959	.00000	1.
.528	1.00046	.83321	.00000	1.05016	.00000	1.00000
.527	1.00208	.83276	.00000	1.05073	.00000	1.00000
.526	1.00371	.83230	.99999	1.05130	.99999	1.00000
.525	1.00533	.83185	.99998	1.05187	.99998	1.00000
.524	1.00696	.83140	.99996	1.05244	.99997	1.00000
.523	1.00859	.83094	.99994	1.05302	.99995	1.00000
.522	1.01022	.83049	.99991	1.05359	.99993	1.00000
.521	1.01185	.83004	.99988	1.05417	.99990	1.00000
.520	1.01348	.82958	.99985	1.05475	.99988	1.00000
.519	1.01512	.82912	.99981	1.05533	.99984	1.00000
.518	1.01675	.82867	.99977	1.05591	.99981	1.00001
.517	1.01839	.82821	.99972	1.05650	.99977	1.00001
.516	1.02003	.82775	.99967	1.05708	.99973	1.00001
.515	1.02167	.82729	.99961	1.05767	.99968	1.00001
.514	1.02331	.82683	.99955	1.05825	.99963	1.00002
.513	1.02496	.82637	.99949	1.05884	.99958	1.00002
.512	1.02660	.82591	.99942	1.05943	.99953	1.00002
.511	1.02825	.82545	.99935	1.06003	.99947	1.00003
.510	1.02989	.82499	.99927	1.06062	.99940	1.00003
.509	1.03154	.82453	.99919	1.06121	.99934	1.00004
.508	1.03319	.82406	.99910	1.06181	.99927	1.00004
.507	1.03485	.82360	.99901	1.06241	.99920	1.00005
.506	1.03650	.82314	.99891	1.06301	.99912	1.00006
.505	1.03815	.82267	.99881	1.06361	.99904	1.00007
.504	1.03981	.82220	.99871	1.06421	.99896	1.00008
.503	1.04147	.82174	.99860	1.06482	.99887	1.00009
.502	1.04313	.82127	.99849	1.06542	.99878	1.00010
.501	1.04479	.82080	.99837	1.06603	.99869	1.00011

TABLE V. Isentropic Exponent = 1.40

p/p_t	M	T/T_t	GAMMA	ISOTHERMAL T_t/T_t^*	RAYLEIGH T_t/T_t^*	SHOCK p_{t1}/p_{t2}
.500	1.04646	.82034	.99825	1.06664	.99859	1.00012
.499	1.04812	.81987	.99812	1.06725	.99849	1.00013
.498	1.04979	.81940	.99799	1.06786	.99839	1.00015
.497	1.05145	.81893	.99786	1.06847	.99828	1.00016
.496	1.05312	.81845	.99772	1.06909	.99817	1.00018
.495	1.05480	.81798	.99757	1.06970	.99806	1.00019
.494	1.05647	.81751	.99743	1.07032	.99795	1.00021
.493	1.05814	.81704	.99727	1.07094	.99783	1.00023
.492	1.05982	.81656	.99712	1.07156	.99771	1.00025
.491	1.06150	.81609	.99696	1.07219	.99758	1.00027
.490	1.06318	.81561	.99679	1.07281	.99745	1.00029
.489	1.06486	.81514	.99662	1.07344	.99732	1.00031
.488	1.06655	.81466	.99645	1.07407	.99719	1.00034
.487	1.06823	.81418	.99627	1.07470	.99705	1.00036
.486	1.06992	.81371	.99609	1.07533	.99691	1.00039
.485	1.07161	.81323	.99590	1.07596	.99676	1.00041
.484	1.07330	.81275	.99571	1.07659	.99662	1.00044
.483	1.07499	.81227	.99551	1.07723	.99647	1.00047
.482	1.07669	.81179	.99531	1.07787	.99631	1.00050
.481	1.07838	.81131	.99511	1.07851	.99616	1.00054
.480	1.08008	.81082	.99490	1.07915	.99600	1.00057
.479	1.08178	.81034	.99468	1.07979	.99584	1.00061
.478	1.08348	.80986	.99447	1.08044	.99567	1.00064
.477	1.08519	.80937	.99424	1.08109	.99550	1.00068
.476	1.08689	.80889	.99402	1.08173	.99533	1.00072
.475	1.08860	.80840	.99378	1.08238	.99516	1.00076
.474	1.09031	.80791	.99355	1.08304	.99498	1.00080
.473	1.09202	.80743	.99331	1.08369	.99480	1.00085
.472	1.09374	.80694	.99307	1.08435	.99462	1.00089
.471	1.09545	.80645	.99282	1.08500	.99443	1.00094
.470	1.09717	.80596	.99256	1.08566	.99424	1.00099
.469	1.09889	.80547	.99231	1.08632	.99405	1.00104
.468	1.10061	.80498	.99204	1.08698	.99385	1.00109
.467	1.10233	.80449	.99178	1.08765	.99366	1.00115
.466	1.10406	.80399	.99151	1.08832	.99346	1.00120
.465	1.10579	.80350	.99123	1.08898	.99325	1.00126
.464	1.10752	.80301	.99095	1.08965	.99305	1.00132
.463	1.10925	.80251	.99067	1.09033	.99284	1.00138
.462	1.11098	.80202	.99038	1.09100	.99262	1.00144
.461	1.11272	.80152	.99009	1.09168	.99241	1.00150
.460	1.11446	.80102	.98979	1.09235	.99219	1.00157
.459	1.11620	.80053	.98949	1.09303	.99197	1.00164
.458	1.11794	.80003	.98918	1.09371	.99175	1.00171
.457	1.11969	.79953	.98887	1.09440	.99152	1.00178
.456	1.12143	.79903	.98856	1.09508	.99129	1.00185
.455	1.12318	.79853	.98824	1.09577	.99106	1.00193
.454	1.12493	.79802	.98791	1.09646	.99083	1.00200
.453	1.12669	.79752	.98758	1.09715	.99059	1.00208
.452	1.12844	.79702	.98725	1.09784	.99035	1.00216
.451	1.13020	.79651	.98691	1.09854	.99011	1.00225

TABLE V. Isentropic Exponent = 1.40

p/p_t	M	T/T_t	GAMMA	ISOTHERMAL T_t/T_t^*	RAYLEIGH T_t/T_t^*	SHOCK p_{t1}/p_{t2}
.450	1.13196	.79601	.98657	1.09923	.98986	
.449	1.13372	.79550	.98623	1.09993	.98961	1.00233
.448	1.13549	.79500	.98588	1.10063	.98936	1.00242
.447	1.13726	.79449	.98552	1.10134	.98911	1.00251
.446	1.13903	.79398	.98516	1.10204	.98885	1.00260
						1.00269
.445	1.14080	.79347	.98480	1.10275	.98859	
.444	1.14257	.79296	.98443	1.10346	.98833	1.00279
.443	1.14435	.79245	.98406	1.10417	.98806	1.00289
.442	1.14613	.79194	.98368	1.10488	.98780	1.00299
.441	1.14791	.79143	.98330	1.10560	.98753	1.00309
						1.00320
.440	1.14970	.79091	.98291	1.10631	.98726	
.439	1.15148	.79040	.98252	1.10703	.98698	1.00330
.438	1.15327	.78989	.98213	1.10776	.98670	1.00341
.437	1.15506	.78937	.98173	1.10848	.98642	1.00352
.436	1.15686	.78885	.98132	1.10921	.98614	1.00364
						1.00375
.435	1.15865	.78834	.98091	1.10993	.98585	
.434	1.16045	.78782	.98050	1.11066	.98557	1.00387
.433	1.16225	.78730	.98008	1.11140	.98528	1.00399
.432	1.16406	.78678	.97966	1.11213	.98498	1.00412
.431	1.16586	.78626	.97923	1.11287	.98469	1.00424
						1.00437
.430	1.16767	.78574	.97880	1.11361	.98439	
.429	1.16948	.78521	.97837	1.11435	.98409	1.00450
.428	1.17130	.78469	.97793	1.11509	.98378	1.00464
.427	1.17312	.78417	.97748	1.11584	.98348	1.00477
.426	1.17494	.78364	.97703	1.11658	.98317	1.00491
						1.00505
.425	1.17676	.78311	.97658	1.11733	.98286	
.424	1.17858	.78259	.97612	1.11809	.98254	1.00520
.423	1.18041	.78206	.97566	1.11884	.98223	1.00535
.422	1.18224	.78153	.97519	1.11960	.98191	1.00549
.421	1.18408	.78100	.97472	1.12036	.98159	1.00565
						1.00580
.420	1.18591	.78047	.97424	1.12112	.98126	
.419	1.18775	.77994	.97376	1.12188	.98094	1.00596
.418	1.18959	.77941	.97328	1.12265	.98061	1.00612
.417	1.19144	.77887	.97278	1.12342	.98028	1.00628
.416	1.19328	.77834	.97229	1.12419	.97995	1.00645
						1.00662
.415	1.19513	.77781	.97179	1.12496	.97961	
.414	1.19699	.77727	.97129	1.12574	.97927	1.00679
.413	1.19884	.77673	.97078	1.12651	.97893	1.00696
.412	1.20070	.77619	.97027	1.12729	.97859	1.00714
.411	1.20256	.77566	.96975	1.12808	.97824	1.00732
						1.00751
.410	1.20443	.77512	.96923	1.12886	.97789	
.409	1.20630	.77458	.96870	1.12965	.97754	1.00769
.408	1.20817	.77403	.96817	1.13044	.97719	1.00788
.407	1.21004	.77349	.96763	1.13123	.97683	1.00807
.406	1.21192	.77295	.96709	1.13203	.97648	1.00827
						1.00847
.405	1.21380	.77240	.96655	1.13283	.97612	
.404	1.21568	.77186	.96600	1.13363	.97575	1.00867
.403	1.21756	.77131	.96544	1.13443	.97539	1.00887
.402	1.21945	.77076	.96488	1.13524	.97502	1.00908
.401	1.22135	.77022	.96432	1.13604	.97465	1.00929
						1.00951

TABLE V. Isentropic Exponent = 1.40

p/p_t	M	T/T_t	GAMMA	ISOTHERMAL T_t/T_t^*	RAYLEIGH T_t/T_t^*	SHOCK p_{t1}/p_{t2}
.400	1.22324	.76967	.96375	1.13686	.97428	1.00973
.399	1.22514	.76912	.96318	1.13767	.97391	1.00995
.398	1.22704	.76857	.96260	1.13848	.97353	1.01017
.397	1.22894	.76801	.96202	1.13930	.97315	1.01040
.396	1.23085	.76746	.96143	1.14012	.97277	1.01063
.395	1.23276	.76691	.96084	1.14095	.97239	1.01086
.394	1.23468	.76635	.96024	1.14178	.97200	1.01110
.393	1.23660	.76579	.95964	1.14260	.97161	1.01134
.392	1.23852	.76524	.95903	1.14344	.97122	1.01159
.391	1.24044	.76468	.95842	1.14427	.97083	1.01184
.390	1.24237	.76412	.95781	1.14511	.97044	1.01209
.389	1.24430	.76356	.95719	1.14595	.97004	1.01234
.388	1.24623	.76300	.95656	1.14679	.96964	1.01260
.387	1.24817	.76244	.95593	1.14764	.96924	1.01286
.386	1.25011	.76187	.95530	1.14849	.96883	1.01313
.385	1.25206	.76131	.95466	1.14934	.96843	1.01340
.384	1.25401	.76074	.95402	1.15019	.96802	1.01367
.383	1.25596	.76018	.95337	1.15105	.96761	1.01395
.382	1.25791	.75961	.95271	1.15191	.96720	1.01423
.381	1.25987	.75904	.95206	1.15277	.96678	1.01451
.380	1.26183	.75847	.95139	1.15364	.96636	1.01480
.379	1.26380	.75790	.95073	1.15451	.96594	1.01509
.378	1.26577	.75733	.95005	1.15538	.96552	1.01539
.377	1.26774	.75675	.94938	1.15625	.96510	1.01569
.376	1.26972	.75618	.94869	1.15713	.96467	1.01599
.375	1.27170	.75560	.94801	1.15801	.96424	1.01630
.374	1.27368	.75503	.94732	1.15890	.96381	1.01661
.373	1.27567	.75445	.94662	1.15978	.96338	1.01692
.372	1.27766	.75387	.94592	1.16067	.96295	1.01724
.371	1.27966	.75329	.94521	1.16157	.96251	1.01757
.370	1.28166	.75271	.94450	1.16246	.96207	1.01789
.369	1.28366	.75213	.94379	1.16336	.96163	1.01823
.368	1.28567	.75155	.94306	1.16426	.96118	1.01856
.367	1.28768	.75096	.94234	1.16517	.96074	1.01890
.366	1.28969	.75038	.94161	1.16608	.96029	1.01925
.365	1.29171	.74979	.94087	1.16699	.95984	1.01959
.364	1.29373	.74920	.94013	1.16791	.95939	1.01995
.363	1.29576	.74862	.93939	1.16882	.95893	1.02030
.362	1.29779	.74803	.93864	1.16975	.95848	1.02066
.361	1.29982	.74744	.93788	1.17067	.95802	1.02103
.360	1.30186	.74684	.93712	1.17160	.95756	1.02140
.359	1.30391	.74625	.93636	1.17253	.95709	1.02177
.358	1.30595	.74566	.93559	1.17346	.95663	1.02215
.357	1.30800	.74506	.93481	1.17440	.95616	1.02253
.356	1.31006	.74446	.93403	1.17534	.95569	1.02292
.355	1.31212	.74386	.93325	1.17629	.95522	1.02331
.354	1.31418	.74327	.93246	1.17724	.95475	1.02371
.353	1.31625	.74266	.93167	1.17819	.95427	1.02411
.352	1.31832	.74206	.93087	1.17915	.95379	1.02452
.351	1.32040	.74146	.93006	1.18010	.95331	1.02493

TABLE V. Isentropic Exponent = 1.40

p/p_t	M	T/T_t	GAMMA	ISOTHERMAL T_t/T_t^*	RAYLEIGH T_t/T_t^*	SHOCK p_{t1}/p_{t2}
.350	1.32248	.74086	.92925	1.18107	.95283	1.02535
.349	1.32456	.74025	.92844	1.18203	.95235	1.02577
.348	1.32665	.73964	.92762	1.18300	.95186	1.02619
.347	1.32875	.73904	.92679	1.18397	.95137	1.02662
.346	1.33085	.73843	.92596	1.18495	.95088	1.02706
.345	1.33295	.73782	.92513	1.18593	.95039	1.02749
.344	1.33506	.73720	.92429	1.18692	.94990	1.02794
.343	1.33717	.73659	.92344	1.18790	.94940	1.02839
.342	1.33929	.73598	.92259	1.18889	.94890	1.02884
.341	1.34141	.73536	.92174	1.18989	.94840	1.02930
.340	1.34353	.73475	.92088	1.19089	.94790	1.02977
.339	1.34566	.73413	.92001	1.19189	.94740	1.03024
.338	1.34780	.73351	.91914	1.19290	.94689	1.03071
.337	1.34994	.73289	.91827	1.19391	.94638	1.03119
.336	1.35208	.73227	.91739	1.19492	.94587	1.03168
.335	1.35423	.73164	.91650	1.19594	.94536	1.03217
.334	1.35639	.73102	.91561	1.19696	.94484	1.03267
.333	1.35855	.73039	.91471	1.19799	.94432	1.03317
.332	1.36071	.72976	.91381	1.19902	.94381	1.03367
.331	1.36288	.72914	.91290	1.20005	.94328	1.03419
.330	1.36505	.72850	.91199	1.20109	.94276	1.03470
.329	1.36723	.72787	.91107	1.20213	.94224	1.03523
.328	1.36942	.72724	.91015	1.20318	.94171	1.03576
.327	1.37161	.72661	.90922	1.20423	.94118	1.03629
.326	1.37380	.72597	.90829	1.20528	.94065	1.03683
.325	1.37600	.72533	.90735	1.20634	.94012	1.03738
.324	1.37820	.72470	.90641	1.20740	.93958	1.03793
.323	1.38041	.72406	.90546	1.20847	.93904	1.03849
.322	1.38263	.72341	.90451	1.20954	.93850	1.03905
.321	1.38485	.72277	.90355	1.21062	.93796	1.03962
.320	1.38708	.72213	.90258	1.21170	.93742	1.04020
.319	1.38931	.72148	.90161	1.21278	.93688	1.04078
.318	1.39154	.72084	.90064	1.21387	.93633	1.04137
.317	1.39379	.72019	.89966	1.21496	.93578	1.04196
.316	1.39603	.71954	.89867	1.21606	.93523	1.04256
.315	1.39829	.71889	.89768	1.21716	.93467	1.04317
.314	1.40054	.71823	.89668	1.21827	.93412	1.04378
.313	1.40281	.71758	.89568	1.21938	.93356	1.04440
.312	1.40508	.71692	.89467	1.22049	.93300	1.04502
.311	1.40735	.71627	.89366	1.22161	.93244	1.04565
.310	1.40964	.71561	.89264	1.22274	.93188	1.04629
.309	1.41192	.71495	.89161	1.22387	.93131	1.04693
.308	1.41422	.71429	.89058	1.22500	.93075	1.04758
.307	1.41651	.71362	.88955	1.22614	.93018	1.04824
.306	1.41882	.71296	.88851	1.22728	.92961	1.04891
.305	1.42113	.71229	.88746	1.22843	.92903	1.04958
.304	1.42345	.71162	.88641	1.22958	.92846	1.05025
.303	1.42577	.71095	.88535	1.23074	.92788	1.05094
.302	1.42810	.71028	.88429	1.23191	.92730	1.05163
.301	1.43043	.70961	.88322	1.23307	.92672	1.05233

TABLE V. Isentropic Exponent = 1.40

p/p_t	M	T/T_t	GAMMA	ISOTHERMAL T_t/T_t^*	RAYLEIGH T_t/T_t^*	SHOCK p_{t1}/p_{t2}
.300	1.43277	.70893	.88214	1.23425	.92614	1.05303
.299	1.43512	.70826	.88106	1.23542	.92556	1.05375
.298	1.43747	.70758	.87998	1.23661	.92497	1.05447
.297	1.43983	.70690	.87889	1.23780	.92438	1.05519
.296	1.44220	.70622	.87779	1.23899	.92379	1.05593
.295	1.44457	.70554	.87669	1.24019	.92320	1.05667
.294	1.44695	.70485	.87558	1.24139	.92260	1.05742
.293	1.44934	.70417	.87446	1.24260	.92200	1.05817
.292	1.45173	.70348	.87334	1.24382	.92141	1.05894
.291	1.45413	.70279	.87222	1.24503	.92081	1.05971
.290	1.45653	.70210	.87109	1.24626	.92020	1.06049
.289	1.45894	.70141	.86995	1.24749	.91960	1.06127
.288	1.46136	.70071	.86881	1.24873	.91899	1.06207
.287	1.46379	.70002	.86766	1.24997	.91838	1.06287
.286	1.46622	.69932	.86650	1.25122	.91777	1.06368
.285	1.46866	.69862	.86534	1.25247	.91716	1.06450
.284	1.47111	.69792	.86418	1.25373	.91655	1.06532
.283	1.47356	.69722	.86301	1.25499	.91593	1.06616
.282	1.47602	.69651	.86183	1.25626	.91531	1.06700
.281	1.47849	.69580	.86064	1.25754	.91469	1.06785
.280	1.48096	.69510	.85945	1.25882	.91407	1.06871
.279	1.48344	.69439	.85826	1.26011	.91344	1.06958
.278	1.48593	.69367	.85706	1.26140	.91282	1.07045
.277	1.48843	.69296	.85585	1.26270	.91219	1.07134
.276	1.49093	.69224	.85464	1.26400	.91156	1.07223
.275	1.49344	.69153	.85342	1.26532	.91093	1.07313
.274	1.49596	.69081	.85219	1.26663	.91029	1.07404
.273	1.49849	.69009	.85096	1.26796	.90966	1.07496
.272	1.50102	.68936	.84972	1.26929	.90902	1.07589
.271	1.50356	.68864	.84848	1.27062	.90838	1.07683
.270	1.50611	.68791	.84723	1.27197	.90774	1.07778
.269	1.50867	.68718	.84597	1.27332	.90709	1.07873
.268	1.51124	.68645	.84471	1.27467	.90645	1.07970
.267	1.51381	.68572	.84344	1.27603	.90580	1.08067
.266	1.51639	.68498	.84216	1.27740	.90515	1.08166
.265	1.51898	.68425	.84088	1.27878	.90450	1.08265
.264	1.52158	.68351	.83959	1.28016	.90384	1.08365
.263	1.52418	.68277	.83830	1.28155	.90319	1.08467
.262	1.52680	.68203	.83700	1.28294	.90253	1.08569
.261	1.52942	.68128	.83569	1.28435	.90187	1.08672
.260	1.53205	.68053	.83438	1.28576	.90120	1.08777
.259	1.53469	.67978	.83306	1.28717	.90054	1.08882
.258	1.53734	.67903	.83174	1.28860	.89987	1.08988
.257	1.53999	.67828	.83041	1.29003	.89921	1.09096
.256	1.54266	.67753	.82907	1.29146	.89854	1.09204
.255	1.54533	.67677	.82772	1.29291	.89786	1.09314
.254	1.54802	.67601	.82637	1.29436	.89719	1.09424
.253	1.55071	.67525	.82502	1.29582	.89651	1.09536
.252	1.55341	.67448	.82365	1.29729	.89584	1.09649
.251	1.55612	.67372	.82228	1.29876	.89516	1.09762

TABLE V.   Isentropic Exponent = 1.40

p/p_t	M	T/T_t	GAMMA	ISOTHERMAL T_t/T*	RAYLEIGH T_t/T*	SHOCK p_t1/p_t2
.250	1.55884	.67295	.82090	1.30025	.89447	1.09877
.249	1.56157	.67218	.81952	1.30173	.89379	1.09993
.248	1.56430	.67141	.81813	1.30323	.89310	1.10111
.247	1.56705	.67063	.81673	1.30474	.89241	1.10229
.246	1.56981	.66986	.81533	1.30625	.89172	1.10348
.245	1.57257	.66908	.81392	1.30777	.89103	1.10469
.244	1.57535	.66830	.81250	1.30930	.89034	1.10591
.243	1.57813	.66751	.81108	1.31084	.88964	1.10714
.242	1.58093	.66673	.80965	1.31238	.88894	1.10838
.241	1.58373	.66594	.80821	1.31394	.88824	1.10964
.240	1.58655	.66515	.80677	1.31550	.88754	1.11091
.239	1.58937	.66435	.80532	1.31707	.88683	1.11218
.238	1.59221	.66356	.80386	1.31865	.88613	1.11348
.237	1.59506	.66276	.80240	1.32024	.88542	1.11478
.236	1.59791	.66196	.80093	1.32183	.88471	1.11610
.235	1.60078	.66116	.79945	1.32344	.88399	1.11743
.234	1.60366	.66035	.79796	1.32505	.88328	1.11877
.233	1.60654	.65955	.79647	1.32667	.88256	1.12013
.232	1.60944	.65874	.79497	1.32830	.88184	1.12150
.231	1.61235	.65792	.79347	1.32994	.88112	1.12289
.230	1.61527	.65711	.79195	1.33159	.88039	1.12428
.229	1.61820	.65629	.79043	1.33325	.87967	1.12570
.228	1.62115	.65547	.78890	1.33492	.87894	1.12712
.227	1.62410	.65465	.78737	1.33660	.87821	1.12856
.226	1.62707	.65382	.78583	1.33828	.87748	1.13002
.225	1.63004	.65299	.78428	1.33998	.87674	1.13149
.224	1.63303	.65216	.78272	1.34169	.87601	1.13297
.223	1.63603	.65133	.78116	1.34340	.87527	1.13447
.222	1.63904	.65049	.77959	1.34513	.87453	1.13598
.221	1.64207	.64966	.77801	1.34687	.87378	1.13751
.220	1.64510	.64881	.77642	1.34861	.87304	1.13906
.219	1.64815	.64797	.77483	1.35037	.87229	1.14062
.218	1.65121	.64712	.77323	1.35214	.87154	1.14219
.217	1.65428	.64627	.77162	1.35391	.87079	1.14378
.216	1.65737	.64542	.77001	1.35570	.87004	1.14539
.215	1.66047	.64457	.76838	1.35750	.86928	1.14702
.214	1.66358	.64371	.76675	1.35931	.86852	1.14866
.213	1.66670	.64285	.76511	1.36113	.86776	1.15031
.212	1.66984	.64198	.76347	1.36296	.86700	1.15199
.211	1.67299	.64112	.76181	1.36480	.86623	1.15368
.210	1.67615	.64025	.76015	1.36666	.86546	1.15539
.209	1.67932	.63938	.75848	1.36852	.86470	1.15711
.208	1.68251	.63850	.75681	1.37040	.86392	1.15886
.207	1.68572	.63762	.75512	1.37229	.86315	1.16062
.206	1.68893	.63674	.75343	1.37419	.86237	1.16240
.205	1.69217	.63586	.75173	1.37610	.86159	1.16420
.204	1.69541	.63497	.75002	1.37802	.86081	1.16602
.203	1.69867	.63408	.74830	1.37996	.86003	1.16785
.202	1.70195	.63318	.74658	1.38191	.85924	1.16971
.201	1.70523	.63229	.74485	1.38387	.85845	1.17158

TABLE V. Isentropic Exponent = 1.40

p/p_t	M	T/T_t	GAMMA	ISOTHERMAL T_t/T_t^*	RAYLEIGH T_t/T_t^*	SHOCK p_{t1}/p_{t2}
.200	1.70854	.63139	.74311	1.38584	.85766	1.17348
.199	1.71185	.63048	.74136	1.38783	.85687	1.17539
.198	1.71519	.62957	.73960	1.38983	.85608	1.17733
.197	1.71854	.62866	.73784	1.39184	.85528	1.17928
.196	1.72190	.62775	.73606	1.39386	.85448	1.18126
.195	1.72528	.62683	.73428	1.39590	.85368	1.18326
.194	1.72867	.62591	.73249	1.39796	.85287	1.18527
.193	1.73209	.62499	.73069	1.40002	.85206	1.18731
.192	1.73551	.62406	.72888	1.40210	.85125	1.18938
.191	1.73896	.62313	.72707	1.40419	.85044	1.19146
.190	1.74241	.62220	.72524	1.40630	.84963	1.19357
.189	1.74589	.62126	.72341	1.40842	.84881	1.19569
.188	1.74938	.62032	.72157	1.41056	.84799	1.19785
.187	1.75289	.61938	.71972	1.41271	.84717	1.20002
.186	1.75642	.61843	.71786	1.41488	.84634	1.20222
.185	1.75996	.61748	.71599	1.41706	.84552	1.20445
.184	1.76353	.61652	.71412	1.41925	.84469	1.20669
.183	1.76710	.61556	.71223	1.42147	.84385	1.20897
.182	1.77070	.61460	.71034	1.42369	.84302	1.21126
.181	1.77432	.61363	.70843	1.42594	.84218	1.21359
.180	1.77795	.61266	.70652	1.42819	.84134	1.21593
.179	1.78160	.61169	.70460	1.43047	.84050	1.21831
.178	1.78527	.61071	.70267	1.43276	.83965	1.22071
.177	1.78896	.60973	.70073	1.43507	.83880	1.22314
.176	1.79267	.60874	.69878	1.43739	.83795	1.22560
.175	1.79640	.60775	.69682	1.43974	.83710	1.22808
.174	1.80015	.60676	.69486	1.44210	.83624	1.23059
.173	1.80392	.60576	.69288	1.44447	.83539	1.23313
.172	1.80771	.60476	.69089	1.44687	.83452	1.23570
.171	1.81152	.60375	.68890	1.44928	.83366	1.23830
.170	1.81535	.60274	.68689	1.45171	.83279	1.24093
.169	1.81920	.60172	.68488	1.45416	.83192	1.24359
.168	1.82307	.60070	.68285	1.45663	.83105	1.24628
.167	1.82696	.59968	.68082	1.45911	.83018	1.24900
.166	1.83088	.59865	.67877	1.46162	.82930	1.25176
.165	1.83481	.59762	.67672	1.46414	.82842	1.25454
.164	1.83877	.59658	.67465	1.46669	.82753	1.25736
.163	1.84275	.59554	.67258	1.46926	.82665	1.26022
.162	1.84676	.59449	.67049	1.47184	.82576	1.26310
.161	1.85079	.59344	.66840	1.47445	.82486	1.26602
.160	1.85484	.59239	.66630	1.47707	.82397	1.26898
.159	1.85891	.59133	.66418	1.47972	.82307	1.27197
.158	1.86301	.59026	.66205	1.48239	.82217	1.27500
.157	1.86714	.58919	.65992	1.48508	.82126	1.27806
.156	1.87128	.58812	.65777	1.48780	.82035	1.28117
.155	1.87546	.58704	.65562	1.49053	.81944	1.28431
.154	1.87965	.58595	.65345	1.49329	.81853	1.28749
.153	1.88388	.58486	.65127	1.49607	.81761	1.29071
.152	1.88813	.58377	.64908	1.49888	.81669	1.29396
.151	1.89241	.58267	.64688	1.50171	.81577	1.29726

TABLE V. Isentropic Exponent = 1.40

p/p_t	M	T/T_t	GAMMA	ISOTHERMAL T_t/T^*	RAYLEIGH T_t/T^*	SHOCK p_{t1}/p_{t2}
.150	1.89671	.58156	.64467	1.50456	.81484	1.30061
.149	1.90104	.58045	.64245	1.50744	.81391	1.30399
.148	1.90540	.57934	.64021	1.51034	.81298	1.30742
.147	1.90978	.57822	.63797	1.51327	.81204	1.31089
.146	1.91420	.57709	.63571	1.51623	.81110	1.31440
.145	1.91864	.57596	.63345	1.51921	.81016	1.31796
.144	1.92311	.57482	.63117	1.52221	.80921	1.32157
.143	1.92762	.57368	.62888	1.52525	.80826	1.32522
.142	1.93215	.57253	.62658	1.52831	.80731	1.32892
.141	1.93671	.57137	.62426	1.53140	.80635	1.33267
.140	1.94130	.57021	.62194	1.53452	.80539	1.33647
.139	1.94593	.56905	.61960	1.53766	.80443	1.34032
.138	1.95059	.56787	.61725	1.54084	.80346	1.34423
.137	1.95528	.56669	.61489	1.54404	.80249	1.34818
.136	1.96000	.56551	.61252	1.54728	.80152	1.35219
.135	1.96475	.56432	.61013	1.55054	.80054	1.35625
.134	1.96954	.56312	.60773	1.55384	.79956	1.36037
.133	1.97436	.56192	.60532	1.55717	.79857	1.36455
.132	1.97922	.56071	.60290	1.56053	.79758	1.36878
.131	1.98412	.55949	.60046	1.56393	.79659	1.37308
.130	1.98905	.55827	.59801	1.56735	.79559	1.37743
.129	1.99401	.55704	.59555	1.57082	.79459	1.38184
.128	1.99902	.55580	.59308	1.57431	.79359	1.38632
.127	2.00406	.55455	.59059	1.57784	.79258	1.39086
.126	2.00914	.55330	.58809	1.58141	.79156	1.39547
.125	2.01426	.55204	.58557	1.58502	.79055	1.40015
.124	2.01942	.55078	.58305	1.58866	.78953	1.40489
.123	2.02462	.54951	.58050	1.59234	.78850	1.40970
.122	2.02986	.54823	.57795	1.59606	.78747	1.41459
.121	2.03514	.54694	.57538	1.59981	.78644	1.41954
.120	2.04046	.54564	.57280	1.60361	.78540	1.42458
.119	2.04583	.54434	.57020	1.60745	.78436	1.42968
.118	2.05124	.54303	.56759	1.61133	.78331	1.43487
.117	2.05670	.54171	.56496	1.61525	.78226	1.44013
.116	2.06220	.54038	.56232	1.61922	.78120	1.44548
.115	2.06775	.53905	.55966	1.62323	.78014	1.45091
.114	2.07335	.53771	.55699	1.62729	.77908	1.45643
.113	2.07899	.53635	.55431	1.63139	.77801	1.46203
.112	2.08469	.53499	.55161	1.63554	.77693	1.46772
.111	2.09043	.53362	.54889	1.63973	.77585	1.47350
.110	2.09622	.53225	.54616	1.64398	.77477	1.47938
.109	2.10207	.53086	.54341	1.64827	.77368	1.48535
.108	2.10797	.52946	.54065	1.65262	.77258	1.49142
.107	2.11392	.52806	.53787	1.65702	.77148	1.49759
.106	2.11993	.52664	.53507	1.66147	.77038	1.50387
.105	2.12599	.52522	.53226	1.66597	.76927	1.51025
.104	2.13212	.52378	.52943	1.67054	.76816	1.51674
.103	2.13829	.52234	.52659	1.67515	.76704	1.52334
.102	2.14453	.52089	.52373	1.67983	.76591	1.53005
.101	2.15083	.51942	.52085	1.68456	.76478	1.53688

TABLE V. Isentropic Exponent = 1.40

p/p$_t$	M	T/T$_t$	GAMMA	ISOTHERMAL T$_t$/T$_t$*	RAYLEIGH T$_t$/T$_t$*	SHOCK p$_{t1}$/p$_{t2}$
.100	2.15719	.51795	.51795	1.68936	.76364	1.54383
.099	2.16362	.51646	.51504	1.69422	.76250	1.55091
.098	2.17011	.51497	.51211	1.69914	.76135	1.55811
.097	2.17666	.51346	.50916	1.70413	.76020	1.56544
.096	2.18329	.51194	.50619	1.70918	.75904	1.57291
.095	2.18998	.51041	.50321	1.71430	.75787	1.58051
.094	2.19674	.50887	.50020	1.71949	.75670	1.58826
.093	2.20357	.50732	.49718	1.72475	.75552	1.59615
.092	2.21048	.50575	.49414	1.73009	.75433	1.60418
.091	2.21746	.50418	.49108	1.73550	.75314	1.61238
.090	2.22452	.50259	.48800	1.74099	.75194	1.62073
.089	2.23166	.50099	.48490	1.74656	.75074	1.62925
.088	2.23888	.49937	.48178	1.75220	.74953	1.63793
.087	2.24618	.49774	.47864	1.75793	.74831	1.64679
.086	2.25357	.49610	.47547	1.76375	.74709	1.65582
.085	2.26104	.49445	.47229	1.76965	.74585	1.66504
.084	2.26860	.49278	.46909	1.77565	.74461	1.67445
.083	2.27626	.49109	.46587	1.78173	.74337	1.68406
.082	2.28400	.48940	.46262	1.78792	.74211	1.69387
.081	2.29184	.48768	.45935	1.79419	.74085	1.70388
.080	2.29978	.48596	.45606	1.80057	.73958	1.71412
.079	2.30782	.48421	.45275	1.80706	.73830	1.72457
.078	2.31596	.48245	.44941	1.81365	.73701	1.73526
.077	2.32421	.48068	.44605	1.82034	.73572	1.74619
.076	2.33257	.47889	.44267	1.82716	.73442	1.75736
.075	2.34104	.47708	.43926	1.83408	.73310	1.76879
.074	2.34963	.47525	.43583	1.84113	.73178	1.78048
.073	2.35833	.47341	.43237	1.84830	.73045	1.79245
.072	2.36716	.47155	.42889	1.85560	.72911	1.80471
.071	2.37611	.46966	.42538	1.86303	.72777	1.81726
.070	2.38519	.46777	.42184	1.87060	.72641	1.83011
.069	2.39440	.46585	.41828	1.87830	.72504	1.84329
.068	2.40375	.46391	.41469	1.88615	.72366	1.85679
.067	2.41324	.46195	.41108	1.89415	.72227	1.87064
.066	2.42288	.45997	.40743	1.90231	.72087	1.88485
.065	2.43267	.45797	.40376	1.91063	.71946	1.89943
.064	2.44261	.45594	.40006	1.91911	.71804	1.91440
.063	2.45271	.45389	.39633	1.92776	.71661	1.92977
.062	2.46298	.45182	.39256	1.93660	.71517	1.94556
.061	2.47342	.44973	.38877	1.94561	.71371	1.96180
.060	2.48403	.44761	.38495	1.95482	.71224	1.97849
.059	2.49483	.44547	.38109	1.96423	.71076	1.99566
.058	2.50582	.44330	.37720	1.97385	.70927	2.01334
.057	2.51701	.44110	.37328	1.98368	.70776	2.03154
.056	2.52840	.43887	.36932	1.99374	.70624	2.05029
.055	2.54000	.43662	.36533	2.00403	.70471	2.06962
.054	2.55182	.43434	.36131	2.01456	.70316	2.08955
.053	2.56387	.43202	.35724	2.02535	.70159	2.11012
.052	2.57616	.42968	.35314	2.03641	.70002	2.13136
.051	2.58870	.42730	.34900	2.04773	.69842	2.15330

TABLE V. Isentropic Exponent = 1.40

p/p_t	M	T/T_t	GAMMA	ISOTHERMAL T_t/T^*	RAYLEIGH T_t/T_t^*	SHOCK p_{t1}/p_{t2}
.050	2.60149	.42489	.34482	2.05935	.69681	2.17598
.049	2.61455	.42245	.34061	2.07127	.69518	2.19945
.048	2.62789	.41996	.33635	2.08351	.69353	2.22373
.047	2.64152	.41745	.33204	2.09608	.69187	2.24889
.046	2.65546	.41489	.32770	2.10900	.69019	2.27496
.045	2.66971	.41229	.32331	2.12229	.68849	2.30202
.044	2.68430	.40965	.31887	2.13596	.68677	2.33010
.043	2.69924	.40697	.31439	2.15004	.68502	2.35928
.042	2.71455	.40424	.30986	2.16454	.68326	2.38963
.041	2.73025	.40147	.30528	2.17949	.68147	2.42123
.040	2.74635	.39865	.30065	2.19492	.67966	2.45414
.039	2.76288	.39577	.29597	2.21086	.67783	2.48848
.038	2.77985	.39285	.29123	2.22733	.67597	2.52433
.037	2.79731	.38987	.28644	2.24436	.67408	2.56180
.036	2.81527	.38683	.28158	2.26200	.67217	2.60101
.035	2.83376	.38372	.27667	2.28028	.67023	2.64210
.034	2.85281	.38056	.27170	2.29925	.66826	2.68522
.033	2.87247	.37733	.26666	2.31894	.66626	2.73051
.032	2.89277	.37402	.26155	2.33942	.66422	2.77818
.031	2.91375	.37065	.25638	2.36074	.66215	2.82841
.030	2.93546	.36719	.25113	2.38296	.66004	2.88144
.029	2.95794	.36365	.24580	2.40615	.65789	2.93752
.028	2.98127	.36002	.24040	2.43040	.65571	2.99694
.027	3.00550	.35630	.23492	2.45578	.65348	3.06004
.026	3.03071	.35248	.22935	2.48241	.65120	3.12718
.025	3.05696	.34855	.22369	2.51038	.64887	3.19879
.024	3.08437	.34451	.21793	2.53983	.64650	3.27538
.023	3.11302	.34035	.21208	2.57090	.64406	3.35751
.022	3.14303	.33605	.20612	2.60376	.64157	3.44585
.021	3.17454	.33161	.20004	2.63860	.63901	3.54119
.020	3.20771	.32702	.19386	2.67564	.63638	3.64446
.019	3.24270	.32227	.18754	2.71514	.63368	3.75676
.018	3.27973	.31733	.18109	2.75741	.63090	3.87943
.017	3.31905	.31219	.17450	2.80281	.62802	4.01407
.016	3.36094	.30683	.16776	2.85178	.62505	4.16268
.015	3.40576	.30122	.16084	2.90486	.62197	4.32772
.014	3.45393	.29534	.15375	2.96269	.61877	4.51229
.013	3.50598	.28915	.14646	3.02609	.61543	4.72038
.012	3.56258	.28261	.13896	3.09609	.61194	4.95717
.011	3.62454	.27568	.13122	3.17402	.60827	5.22954
.010	3.69296	.26827	.12321	3.26165	.60440	5.54690
.009	3.76929	.26031	.11489	3.36132	.60029	5.92241
.008	3.85551	.25170	.10624	3.47637	.59590	6.37522
.007	3.95442	.24228	.09718	3.61156	.59117	6.93433
.006	4.07022	.23184	.08764	3.77418	.58600	7.64614
.005	4.20951	.22007	.07753	3.97599	.58028	8.59038
.004	4.38357	.20648	.06668	4.23774	.57379	9.91784
.003	4.61413	.19019	.05485	4.60078	.56620	11.95707
.002	4.95169	.16938	.04158	5.16587	.55676	15.60825
.001	5.56637	.13895	.02580	6.29728	.54349	24.77866

TABLE VI.   Isentropic Exponent = 1.67

p/p_t	M	T/T_t	GAMMA	ISOTHERMAL T_t/T_t^*	RAYLEIGH T_t/T_t^*	SHOCK p_{t1}/p_{t2}
1.0000	.00000	1.00000	.00000	.83292	.00000	1.
.9995	.02447	.99980	.04351	.83308	.00319	1.
.9990	.03462	.99960	.06152	.83325	.00638	1.
.9985	.04241	.99940	.07532	.83342	.00955	1.
.9980	.04898	.99920	.08696	.83359	.01272	1.
.9975	.05477	.99900	.09720	.83375	.01587	1.
.9970	.06000	.99880	.10645	.83392	.01902	1.
.9965	.06482	.99859	.11496	.83409	.02216	1.
.9960	.06931	.99839	.12287	.83426	.02529	1.
.9955	.07353	.99819	.13029	.83443	.02841	1.
.9950	.07752	.99799	.13731	.83459	.03152	1.
.9945	.08132	.99779	.14398	.83476	.03462	1.
.9940	.08495	.99759	.15034	.83493	.03771	1.
.9935	.08843	.99739	.15645	.83510	.04080	1.
.9930	.09179	.99719	.16232	.83527	.04387	1.
.9925	.09502	.99698	.16798	.83544	.04694	1.
.9920	.09816	.99678	.17344	.83561	.04999	1.
.9915	.10120	.99658	.17874	.83578	.05304	1.
.9910	.10415	.99638	.18388	.83594	.05608	1.
.9905	.10702	.99618	.18888	.83611	.05911	1.
.9900	.10982	.99598	.19374	.83628	.06214	1.
.9895	.11255	.99577	.19848	.83645	.06515	1.
.9890	.11522	.99557	.20311	.83662	.06815	1.
.9885	.11783	.99537	.20762	.83679	.07115	1.
.9880	.12039	.99517	.21204	.83696	.07414	1.
.9875	.12289	.99497	.21636	.83713	.07712	1.
.9870	.12535	.99476	.22060	.83730	.08009	1.
.9865	.12776	.99456	.22475	.83747	.08305	1.
.9860	.13013	.99436	.22882	.83764	.08600	1.
.9855	.13245	.99416	.23282	.83781	.08895	1.
.9850	.13474	.99395	.23674	.83798	.09188	1.
.9845	.13699	.99375	.24060	.83815	.09481	1.
.9840	.13921	.99355	.24440	.83833	.09773	1.
.9835	.14139	.99335	.24813	.83850	.10064	1.
.9830	.14354	.99314	.25180	.83867	.10354	1.
.9825	.14567	.99294	.25542	.83884	.10644	1.
.9820	.14776	.99274	.25898	.83901	.10932	1.
.9815	.14982	.99254	.26250	.83918	.11220	1.
.9810	.15186	.99233	.26596	.83935	.11507	1.
.9805	.15387	.99213	.26937	.83952	.11793	1.
.9800	.15586	.99193	.27274	.83970	.12078	1.
.9795	.15783	.99172	.27607	.83987	.12363	1.
.9790	.15977	.99152	.27935	.84004	.12646	1.
.9785	.16169	.99132	.28259	.84021	.12929	1.
.9780	.16359	.99111	.28579	.84038	.13211	1.
.9775	.16547	.99091	.28896	.84056	.13492	1.
.9770	.16732	.99071	.29208	.84073	.13773	1.
.9765	.16916	.99050	.29517	.84090	.14052	1.
.9760	.17098	.99030	.29823	.84108	.14331	1.
.9755	.17279	.99010	.30125	.84125	.14609	1.

TABLE VI. Isentropic Exponent = 1.67

p/p_t	M	T/T_t	GAMMA	ISOTHERMAL T_t/T^*	RAYLEIGH T_t/T^*	SHOCK p_{t1}/p_{t2}
.9750	.17457	.98989	.30424	.84142	.14886	1.
.9745	.17634	.98969	.30719	.84159	.15163	1.
.9740	.17809	.98949	.31012	.84177	.15438	1.
.9735	.17983	.98928	.31301	.84194	.15713	1.
.9730	.18155	.98908	.31588	.84211	.15987	1.
.9725	.18326	.98887	.31872	.84229	.16260	1.
.9720	.18495	.98867	.32153	.84246	.16533	1.
.9715	.18663	.98847	.32431	.84264	.16804	1.
.9710	.18829	.98826	.32707	.84281	.17075	1.
.9705	.18994	.98806	.32980	.84298	.17345	1.
.9700	.19158	.98785	.33250	.84316	.17614	1.
.9695	.19320	.98765	.33518	.84333	.17883	1.
.9690	.19481	.98745	.33784	.84351	.18151	1.
.9685	.19641	.98724	.34048	.84368	.18418	1.
.9680	.19800	.98704	.34309	.84386	.18684	1.
.9675	.19958	.98683	.34568	.84403	.18949	1.
.9670	.20115	.98663	.34825	.84421	.19214	1.
.9665	.20270	.98642	.35079	.84438	.19478	1.
.9660	.20424	.98622	.35332	.84456	.19741	1.
.9655	.20578	.98601	.35582	.84473	.20004	1.
.9650	.20730	.98581	.35831	.84491	.20265	1.
.9645	.20881	.98560	.36078	.84508	.20526	1.
.9640	.21032	.98540	.36322	.84526	.20786	1.
.9635	.21181	.98519	.36565	.84544	.21046	1.
.9630	.21330	.98499	.36806	.84561	.21304	1.
.9625	.21477	.98478	.37045	.84579	.21562	1.
.9620	.21624	.98458	.37283	.84596	.21820	1.
.9615	.21770	.98437	.37518	.84614	.22076	1.
.9610	.21914	.98417	.37752	.84632	.22332	1.
.9605	.22058	.98396	.37985	.84649	.22587	1.
.9600	.22202	.98376	.38215	.84667	.22841	1.
.9595	.22344	.98355	.38444	.84685	.23095	1.
.9590	.22486	.98334	.38672	.84703	.23347	1.
.9585	.22626	.98314	.38898	.84720	.23599	1.
.9580	.22766	.98293	.39122	.84738	.23851	1.
.9575	.22906	.98273	.39345	.84756	.24101	1.
.9570	.23044	.98252	.39567	.84774	.24351	1.
.9565	.23182	.98232	.39787	.84791	.24601	1.
.9560	.23319	.98211	.40005	.84809	.24849	1.
.9555	.23456	.98190	.40222	.84827	.25097	1.
.9550	.23591	.98170	.40438	.84845	.25344	1.
.9545	.23726	.98149	.40653	.84863	.25590	1.
.9540	.23861	.98128	.40866	.84880	.25836	1.
.9535	.23994	.98108	.41078	.84898	.26081	1.
.9530	.24128	.98087	.41288	.84916	.26325	1.
.9525	.24260	.98066	.41497	.84934	.26569	1.
.9520	.24392	.98046	.41705	.84952	.26811	1.
.9515	.24523	.98025	.41912	.84970	.27054	1.
.9510	.24654	.98005	.42118	.84988	.27295	1.
.9505	.24784	.97984	.42322	.85006	.27536	1.

TABLE VI. Isentropic Exponent = 1.67

p/p_t	M	T/T_t	GAMMA	ISOTHERMAL T_t/T_t^*	RAYLEIGH T_t/T_t^*	SHOCK p_{t1}/p_{t2}
.950	.24913	.97963	.42525	.85024	.27776	1.
.949	.25170	.97922	.42928	.85060	.28254	1.
.948	.25425	.97880	.43326	.85095	.28729	1.
.947	.25678	.97839	.43720	.85132	.29202	1.
.946	.25928	.97797	.44109	.85168	.29672	1.
.945	.26177	.97756	.44495	.85204	.30139	1.
.944	.26424	.97714	.44876	.85240	.30604	1.
.943	.26668	.97673	.45253	.85276	.31065	1.
.942	.26911	.97631	.45626	.85313	.31525	1.
.941	.27152	.97590	.45996	.85349	.31981	1.
.940	.27392	.97548	.46362	.85385	.32435	1.
.939	.27629	.97506	.46724	.85422	.32887	1.
.938	.27865	.97465	.47083	.85458	.33336	1.
.937	.28099	.97423	.47438	.85495	.33782	1.
.936	.28332	.97381	.47790	.85532	.34226	1.
.935	.28563	.97340	.48139	.85568	.34667	1.
.934	.28793	.97298	.48484	.85605	.35106	1.
.933	.29021	.97256	.48827	.85642	.35542	1.
.932	.29247	.97214	.49166	.85679	.35975	1.
.931	.29473	.97172	.49502	.85716	.36407	1.
.930	.29697	.97130	.49835	.85752	.36835	1.
.929	.29919	.97089	.50166	.85789	.37261	1.
.928	.30140	.97047	.50493	.85827	.37685	1.
.927	.30360	.97005	.50818	.85864	.38107	1.
.926	.30579	.96963	.51140	.85901	.38526	1.
.925	.30797	.96921	.51459	.85938	.38942	1.
.924	.31013	.96879	.51776	.85975	.39356	1.
.923	.31228	.96836	.52090	.86013	.39768	1.
.922	.31442	.96794	.52401	.86050	.40178	1.
.921	.31655	.96752	.52710	.86088	.40585	1.
.920	.31866	.96710	.53017	.86125	.40990	1.
.919	.32077	.96668	.53321	.86163	.41392	1.
.918	.32287	.96626	.53622	.86200	.41792	1.
.917	.32495	.96583	.53922	.86238	.42190	1.
.916	.32703	.96541	.54219	.86276	.42586	1.
.915	.32909	.96499	.54514	.86314	.42979	1.
.914	.33115	.96457	.54806	.86352	.43370	1.
.913	.33320	.96414	.55097	.86390	.43759	1.
.912	.33523	.96372	.55385	.86428	.44146	1.
.911	.33726	.96329	.55671	.86466	.44530	1.
.910	.33928	.96287	.55955	.86504	.44912	1.
.909	.34129	.96244	.56237	.86542	.45292	1.
.908	.34329	.96202	.56517	.86580	.45670	1.
.907	.34528	.96159	.56795	.86618	.46046	1.
.906	.34727	.96117	.57071	.86657	.46419	1.
.905	.34924	.96074	.57346	.86695	.46791	1.
.904	.35121	.96032	.57618	.86734	.47160	1.
.903	.35317	.95989	.57888	.86772	.47527	1.
.902	.35512	.95946	.58157	.86811	.47892	1.
.901	.35707	.95904	.58423	.86849	.48255	1.

TABLE VI. Isentropic Exponent = 1.67

p/p_t	M	T/T_t	GAMMA	ISOTHERMAL T_t/T_t^*	RAYLEIGH T_t/T_t^*	SHOCK p_{t1}/p_{t2}
.900	.35901	.95861	.58688	.86888	.48616	1.
.899	.36094	.95818	.58951	.86927	.48975	1.
.898	.36286	.95776	.59212	.86966	.49331	1.
.897	.36477	.95733	.59472	.87004	.49686	1.
.896	.36668	.95690	.59730	.87043	.50039	1.
.895	.36858	.95647	.59986	.87082	.50389	1.
.894	.37048	.95604	.60240	.87122	.50738	1.
.893	.37236	.95561	.60493	.87161	.51084	1.
.892	.37425	.95518	.60744	.87200	.51429	1.
.891	.37612	.95475	.60994	.87239	.51772	1.
.890	.37799	.95432	.61242	.87278	.52112	1.
.889	.37985	.95389	.61488	.87318	.52451	1.
.888	.38171	.95346	.61733	.87357	.52788	1.
.887	.38356	.95303	.61976	.87397	.53122	1.
.886	.38540	.95260	.62218	.87436	.53455	1.
.885	.38724	.95217	.62458	.87476	.53786	1.
.884	.38907	.95174	.62697	.87516	.54115	1.
.883	.39090	.95130	.62935	.87555	.54442	1.
.882	.39272	.95087	.63170	.87595	.54768	1.
.881	.39453	.95044	.63405	.87635	.55091	1.
.880	.39634	.95001	.63638	.87675	.55412	1.
.879	.39815	.94957	.63870	.87715	.55732	1.
.878	.39995	.94914	.64100	.87755	.56050	1.
.877	.40174	.94871	.64329	.87795	.56366	1.
.876	.40353	.94827	.64556	.87835	.56680	1.
.875	.40531	.94784	.64783	.87876	.56992	1.
.874	.40709	.94740	.65007	.87916	.57303	1.
.873	.40887	.94697	.65231	.87956	.57611	1.
.872	.41064	.94653	.65453	.87997	.57918	1.
.871	.41240	.94610	.65674	.88037	.58223	1.
.870	.41416	.94566	.65894	.88078	.58527	1.
.869	.41591	.94522	.66112	.88119	.58828	1.
.868	.41767	.94479	.66330	.88159	.59128	1.
.867	.41941	.94435	.66545	.88200	.59426	1.
.866	.42115	.94391	.66760	.88241	.59723	1.
.865	.42289	.94348	.66974	.88282	.60017	1.
.864	.42462	.94304	.67186	.88323	.60310	1.
.863	.42635	.94260	.67397	.88364	.60602	1.
.862	.42808	.94216	.67607	.88405	.60891	1.
.861	.42980	.94172	.67816	.88446	.61179	1.
.860	.43151	.94128	.68023	.88487	.61465	1.
.859	.43322	.94085	.68230	.88529	.61750	1.
.858	.43493	.94041	.68435	.88570	.62033	1.
.857	.43664	.93997	.68639	.88611	.62314	1.
.856	.43834	.93953	.68842	.88653	.62593	1.
.855	.44003	.93909	.69044	.88695	.62871	1.
.854	.44173	.93864	.69245	.88736	.63148	1.
.853	.44342	.93820	.69445	.88778	.63422	1.
.852	.44510	.93776	.69643	.88820	.63695	1.
.851	.44678	.93732	.69841	.88862	.63967	1.

TABLE VI. Isentropic Exponent = 1.67

p/p_t	M	T/T_t	GAMMA	ISOTHERMAL T_t/T_t^*	RAYLEIGH T_t/T_t^*	SHOCK p_{t1}/p_{t2}
.850	.44846	.93688	.70037	.88904	.64237	1.
.849	.45014	.93644	.70233	.88946	.64505	1.
.848	.45181	.93599	.70427	.88988	.64772	1.
.847	.45348	.93555	.70621	.89030	.65037	1.
.846	.45514	.93511	.70813	.89072	.65301	1.
.845	.45680	.93466	.71004	.89114	.65563	1.
.844	.45846	.93422	.71195	.89157	.65823	1.
.843	.46012	.93377	.71384	.89199	.66082	1.
.842	.46177	.93333	.71572	.89241	.66340	1.
.841	.46342	.93289	.71760	.89284	.66596	1.
.840	.46506	.93244	.71946	.89327	.66850	1.
.839	.46670	.93199	.72131	.89369	.67103	1.
.838	.46834	.93155	.72316	.89412	.67355	1.
.837	.46998	.93110	.72499	.89455	.67605	1.
.836	.47161	.93066	.72682	.89498	.67853	1.
.835	.47324	.93021	.72863	.89541	.68100	1.
.834	.47487	.92976	.73044	.89584	.68346	1.
.833	.47650	.92931	.73223	.89627	.68590	1.
.832	.47812	.92887	.73402	.89670	.68833	1.
.831	.47974	.92842	.73580	.89714	.69074	1.
.830	.48135	.92797	.73757	.89757	.69314	1.
.829	.48297	.92752	.73933	.89800	.69552	1.
.828	.48458	.92707	.74108	.89844	.69789	1.
.827	.48619	.92662	.74282	.89887	.70025	1.
.826	.48779	.92617	.74455	.89931	.70259	1.
.825	.48940	.92572	.74628	.89975	.70492	1.
.824	.49100	.92527	.74799	.90019	.70723	1.
.823	.49260	.92482	.74970	.90062	.70953	1.
.822	.49419	.92437	.75140	.90106	.71182	1.
.821	.49579	.92392	.75309	.90150	.71409	1.
.820	.49738	.92347	.75477	.90194	.71635	1.
.819	.49897	.92302	.75644	.90239	.71860	1.
.818	.50055	.92256	.75811	.90283	.72083	1.
.817	.50214	.92211	.75976	.90327	.72305	1.
.816	.50372	.92166	.76141	.90372	.72525	1.
.815	.50530	.92121	.76305	.90416	.72744	1.
.814	.50687	.92075	.76468	.90461	.72962	1.
.813	.50845	.92030	.76630	.90505	.73179	1.
.812	.51002	.91984	.76792	.90550	.73394	1.
.811	.51159	.91939	.76952	.90595	.73608	1.
.810	.51316	.91893	.77112	.90640	.73821	1.
.809	.51473	.91848	.77271	.90684	.74032	1.
.808	.51629	.91802	.77430	.90729	.74243	1.
.807	.51786	.91757	.77587	.90775	.74451	1.
.806	.51942	.91711	.77744	.90820	.74659	1.
.805	.52098	.91665	.77900	.90865	.74865	1.
.804	.52253	.91620	.78055	.90910	.75071	1.
.803	.52409	.91574	.78209	.90956	.75274	1.
.802	.52564	.91528	.78363	.91001	.75477	1.
.801	.52719	.91482	.78516	.91047	.75678	1.

TABLE VI.   Isentropic Exponent = 1.67

p/p_t	M	T/T_t	GAMMA	ISOTHERMAL T_t/T_t^*	RAYLEIGH T_t/T_t^*	SHOCK p_{t1}/p_{t2}
.800	.52874	.91437	.78668	.91092	.75879	1.
.799	.53029	.91391	.78819	.91138	.76077	1.
.798	.53183	.91345	.78970	.91184	.76275	1.
.797	.53338	.91299	.79120	.91230	.76472	1.
.796	.53492	.91253	.79269	.91276	.76667	1.
.795	.53646	.91207	.79417	.91322	.76861	1.
.794	.53800	.91161	.79565	.91368	.77054	1.
.793	.53953	.91115	.79712	.91414	.77246	1.
.792	.54107	.91069	.79858	.91460	.77436	1.
.791	.54260	.91022	.80004	.91507	.77626	1.
.790	.54413	.90976	.80148	.91553	.77814	1.
.789	.54566	.90930	.80293	.91600	.78001	1.
.788	.54719	.90884	.80436	.91646	.78187	1.
.787	.54872	.90838	.80579	.91693	.78372	1.
.786	.55025	.90791	.80721	.91740	.78556	1.
.785	.55177	.90745	.80862	.91787	.78738	1.
.784	.55329	.90698	.81002	.91834	.78919	1.
.783	.55481	.90652	.81142	.91881	.79100	1.
.782	.55633	.90606	.81281	.91928	.79279	1.
.781	.55785	.90559	.81420	.91975	.79457	1.
.780	.55937	.90513	.81558	.92022	.79634	1.
.779	.56089	.90466	.81695	.92070	.79809	1.
.778	.56240	.90419	.81831	.92117	.79984	1.
.777	.56391	.90373	.81967	.92165	.80158	1.
.776	.56542	.90326	.82102	.92212	.80330	1.
.775	.56693	.90279	.82237	.92260	.80502	1.
.774	.56844	.90233	.82371	.92308	.80672	1.
.773	.56995	.90186	.82504	.92356	.80841	1.
.772	.57146	.90139	.82636	.92404	.81010	1.
.771	.57296	.90092	.82768	.92452	.81177	1.
.770	.57447	.90045	.82899	.92500	.81343	1.
.769	.57597	.89998	.83030	.92548	.81508	1.
.768	.57747	.89951	.83160	.92597	.81672	1.
.767	.57897	.89904	.83289	.92645	.81835	1.
.766	.58047	.89857	.83418	.92693	.81997	1.
.765	.58197	.89810	.83546	.92742	.82158	1.
.764	.58347	.89763	.83673	.92791	.82318	1.
.763	.58496	.89716	.83800	.92840	.82477	1.
.762	.58646	.89669	.83926	.92888	.82635	1.
.761	.58795	.89621	.84052	.92937	.82791	1.
.760	.58944	.89574	.84176	.92986	.82947	1.
.759	.59094	.89527	.84301	.93036	.83102	1.
.758	.59243	.89480	.84424	.93085	.83256	1.
.757	.59392	.89432	.84547	.93134	.83409	1.
.756	.59540	.89385	.84670	.93183	.83561	1.
.755	.59689	.89337	.84792	.93233	.83712	1.
.754	.59838	.89290	.84913	.93283	.83862	1.
.753	.59986	.89242	.85033	.93332	.84010	1.
.752	.60135	.89195	.85153	.93382	.84158	1.
.751	.60283	.89147	.85273	.93432	.84306	1.

TABLE VI. Isentropic Exponent = 1.67

p/p_t	M	T/T_t	GAMMA	ISOTHERMAL T_t/T^*	RAYLEIGH T_t/T^*	SHOCK p_{t1}/p_{t2}
.750	.60432	.89099	.85392	.93482	.84452	1.
.749	.60580	.89052	.85510	.93532	.84597	1.
.748	.60728	.89004	.85627	.93582	.84741	1.
.747	.60876	.88956	.85744	.93632	.84884	1.
.746	.61024	.88908	.85861	.93683	.85026	1.
.745	.61172	.88861	.85977	.93733	.85168	1.
.744	.61320	.88813	.86092	.93784	.85308	1.
.743	.61468	.88765	.86207	.93834	.85448	1.
.742	.61615	.88717	.86321	.93885	.85586	1.
.741	.61763	.88669	.86434	.93936	.85724	1.
.740	.61910	.88621	.86547	.93987	.85861	1.
.739	.62058	.88573	.86660	.94038	.85997	1.
.738	.62205	.88525	.86772	.94089	.86132	1.
.737	.62353	.88477	.86883	.94140	.86266	1.
.736	.62500	.88428	.86994	.94191	.86399	1.
.735	.62647	.88380	.87104	.94243	.86531	1.
.734	.62794	.88332	.87213	.94294	.86663	1.
.733	.62941	.88284	.87322	.94346	.86793	1.
.732	.63088	.88235	.87431	.94397	.86923	1.
.731	.63235	.88187	.87539	.94449	.87052	1.
.730	.63382	.88138	.87646	.94501	.87179	1.
.729	.63529	.88090	.87753	.94553	.87307	1.
.728	.63675	.88041	.87859	.94605	.87433	1.
.727	.63822	.87993	.87965	.94657	.87558	1.
.726	.63969	.87944	.88070	.94710	.87683	1.
.725	.64115	.87896	.88174	.94762	.87806	1.
.724	.64262	.87847	.88278	.94814	.87929	1.
.723	.64408	.87798	.88382	.94867	.88051	1.
.722	.64555	.87750	.88485	.94920	.88172	1.
.721	.64701	.87701	.88587	.94973	.88292	1.
.720	.64848	.87652	.88689	.95025	.88412	1.
.719	.64994	.87603	.88791	.95078	.88530	1.
.718	.65140	.87554	.88892	.95132	.88648	1.
.717	.65286	.87505	.88992	.95185	.88765	1.
.716	.65433	.87456	.89092	.95238	.88881	1.
.715	.65579	.87407	.89191	.95292	.88997	1.
.714	.65725	.87358	.89290	.95345	.89111	1.
.713	.65871	.87309	.89388	.95399	.89225	1.
.712	.66017	.87260	.89486	.95452	.89338	1.
.711	.66163	.87211	.89583	.95506	.89450	1.
.710	.66309	.87162	.89679	.95560	.89562	1.
.709	.66455	.87112	.89775	.95614	.89672	1.
.708	.66600	.87063	.89871	.95668	.89782	1.
.707	.66746	.87014	.89966	.95723	.89891	1.
.706	.66892	.86964	.90061	.95777	.89999	1.
.705	.67038	.86915	.90155	.95832	.90107	1.
.704	.67184	.86865	.90248	.95886	.90214	1.
.703	.67329	.86816	.90341	.95941	.90320	1.
.702	.67475	.86766	.90434	.95996	.90425	1.
.701	.67621	.86717	.90526	.96051	.90529	1.

TABLE VI.   Isentropic Exponent = 1.67

p/p_t	M	T/T_t	GAMMA	ISOTHERMAL T_t/T^*	RAYLEIGH T_t/T_t^*	SHOCK p_{t1}/p_{t2}
.700	.67766	.86667	.90618	.96106	.90633	1.
.699	.67912	.86617	.90709	.96161	.90736	1.
.698	.68058	.86568	.90799	.96216	.90838	1.
.697	.68203	.86518	.90889	.96271	.90940	1.
.696	.68349	.86468	.90979	.96327	.91041	1.
.695	.68494	.86418	.91068	.96382	.91141	1.
.694	.68640	.86368	.91156	.96438	.91240	1.
.693	.68786	.86318	.91244	.96494	.91338	1.
.692	.68931	.86268	.91332	.96550	.91436	1.
.691	.69077	.86218	.91419	.96606	.91533	1.
.690	.69222	.86168	.91505	.96662	.91630	1.
.689	.69368	.86118	.91591	.96718	.91725	1.
.688	.69513	.86068	.91677	.96775	.91820	1.
.687	.69659	.86018	.91762	.96831	.91915	1.
.686	.69804	.85967	.91847	.96888	.92008	1.
.685	.69949	.85917	.91931	.96944	.92101	1.
.684	.70095	.85867	.92015	.97001	.92193	1.
.683	.70240	.85816	.92098	.97058	.92284	1.
.682	.70386	.85766	.92180	.97115	.92375	1.
.681	.70531	.85715	.92263	.97172	.92465	1.
.680	.70677	.85665	.92344	.97230	.92555	1.
.679	.70822	.85614	.92425	.97287	.92643	1.
.678	.70968	.85564	.92506	.97345	.92731	1.
.677	.71113	.85513	.92587	.97402	.92819	1.
.676	.71258	.85462	.92666	.97460	.92905	1.
.675	.71404	.85412	.92746	.97518	.92991	1.
.674	.71549	.85361	.92825	.97576	.93077	1.
.673	.71695	.85310	.92903	.97634	.93161	1.
.672	.71840	.85259	.92981	.97692	.93245	1.
.671	.71986	.85208	.93058	.97751	.93329	1.
.670	.72131	.85157	.93135	.97809	.93412	1.
.669	.72277	.85106	.93212	.97868	.93494	1.
.668	.72422	.85055	.93288	.97927	.93575	1.
.667	.72568	.85004	.93363	.97986	.93656	1.
.666	.72713	.84953	.93439	.98045	.93736	1.
.665	.72859	.84902	.93513	.98104	.93815	1.
.664	.73005	.84850	.93587	.98163	.93894	1.
.663	.73150	.84799	.93661	.98222	.93972	1.
.662	.73296	.84748	.93734	.98282	.94050	1.
.661	.73441	.84696	.93807	.98342	.94127	1.
.660	.73587	.84645	.93879	.98401	.94203	1.
.659	.73733	.84594	.93951	.98461	.94279	1.
.658	.73878	.84542	.94023	.98521	.94354	1.
.657	.74024	.84490	.94094	.98581	.94428	1.
.656	.74170	.84439	.94164	.98642	.94502	1.
.655	.74316	.84387	.94234	.98702	.94575	1.
.654	.74461	.84335	.94304	.98762	.94648	1.
.653	.74607	.84284	.94373	.98823	.94720	1.
.652	.74753	.84232	.94442	.98884	.94791	1.
.651	.74899	.84180	.94510	.98945	.94862	1.

TABLE VI. Isentropic Exponent = 1.67

p/p_t	M	T/T_t	GAMMA	ISOTHERMAL T_t/T_t^*	RAYLEIGH T_t/T_t^*	SHOCK p_{t1}/p_{t2}
.650	.75045	.84128	.94578	.99006	.94933	1.
.649	.75191	.84076	.94645	.99067	.95002	1.
.648	.75337	.84024	.94712	.99128	.95071	1.
.647	.75483	.83972	.94779	.99190	.95140	1.
.646	.75629	.83920	.94845	.99251	.95207	1.
.645	.75775	.83868	.94910	.99313	.95275	1.
.644	.75921	.83816	.94975	.99375	.95341	1.
.643	.76067	.83763	.95040	.99437	.95407	1.
.642	.76213	.83711	.95104	.99499	.95473	1.
.641	.76359	.83659	.95168	.99561	.95538	1.
.640	.76506	.83606	.95231	.99624	.95602	1.
.639	.76652	.83554	.95294	.99686	.95666	1.
.638	.76798	.83502	.95357	.99749	.95729	1.
.637	.76945	.83449	.95419	.99812	.95792	1.
.636	.77091	.83396	.95480	.99875	.95854	1.
.635	.77238	.83344	.95542	.99938	.95916	1.
.634	.77384	.83291	.95602	1.00001	.95977	1.
.633	.77531	.83238	.95663	1.00064	.96037	1.
.632	.77677	.83186	.95722	1.00128	.96097	1.
.631	.77824	.83133	.95782	1.00191	.96156	1.
.630	.77971	.83080	.95841	1.00255	.96215	1.
.629	.78117	.83027	.95899	1.00319	.96274	1.
.628	.78264	.82974	.95957	1.00383	.96331	1.
.627	.78411	.82921	.96015	1.00447	.96388	1.
.626	.78558	.82868	.96072	1.00512	.96445	1.
.625	.78705	.82815	.96129	1.00576	.96501	1.
.624	.78852	.82762	.96185	1.00641	.96557	1.
.623	.78999	.82708	.96241	1.00705	.96612	1.
.622	.79146	.82655	.96297	1.00770	.96666	1.
.621	.79293	.82602	.96352	1.00835	.96720	1.
.620	.79441	.82548	.96407	1.00901	.96774	1.
.619	.79588	.82495	.96461	1.00966	.96827	1.
.618	.79735	.82441	.96515	1.01032	.96879	1.
.617	.79883	.82388	.96568	1.01097	.96931	1.
.616	.80030	.82334	.96621	1.01163	.96982	1.
.615	.80178	.82281	.96673	1.01229	.97033	1.
.614	.80325	.82227	.96725	1.01295	.97084	1.
.613	.80473	.82173	.96777	1.01361	.97134	1.
.612	.80621	.82119	.96828	1.01428	.97183	1.
.611	.80769	.82065	.96879	1.01494	.97232	1.
.610	.80917	.82012	.96930	1.01561	.97280	1.
.609	.81065	.81958	.96979	1.01628	.97328	1.
.608	.81213	.81904	.97029	1.01695	.97375	1.
.607	.81361	.81849	.97078	1.01762	.97422	1.
.606	.81509	.81795	.97127	1.01829	.97469	1.
.605	.81657	.81741	.97175	1.01897	.97514	1.
.604	.81805	.81687	.97223	1.01965	.97560	1.
.603	.81954	.81633	.97270	1.02032	.97605	1.
.602	.82102	.81578	.97317	1.02100	.97649	1.
.601	.82251	.81524	.97364	1.02169	.97693	1.

TABLE VI. Isentropic Exponent = 1.67

p/p_t	M	T/T_t	GAMMA	ISOTHERMAL T_t/T^*	RAYLEIGH T_t/T^*	SHOCK p_{t1}/p_{t2}
.600	.82399	.81469	.97410	1.02237	.97737	1.
.599	.82548	.81415	.97456	1.02305	.97780	1.
.598	.82697	.81360	.97501	1.02374	.97822	1.
.597	.82846	.81306	.97546	1.02443	.97864	1.
.596	.82995	.81251	.97591	1.02512	.97906	1.
.595	.83144	.81196	.97635	1.02581	.97947	1.
.594	.83293	.81142	.97678	1.02650	.97987	1.
.593	.83442	.81087	.97722	1.02719	.98028	1.
.592	.83591	.81032	.97765	1.02789	.98067	1.
.591	.83741	.80977	.97807	1.02859	.98107	1.
.590	.83890	.80922	.97849	1.02929	.98145	1.
.589	.84040	.80867	.97891	1.02999	.98184	1.
.588	.84189	.80812	.97932	1.03069	.98222	1.
.587	.84339	.80757	.97973	1.03139	.98259	1.
.586	.84489	.80701	.98013	1.03210	.98296	1.
.585	.84639	.80646	.98053	1.03281	.98332	1.
.584	.84789	.80591	.98092	1.03351	.98368	1.
.583	.84939	.80535	.98131	1.03423	.98404	1.
.582	.85089	.80480	.98170	1.03494	.98439	1.
.581	.85239	.80424	.98208	1.03565	.98474	1.
.580	.85390	.80369	.98246	1.03637	.98508	1.
.579	.85540	.80313	.98284	1.03709	.98542	1.
.578	.85691	.80258	.98321	1.03781	.98576	1.
.577	.85842	.80202	.98358	1.03853	.98609	1.
.576	.85992	.80146	.98394	1.03925	.98641	1.
.575	.86143	.80090	.98430	1.03997	.98673	1.
.574	.86294	.80034	.98465	1.04070	.98705	1.
.573	.86445	.79978	.98500	1.04143	.98736	1.
.572	.86597	.79922	.98535	1.04216	.98767	1.
.571	.86748	.79866	.98569	1.04289	.98797	1.
.570	.86899	.79810	.98603	1.04363	.98827	1.
.569	.87051	.79754	.98636	1.04436	.98857	1.
.568	.87203	.79698	.98669	1.04510	.98886	1.
.567	.87354	.79641	.98702	1.04584	.98915	1.
.566	.87506	.79585	.98734	1.04658	.98943	1.
.565	.87658	.79528	.98765	1.04732	.98971	1.
.564	.87810	.79472	.98797	1.04807	.98999	1.
.563	.87962	.79415	.98828	1.04881	.99026	1.
.562	.88115	.79359	.98858	1.04956	.99052	1.
.561	.88267	.79302	.98888	1.05031	.99079	1.
.560	.88420	.79245	.98918	1.05106	.99104	1.
.559	.88572	.79189	.98947	1.05182	.99130	1.
.558	.88725	.79132	.98976	1.05257	.99155	1.
.557	.88878	.79075	.99005	1.05333	.99180	1.
.556	.89031	.79018	.99033	1.05409	.99204	1.
.555	.89184	.78961	.99061	1.05485	.99228	1.
.554	.89337	.78904	.99088	1.05561	.99251	1.
.553	.89491	.78846	.99115	1.05638	.99274	1.
.552	.89644	.78789	.99141	1.05715	.99297	1.
.551	.89798	.78732	.99167	1.05792	.99319	1.

TABLE VI. Isentropic Exponent = 1.67

p/p_t	M	T/T_t	GAMMA	ISOTHERMAL T_t/T_t*	RAYLEIGH T_t/T_t*	SHOCK p_t1/p_t2
.550	.89952	.78675	.99193	1.05869	.99341	1.
.549	.90106	.78617	.99218	1.05946	.99363	1.
.548	.90260	.78560	.99243	1.06024	.99384	1.
.547	.90414	.78502	.99268	1.06101	.99404	1.
.546	.90568	.78444	.99292	1.06179	.99425	1.
.545	.90723	.78387	.99315	1.06257	.99445	1.
.544	.90877	.78329	.99339	1.06336	.99464	1.
.543	.91032	.78271	.99362	1.06414	.99484	1.
.542	.91187	.78213	.99384	1.06493	.99502	1.
.541	.91342	.78155	.99406	1.06572	.99521	1.
.540	.91497	.78097	.99428	1.06651	.99539	1.
.539	.91652	.78039	.99449	1.06730	.99557	1.
.538	.91808	.77981	.99470	1.06810	.99574	1.
.537	.91963	.77923	.99490	1.06890	.99591	1.
.536	.92119	.77865	.99510	1.06970	.99608	1.
.535	.92275	.77807	.99530	1.07050	.99624	1.
.534	.92431	.77748	.99549	1.07130	.99640	1.
.533	.92587	.77690	.99568	1.07211	.99655	1.
.532	.92743	.77631	.99586	1.07292	.99670	1.
.531	.92899	.77573	.99604	1.07373	.99685	1.
.530	.93056	.77514	.99622	1.07454	.99700	1.
.529	.93213	.77455	.99639	1.07535	.99714	1.
.528	.93369	.77397	.99656	1.07617	.99727	1.
.527	.93526	.77338	.99673	1.07699	.99741	1.
.526	.93684	.77279	.99689	1.07781	.99754	1.
.525	.93841	.77220	.99704	1.07863	.99766	1.
.524	.93998	.77161	.99719	1.07946	.99779	1.
.523	.94156	.77102	.99734	1.08029	.99791	1.
.522	.94314	.77042	.99749	1.08112	.99802	1.
.521	.94472	.76983	.99763	1.08195	.99814	1.
.520	.94630	.76924	.99776	1.08278	.99825	1.
.519	.94788	.76864	.99789	1.08362	.99835	1.
.518	.94947	.76805	.99802	1.08446	.99845	1.
.517	.95105	.76746	.99815	1.08530	.99855	1.
.516	.95264	.76686	.99827	1.08614	.99865	1.
.515	.95423	.76626	.99838	1.08699	.99874	1.
.514	.95582	.76567	.99849	1.08784	.99883	1.
.513	.95741	.76507	.99860	1.08869	.99892	1.
.512	.95901	.76447	.99871	1.08954	.99900	1.
.511	.96060	.76387	.99881	1.09039	.99908	1.
.510	.96220	.76327	.99890	1.09125	.99915	1.
.509	.96380	.76267	.99899	1.09211	.99922	1.
.508	.96540	.76207	.99908	1.09297	.99929	1.
.507	.96701	.76146	.99917	1.09384	.99936	1.
.506	.96861	.76086	.99925	1.09470	.99942	1.
.505	.97022	.76026	.99932	1.09557	.99948	1.
.504	.97183	.75965	.99939	1.09644	.99954	1.
.503	.97344	.75905	.99946	1.09732	.99959	1.
.502	.97505	.75844	.99953	1.09819	.99964	1.
.501	.97666	.75784	.99959	1.09907	.99968	1.

TABLE VI. Isentropic Exponent = 1.67

p/p_t	M	T/T_t	GAMMA	ISOTHERMAL T_t/T_t^*	RAYLEIGH T_t/T_t^*	SHOCK p_{t1}/p_{t2}
.500	.97828	.75723	.99964	1.09995	.99973	1.
.499	.97989	.75662	.99969	1.10084	.99977	1.
.498	.98151	.75601	.99974	1.10172	.99980	1.
.497	.98314	.75540	.99978	1.10261	.99984	1.
.496	.98476	.75479	.99982	1.10350	.99987	1.
.495	.98638	.75418	.99986	1.10440	.99989	1.
.494	.98801	.75357	.99989	1.10529	.99992	1.
.493	.98964	.75296	.99992	1.10619	.99994	1.
.492	.99127	.75235	.99994	1.10710	.99996	1.
.491	.99290	.75173	.99996	1.10800	.99997	1.
.490	.99454	.75112	.99998	1.10891	.99998	1.
.489	.99618	.75050	.99999	1.10982	.99999	1.
.488	.99781	.74989	.00000	1.11073	.00000	1.
.487	.99946	.74927	.00000	1.11164	.00000	1.
.486	1.00110	.74865	.00000	1.11256	.00000	1.00000
.485	1.00274	.74803	.99999	1.11348	.00000	1.00000
.484	1.00439	.74741	.99999	1.11440	.99999	1.00000
.483	1.00604	.74679	.99997	1.11533	.99998	1.00000
.482	1.00769	.74617	.99996	1.11625	.99997	1.00000
.481	1.00934	.74555	.99993	1.11718	.99995	1.00000
.480	1.01100	.74493	.99991	1.11812	.99993	1.00000
.479	1.01266	.74431	.99988	1.11905	.99991	1.00000
.478	1.01432	.74368	.99985	1.11999	.99989	1.00000
.477	1.01598	.74306	.99981	1.12093	.99986	1.00000
.476	1.01764	.74243	.99977	1.12188	.99983	1.00001
.475	1.01931	.74181	.99972	1.12282	.99980	1.00001
.474	1.02098	.74118	.99968	1.12377	.99976	1.00001
.473	1.02265	.74055	.99962	1.12473	.99972	1.00001
.472	1.02432	.73992	.99956	1.12568	.99968	1.00002
.471	1.02600	.73929	.99950	1.12664	.99964	1.00002
.470	1.02767	.73866	.99944	1.12760	.99959	1.00002
.469	1.02935	.73803	.99937	1.12857	.99954	1.00003
.468	1.03103	.73740	.99929	1.12953	.99948	1.00003
.467	1.03272	.73677	.99922	1.13050	.99943	1.00004
.466	1.03440	.73613	.99913	1.13147	.99937	1.00005
.465	1.03609	.73550	.99905	1.13245	.99931	1.00005
.464	1.03778	.73487	.99896	1.13343	.99924	1.00006
.463	1.03948	.73423	.99886	1.13441	.99918	1.00007
.462	1.04117	.73359	.99877	1.13539	.99911	1.00008
.461	1.04287	.73296	.99866	1.13638	.99903	1.00009
.460	1.04457	.73232	.99856	1.13737	.99896	1.00010
.459	1.04627	.73168	.99845	1.13837	.99888	1.00011
.458	1.04798	.73104	.99833	1.13936	.99880	1.00012
.457	1.04969	.73040	.99821	1.14036	.99871	1.00014
.456	1.05140	.72976	.99809	1.14137	.99863	1.00015
.455	1.05311	.72911	.99796	1.14237	.99854	1.00017
.454	1.05483	.72847	.99783	1.14338	.99845	1.00018
.453	1.05654	.72783	.99770	1.14439	.99835	1.00020
.452	1.05826	.72718	.99756	1.14541	.99825	1.00022
.451	1.05999	.72653	.99741	1.14643	.99815	1.00024

TABLE VI.   Isentropic Exponent = 1.67

p/p_t	M	T/T_t	GAMMA	ISOTHERMAL T_t/T_t^*	RAYLEIGH T_t/T_t^*	SHOCK p_{t1}/p_{t2}
.450	1.06171	.72589	.99726	1.14745	.99805	1.00026
.449	1.06344	.72524	.99711	1.14847	.99795	1.00028
.448	1.06517	.72459	.99696	1.14950	.99784	1.00030
.447	1.06690	.72394	.99680	1.15053	.99773	1.00032
.446	1.06864	.72329	.99663	1.15156	.99761	1.00035
.445	1.07038	.72264	.99646	1.15260	.99750	1.00037
.444	1.07212	.72199	.99629	1.15364	.99738	1.00040
.443	1.07386	.72134	.99611	1.15469	.99726	1.00043
.442	1.07561	.72068	.99593	1.15573	.99714	1.00046
.441	1.07736	.72003	.99575	1.15678	.99701	1.00049
.440	1.07911	.71937	.99556	1.15784	.99688	1.00052
.439	1.08086	.71872	.99537	1.15890	.99675	1.00056
.438	1.08262	.71806	.99517	1.15996	.99661	1.00059
.437	1.08438	.71740	.99497	1.16102	.99648	1.00063
.436	1.08614	.71674	.99476	1.16209	.99634	1.00066
.435	1.08791	.71608	.99455	1.16316	.99620	1.00070
.434	1.08968	.71542	.99434	1.16423	.99605	1.00074
.433	1.09145	.71476	.99412	1.16531	.99591	1.00079
.432	1.09322	.71410	.99389	1.16639	.99576	1.00083
.431	1.09500	.71343	.99367	1.16748	.99561	1.00087
.430	1.09678	.71277	.99344	1.16857	.99545	1.00092
.429	1.09856	.71210	.99320	1.16966	.99529	1.00097
.428	1.10035	.71144	.99296	1.17076	.99514	1.00102
.427	1.10214	.71077	.99272	1.17185	.99497	1.00107
.426	1.10393	.71010	.99247	1.17296	.99481	1.00113
.425	1.10572	.70943	.99222	1.17406	.99464	1.00118
.424	1.10752	.70876	.99196	1.17517	.99447	1.00124
.423	1.10932	.70809	.99170	1.17629	.99430	1.00130
.422	1.11113	.70742	.99143	1.17741	.99413	1.00136
.421	1.11293	.70675	.99116	1.17853	.99395	1.00142
.420	1.11474	.70607	.99089	1.17965	.99377	1.00148
.419	1.11656	.70540	.99061	1.18078	.99359	1.00155
.418	1.11837	.70472	.99033	1.18191	.99341	1.00162
.417	1.12019	.70404	.99004	1.18305	.99322	1.00169
.416	1.12201	.70337	.98975	1.18419	.99304	1.00176
.415	1.12384	.70269	.98946	1.18533	.99284	1.00183
.414	1.12567	.70201	.98916	1.18648	.99265	1.00191
.413	1.12750	.70133	.98885	1.18763	.99246	1.00198
.412	1.12933	.70064	.98855	1.18879	.99226	1.00206
.411	1.13117	.69996	.98823	1.18995	.99206	1.00214
.410	1.13301	.69928	.98792	1.19111	.99186	1.00223
.409	1.13486	.69859	.98760	1.19228	.99165	1.00231
.408	1.13671	.69791	.98727	1.19345	.99144	1.00240
.407	1.13856	.69722	.98694	1.19463	.99123	1.00249
.406	1.14041	.69653	.98661	1.19581	.99102	1.00258
.405	1.14227	.69584	.98627	1.19699	.99081	1.00268
.404	1.14413	.69515	.98593	1.19818	.99059	1.00278
.403	1.14600	.69446	.98558	1.19937	.99037	1.00287
.402	1.14787	.69377	.98523	1.20057	.99015	1.00298
.401	1.14974	.69308	.98487	1.20177	.98993	1.00308

TABLE VI. Isentropic Exponent = 1.67

p/p_t	M	T/T_t	GAMMA	ISOTHERMAL T_t/T_t^*	RAYLEIGH T_t/T_t^*	SHOCK p_{t1}/p_{t2}
.400	1.15162	.69238	.98451	1.20297	.98970	1.00319
.399	1.15350	.69169	.98415	1.20418	.98947	1.00329
.398	1.15538	.69099	.98378	1.20539	.98924	1.00340
.397	1.15726	.69030	.98340	1.20661	.98901	1.00352
.396	1.15915	.68960	.98302	1.20783	.98878	1.00363
.395	1.16105	.68890	.98264	1.20906	.98854	1.00375
.394	1.16295	.68820	.98226	1.21029	.98830	1.00387
.393	1.16485	.68750	.98186	1.21152	.98806	1.00399
.392	1.16675	.68680	.98147	1.21276	.98782	1.00412
.391	1.16866	.68609	.98107	1.21400	.98757	1.00425
.390	1.17057	.68539	.98066	1.21525	.98732	1.00438
.389	1.17249	.68468	.98026	1.21650	.98707	1.00451
.388	1.17441	.68398	.97984	1.21776	.98682	1.00465
.387	1.17633	.68327	.97942	1.21902	.98657	1.00479
.386	1.17826	.68256	.97900	1.22029	.98631	1.00493
.385	1.18019	.68185	.97857	1.22156	.98605	1.00507
.384	1.18212	.68114	.97814	1.22283	.98579	1.00522
.383	1.18406	.68043	.97771	1.22411	.98553	1.00537
.382	1.18600	.67971	.97727	1.22540	.98526	1.00552
.381	1.18795	.67900	.97682	1.22669	.98499	1.00567
.380	1.18990	.67828	.97637	1.22798	.98472	1.00583
.379	1.19185	.67757	.97592	1.22928	.98445	1.00599
.378	1.19381	.67685	.97546	1.23058	.98418	1.00616
.377	1.19578	.67613	.97499	1.23189	.98390	1.00632
.376	1.19774	.67541	.97453	1.23321	.98362	1.00649
.375	1.19971	.67469	.97405	1.23452	.98334	1.00666
.374	1.20169	.67396	.97358	1.23585	.98306	1.00684
.373	1.20367	.67324	.97310	1.23718	.98277	1.00702
.372	1.20565	.67252	.97261	1.23851	.98249	1.00720
.371	1.20764	.67179	.97212	1.23985	.98220	1.00739
.370	1.20963	.67106	.97162	1.24119	.98191	1.00757
.369	1.21162	.67034	.97112	1.24254	.98161	1.00776
.368	1.21362	.66961	.97062	1.24389	.98132	1.00796
.367	1.21563	.66888	.97011	1.24525	.98102	1.00816
.366	1.21764	.66814	.96960	1.24662	.98072	1.00836
.365	1.21965	.66741	.96908	1.24798	.98042	1.00856
.364	1.22167	.66668	.96855	1.24936	.98012	1.00877
.363	1.22369	.66594	.96802	1.25074	.97981	1.00898
.362	1.22572	.66520	.96749	1.25212	.97950	1.00919
.361	1.22775	.66447	.96695	1.25351	.97919	1.00941
.360	1.22978	.66373	.96641	1.25491	.97888	1.00963
.359	1.23182	.66299	.96587	1.25631	.97857	1.00985
.358	1.23387	.66225	.96531	1.25772	.97825	1.01008
.357	1.23592	.66150	.96476	1.25913	.97793	1.01031
.356	1.23797	.66076	.96420	1.26055	.97761	1.01055
.355	1.24003	.66001	.96363	1.26197	.97729	1.01078
.354	1.24209	.65927	.96306	1.26340	.97697	1.01102
.353	1.24416	.65852	.96248	1.26483	.97664	1.01127
.352	1.24623	.65777	.96190	1.26628	.97631	1.01152
.351	1.24831	.65702	.96132	1.26772	.97598	1.01177

TABLE VI. Isentropic Exponent = 1.67

p/p_t	M	T/T_t	GAMMA	ISOTHERMAL T_t/T_t^*	RAYLEIGH T_t/T_t^*	SHOCK p_{t1}/p_{t2}
.350	1.25039	.65627	.96073	1.26917	.97565	1.01203
.349	1.25248	.65551	.96013	1.27063	.97532	1.01229
.348	1.25457	.65476	.95953	1.27209	.97498	1.01255
.347	1.25667	.65401	.95893	1.27356	.97464	1.01282
.346	1.25877	.65325	.95832	1.27504	.97430	1.01309
.345	1.26088	.65249	.95771	1.27652	.97396	1.01336
.344	1.26299	.65173	.95709	1.27801	.97361	1.01364
.343	1.26511	.65097	.95646	1.27950	.97327	1.01393
.342	1.26723	.65021	.95583	1.28100	.97292	1.01421
.341	1.26936	.64944	.95520	1.28251	.97257	1.01450
.340	1.27149	.64868	.95456	1.28402	.97222	1.01480
.339	1.27363	.64791	.95392	1.28554	.97186	1.01510
.338	1.27577	.64715	.95327	1.28706	.97151	1.01540
.337	1.27792	.64638	.95261	1.28859	.97115	1.01571
.336	1.28008	.64561	.95196	1.29013	.97079	1.01602
.335	1.28224	.64484	.95129	1.29167	.97043	1.01634
.334	1.28440	.64406	.95062	1.29322	.97006	1.01665
.333	1.28657	.64329	.94995	1.29478	.96970	1.01698
.332	1.28875	.64251	.94927	1.29634	.96933	1.01731
.331	1.29093	.64174	.94859	1.29791	.96896	1.01764
.330	1.29311	.64096	.94790	1.29949	.96859	1.01798
.329	1.29530	.64018	.94720	1.30107	.96822	1.01832
.328	1.29750	.63940	.94651	1.30266	.96784	1.01866
.327	1.29971	.63861	.94580	1.30426	.96746	1.01901
.326	1.30192	.63783	.94509	1.30586	.96708	1.01937
.325	1.30413	.63704	.94438	1.30747	.96670	1.01973
.324	1.30635	.63626	.94366	1.30909	.96632	1.02009
.323	1.30858	.63547	.94293	1.31072	.96593	1.02046
.322	1.31081	.63468	.94220	1.31235	.96555	1.02083
.321	1.31305	.63389	.94147	1.31399	.96516	1.02121
.320	1.31529	.63309	.94073	1.31563	.96477	1.02159
.319	1.31754	.63230	.93998	1.31729	.96437	1.02198
.318	1.31980	.63150	.93923	1.31895	.96398	1.02237
.317	1.32206	.63070	.93848	1.32061	.96358	1.02277
.316	1.32433	.62991	.93772	1.32229	.96319	1.02317
.315	1.32660	.62911	.93695	1.32397	.96279	1.02358
.314	1.32888	.62830	.93618	1.32566	.96238	1.02399
.313	1.33117	.62750	.93540	1.32736	.96198	1.02441
.312	1.33347	.62669	.93462	1.32906	.96157	1.02483
.311	1.33577	.62589	.93383	1.33078	.96117	1.02526
.310	1.33807	.62508	.93304	1.33250	.96076	1.02569
.309	1.34038	.62427	.93224	1.33423	.96034	1.02613
.308	1.34270	.62346	.93144	1.33596	.95993	1.02657
.307	1.34503	.62265	.93063	1.33771	.95951	1.02702
.306	1.34736	.62183	.92981	1.33946	.95910	1.02747
.305	1.34970	.62102	.92899	1.34122	.95868	1.02793
.304	1.35205	.62020	.92817	1.34299	.95826	1.02839
.303	1.35440	.61938	.92734	1.34476	.95783	1.02886
.302	1.35676	.61856	.92650	1.34655	.95741	1.02934
.301	1.35912	.61773	.92566	1.34834	.95698	1.02982

TABLE VI.   Isentropic Exponent = 1.67

p/p_t	M	T/T_t	GAMMA	ISOTHERMAL T_t/T_t^*	RAYLEIGH T_t/T_t^*	SHOCK p_{t1}/p_{t2}
.300	1.36150	.61691	.92481	1.35014	.95655	1.03031
.299	1.36388	.61608	.92396	1.35195	.95612	1.03080
.298	1.36626	.61526	.92310	1.35377	.95569	1.03130
.297	1.36866	.61443	.92224	1.35560	.95526	1.03180
.296	1.37106	.61360	.92137	1.35743	.95482	1.03231
.295	1.37347	.61276	.92049	1.35928	.95438	1.03282
.294	1.37588	.61193	.91961	1.36113	.95395	1.03334
.293	1.37830	.61109	.91872	1.36299	.95350	1.03387
.292	1.38073	.61026	.91783	1.36486	.95306	1.03440
.291	1.38317	.60942	.91693	1.36674	.95262	1.03494
.290	1.38562	.60858	.91603	1.36863	.95217	1.03549
.289	1.38807	.60773	.91512	1.37053	.95172	1.03604
.288	1.39053	.60689	.91421	1.37244	.95127	1.03660
.287	1.39300	.60604	.91329	1.37435	.95082	1.03716
.286	1.39547	.60520	.91236	1.37628	.95036	1.03773
.285	1.39796	.60435	.91143	1.37822	.94991	1.03831
.284	1.40045	.60349	.91049	1.38016	.94945	1.03889
.283	1.40294	.60264	.90954	1.38211	.94899	1.03948
.282	1.40545	.60178	.90860	1.38408	.94853	1.04008
.281	1.40797	.60093	.90764	1.38605	.94806	1.04068
.280	1.41049	.60007	.90668	1.38804	.94760	1.04129
.279	1.41302	.59921	.90571	1.39003	.94713	1.04190
.278	1.41556	.59835	.90474	1.39203	.94666	1.04253
.277	1.41811	.59748	.90376	1.39405	.94619	1.04316
.276	1.42066	.59661	.90277	1.39607	.94572	1.04379
.275	1.42323	.59575	.90178	1.39811	.94524	1.04444
.274	1.42580	.59488	.90078	1.40015	.94477	1.04509
.273	1.42838	.59400	.89978	1.40221	.94429	1.04574
.272	1.43097	.59313	.89877	1.40427	.94381	1.04641
.271	1.43357	.59225	.89776	1.40635	.94333	1.04708
.270	1.43617	.59138	.89674	1.40844	.94284	1.04776
.269	1.43879	.59050	.89571	1.41054	.94236	1.04845
.268	1.44141	.58962	.89467	1.41265	.94187	1.04914
.267	1.44405	.58873	.89363	1.41477	.94138	1.04984
.266	1.44669	.58785	.89259	1.41690	.94089	1.05055
.265	1.44934	.58696	.89154	1.41904	.94039	1.05127
.264	1.45200	.58607	.89048	1.42119	.93990	1.05199
.263	1.45467	.58518	.88941	1.42336	.93940	1.05273
.262	1.45735	.58428	.88834	1.42554	.93890	1.05347
.261	1.46004	.58339	.88726	1.42773	.93840	1.05422
.260	1.46274	.58249	.88618	1.42993	.93790	1.05497
.259	1.46545	.58159	.88509	1.43214	.93740	1.05574
.258	1.46816	.58069	.88399	1.43436	.93689	1.05651
.257	1.47089	.57978	.88289	1.43660	.93638	1.05729
.256	1.47363	.57888	.88178	1.43885	.93587	1.05808
.255	1.47637	.57797	.88066	1.44111	.93536	1.05888
.254	1.47913	.57706	.87954	1.44338	.93485	1.05969
.253	1.48190	.57615	.87841	1.44567	.93433	1.06050
.252	1.48468	.57523	.87728	1.44797	.93382	1.06132
.251	1.48746	.57432	.87613	1.45028	.93330	1.06216

TABLE VI. Isentropic Exponent = 1.67

p/p_t	M	T/T_t	GAMMA	ISOTHERMAL T_t/T_t^*	RAYLEIGH T_t/T_t^*	SHOCK p_{t1}/p_{t2}
.250	1.49026	.57340	.87499	1.45260	.93278	1.06300
.249	1.49307	.57248	.87383	1.45494	.93226	1.06385
.248	1.49589	.57155	.87267	1.45729	.93173	1.06471
.247	1.49872	.57063	.87150	1.45966	.93120	1.06558
.246	1.50156	.56970	.87032	1.46203	.93068	1.06645
.245	1.50441	.56877	.86914	1.46443	.93015	1.06734
.244	1.50727	.56784	.86795	1.46683	.92961	1.06824
.243	1.51014	.56690	.86676	1.46925	.92908	1.06914
.242	1.51303	.56596	.86555	1.47168	.92855	1.07006
.241	1.51592	.56502	.86434	1.47413	.92801	1.07098
.240	1.51883	.56408	.86313	1.47659	.92747	1.07192
.239	1.52175	.56314	.86190	1.47907	.92693	1.07286
.238	1.52468	.56219	.86067	1.48156	.92638	1.07382
.237	1.52762	.56124	.85944	1.48406	.92584	1.07479
.236	1.53057	.56029	.85819	1.48658	.92529	1.07576
.235	1.53354	.55934	.85694	1.48911	.92474	1.07675
.234	1.53651	.55838	.85568	1.49166	.92419	1.07774
.233	1.53950	.55742	.85442	1.49423	.92364	1.07875
.232	1.54250	.55646	.85314	1.49681	.92309	1.07977
.231	1.54551	.55550	.85186	1.49941	.92253	1.08079
.230	1.54854	.55453	.85058	1.50202	.92197	1.08183
.229	1.55158	.55356	.84928	1.50465	.92141	1.08288
.228	1.55463	.55259	.84798	1.50729	.92085	1.08395
.227	1.55769	.55162	.84667	1.50995	.92028	1.08502
.226	1.56077	.55064	.84535	1.51263	.91972	1.08610
.225	1.56386	.54966	.84403	1.51532	.91915	1.08720
.224	1.56696	.54868	.84270	1.51803	.91858	1.08830
.223	1.57008	.54770	.84136	1.52076	.91801	1.08942
.222	1.57321	.54671	.84001	1.52350	.91744	1.09055
.221	1.57635	.54572	.83866	1.52627	.91686	1.09170
.220	1.57951	.54473	.83730	1.52905	.91628	1.09285
.219	1.58268	.54374	.83593	1.53184	.91570	1.09402
.218	1.58586	.54274	.83455	1.53466	.91512	1.09520
.217	1.58906	.54174	.83317	1.53749	.91454	1.09639
.216	1.59227	.54073	.83177	1.54034	.91395	1.09760
.215	1.59550	.53973	.83037	1.54321	.91337	1.09882
.214	1.59874	.53872	.82896	1.54610	.91278	1.10005
.213	1.60200	.53771	.82755	1.54901	.91219	1.10129
.212	1.60527	.53669	.82612	1.55194	.91159	1.10255
.211	1.60855	.53568	.82469	1.55489	.91100	1.10382
.210	1.61185	.53466	.82325	1.55785	.91040	1.10511
.209	1.61517	.53363	.82180	1.56084	.90980	1.10641
.208	1.61850	.53261	.82035	1.56384	.90920	1.10772
.207	1.62185	.53158	.81888	1.56687	.90860	1.10905
.206	1.62521	.53055	.81741	1.56992	.90799	1.11039
.205	1.62859	.52951	.81593	1.57299	.90738	1.11174
.204	1.63199	.52848	.81444	1.57608	.90677	1.11311
.203	1.63540	.52744	.81295	1.57919	.90616	1.11450
.202	1.63883	.52639	.81144	1.58232	.90555	1.11590
.201	1.64227	.52534	.80993	1.58547	.90493	1.11732

TABLE VI.   Isentropic Exponent = 1.67

p/p_t	M	T/T_t	GAMMA	ISOTHERMAL T_t/T_t^*	RAYLEIGH T_t/T_t^*	SHOCK p_{t1}/p_{t2}
.200	1.64573	.52429	.80840	1.58865	.90432	1.11875
.199	1.64921	.52324	.80687	1.59184	.90370	1.12019
.198	1.65271	.52218	.80533	1.59506	.90308	1.12166
.197	1.65622	.52112	.80378	1.59831	.90245	1.12314
.196	1.65975	.52006	.80223	1.60157	.90183	1.12463
.195	1.66330	.51900	.80066	1.60486	.90120	1.12614
.194	1.66687	.51793	.79909	1.60818	.90057	1.12767
.193	1.67045	.51685	.79750	1.61152	.89994	1.12922
.192	1.67405	.51578	.79591	1.61488	.89930	1.13078
.191	1.67767	.51470	.79431	1.61827	.89867	1.13236
.190	1.68131	.51361	.79270	1.62168	.89803	1.13396
.189	1.68497	.51253	.79108	1.62511	.89739	1.13557
.188	1.68865	.51144	.78945	1.62858	.89675	1.13721
.187	1.69235	.51035	.78782	1.63207	.89610	1.13886
.186	1.69607	.50925	.78617	1.63558	.89545	1.14053
.185	1.69980	.50815	.78451	1.63912	.89480	1.14222
.184	1.70356	.50705	.78285	1.64269	.89415	1.14393
.183	1.70734	.50594	.78118	1.64628	.89350	1.14566
.182	1.71114	.50483	.77949	1.64991	.89284	1.14740
.181	1.71496	.50371	.77780	1.65356	.89219	1.14917
.180	1.71880	.50259	.77609	1.65724	.89153	1.15096
.179	1.72266	.50147	.77438	1.66095	.89086	1.15277
.178	1.72654	.50035	.77266	1.66468	.89020	1.15459
.177	1.73045	.49922	.77093	1.66845	.88953	1.15644
.176	1.73437	.49808	.76919	1.67225	.88886	1.15831
.175	1.73832	.49695	.76743	1.67607	.88819	1.16021
.174	1.74230	.49580	.76567	1.67993	.88752	1.16212
.173	1.74629	.49466	.76390	1.68382	.88684	1.16406
.172	1.75031	.49351	.76212	1.68774	.88616	1.16602
.171	1.75435	.49236	.76032	1.69170	.88548	1.16800
.170	1.75842	.49120	.75852	1.69568	.88480	1.17001
.169	1.76251	.49004	.75671	1.69970	.88411	1.17204
.168	1.76662	.48887	.75489	1.70375	.88343	1.17409
.167	1.77076	.48770	.75305	1.70784	.88274	1.17617
.166	1.77493	.48653	.75121	1.71196	.88204	1.17827
.165	1.77912	.48535	.74935	1.71611	.88135	1.18040
.164	1.78333	.48417	.74749	1.72030	.88065	1.18255
.163	1.78758	.48298	.74561	1.72453	.87995	1.18473
.162	1.79184	.48179	.74372	1.72879	.87925	1.18693
.161	1.79614	.48060	.74183	1.73309	.87854	1.18917
.160	1.80046	.47940	.73992	1.73743	.87784	1.19142
.159	1.80481	.47819	.73800	1.74181	.87713	1.19371
.158	1.80919	.47698	.73606	1.74622	.87642	1.19603
.157	1.81359	.47577	.73412	1.75067	.87570	1.19837
.156	1.81803	.47455	.73217	1.75517	.87498	1.20074
.155	1.82249	.47333	.73020	1.75970	.87426	1.20314
.154	1.82699	.47210	.72822	1.76428	.87354	1.20557
.153	1.83151	.47087	.72623	1.76889	.87282	1.20804
.152	1.83606	.46963	.72423	1.77355	.87209	1.21053
.151	1.84065	.46839	.72222	1.77826	.87136	1.21305

TABLE VI. Isentropic Exponent = 1.67

p/p_t	M	T/T_t	GAMMA	ISOTHERMAL T_t/T_t^*	RAYLEIGH T_t/T_t^*	SHOCK p_{t1}/p_{t2}
.150	1.84526	.46714	.72020	1.78300	.87062	1.21561
.149	1.84991	.46589	.71816	1.78780	.86989	1.21820
.148	1.85459	.46463	.71611	1.79263	.86915	1.22082
.147	1.85930	.46337	.71405	1.79751	.86841	1.22348
.146	1.86404	.46210	.71198	1.80244	.86767	1.22617
.145	1.86882	.46083	.70989	1.80742	.86692	1.22889
.144	1.87364	.45955	.70779	1.81245	.86617	1.23165
.143	1.87848	.45827	.70568	1.81752	.86542	1.23445
.142	1.88336	.45698	.70356	1.82264	.86466	1.23728
.141	1.88828	.45569	.70142	1.82782	.86391	1.24015
.140	1.89323	.45439	.69927	1.83305	.86314	1.24306
.139	1.89823	.45308	.69711	1.83833	.86238	1.24601
.138	1.90325	.45177	.69493	1.84366	.86162	1.24899
.137	1.90832	.45046	.69274	1.84905	.86085	1.25202
.136	1.91342	.44914	.69054	1.85449	.86007	1.25509
.135	1.91857	.44781	.68832	1.85999	.85930	1.25820
.134	1.92375	.44647	.68609	1.86554	.85852	1.26135
.133	1.92897	.44513	.68385	1.87116	.85774	1.26455
.132	1.93424	.44379	.68159	1.87683	.85695	1.26779
.131	1.93954	.44244	.67931	1.88257	.85617	1.27107
.130	1.94489	.44108	.67703	1.88836	.85538	1.27440
.129	1.95028	.43971	.67473	1.89422	.85458	1.27778
.128	1.95571	.43834	.67241	1.90015	.85378	1.28121
.127	1.96119	.43697	.67008	1.90614	.85298	1.28468
.126	1.96672	.43558	.66773	1.91219	.85218	1.28821
.125	1.97229	.43419	.66537	1.91831	.85137	1.29178
.124	1.97791	.43280	.66300	1.92451	.85056	1.29541
.123	1.98357	.43139	.66061	1.93077	.84975	1.29909
.122	1.98929	.42998	.65820	1.93710	.84893	1.30282
.121	1.99505	.42856	.65578	1.94351	.84811	1.30661
.120	2.00086	.42714	.65334	1.94999	.84729	1.31046
.119	2.00673	.42571	.65089	1.95655	.84646	1.31436
.118	2.01264	.42427	.64842	1.96318	.84563	1.31833
.117	2.01861	.42282	.64593	1.96990	.84480	1.32235
.116	2.02464	.42137	.64343	1.97669	.84396	1.32643
.115	2.03071	.41991	.64091	1.98357	.84312	1.33058
.114	2.03685	.41844	.63837	1.99053	.84227	1.33479
.113	2.04304	.41696	.63582	1.99758	.84142	1.33907
.112	2.04929	.41548	.63325	2.00472	.84057	1.34341
.111	2.05560	.41399	.63066	2.01195	.83972	1.34783
.110	2.06197	.41249	.62805	2.01926	.83886	1.35231
.109	2.06840	.41098	.62543	2.02668	.83799	1.35686
.108	2.07490	.40946	.62279	2.03418	.83712	1.36149
.107	2.08145	.40794	.62013	2.04179	.83625	1.36620
.106	2.08808	.40640	.61745	2.04950	.83537	1.37098
.105	2.09477	.40486	.61475	2.05730	.83449	1.37584
.104	2.10153	.40331	.61203	2.06522	.83361	1.38078
.103	2.10835	.40175	.60930	2.07324	.83272	1.38581
.102	2.11525	.40018	.60654	2.08137	.83183	1.39092
.101	2.12223	.39860	.60377	2.08961	.83093	1.39612

TABLE VI. Isentropic Exponent = 1.67

p/p_t	M	T/T_t	GAMMA	ISOTHERMAL T_t/T_t*	RAYLEIGH T_t/T_t*	SHOCK p_{t1}/p_{t2}
.100	2.12927	.39701	.60097	2.09797	.83003	1.40141
.099	2.13639	.39541	.59816	2.10645	.82912	1.40679
.098	2.14359	.39381	.59532	2.11505	.82821	1.41226
.097	2.15087	.39219	.59247	2.12377	.82730	1.41784
.096	2.15823	.39056	.58959	2.13261	.82638	1.42351
.095	2.16567	.38892	.58669	2.14159	.82545	1.42928
.094	2.17320	.38728	.58377	2.15070	.82452	1.43516
.093	2.18081	.38562	.58083	2.15995	.82359	1.44115
.092	2.18851	.38395	.57786	2.16934	.82265	1.44725
.091	2.19630	.38227	.57488	2.17887	.82171	1.45346
.090	2.20419	.38058	.57187	2.18856	.82076	1.45979
.089	2.21217	.37888	.56883	2.19839	.81980	1.46624
.088	2.22024	.37716	.56578	2.20838	.81884	1.47281
.087	2.22842	.37544	.56270	2.21853	.81788	1.47952
.086	2.23670	.37370	.55959	2.22884	.81691	1.48635
.085	2.24508	.37195	.55646	2.23932	.81593	1.49332
.084	2.25357	.37019	.55331	2.24998	.81495	1.50043
.083	2.26217	.36841	.55012	2.26082	.81396	1.50769
.082	2.27088	.36663	.54692	2.27184	.81297	1.51509
.081	2.27971	.36483	.54369	2.28305	.81197	1.52265
.080	2.28866	.36301	.54043	2.29446	.81097	1.53037
.079	2.29773	.36119	.53714	2.30606	.80996	1.53825
.078	2.30693	.35934	.53383	2.31788	.80894	1.54629
.077	2.31626	.35749	.53048	2.32991	.80792	1.55452
.076	2.32571	.35562	.52711	2.34216	.80689	1.56292
.075	2.33531	.35373	.52371	2.35464	.80585	1.57151
.074	2.34505	.35183	.52028	2.36736	.80481	1.58030
.073	2.35493	.34992	.51682	2.38031	.80376	1.58928
.072	2.36496	.34799	.51333	2.39352	.80270	1.59848
.071	2.37514	.34604	.50981	2.40699	.80164	1.60788
.070	2.38548	.34408	.50626	2.42073	.80057	1.61751
.069	2.39598	.34210	.50267	2.43474	.79949	1.62738
.068	2.40666	.34010	.49905	2.44904	.79840	1.63748
.067	2.41750	.33808	.49540	2.46364	.79731	1.64783
.066	2.42853	.33605	.49171	2.47855	.79620	1.65844
.065	2.43974	.33400	.48799	2.49378	.79509	1.66932
.064	2.45114	.33193	.48423	2.50934	.79398	1.68047
.063	2.46274	.32984	.48044	2.52525	.79285	1.69192
.062	2.47455	.32773	.47660	2.54151	.79171	1.70368
.061	2.48656	.32559	.47273	2.55814	.79057	1.71575
.060	2.49880	.32344	.46882	2.57516	.78942	1.72814
.059	2.51126	.32127	.46487	2.59259	.78825	1.74089
.058	2.52396	.31907	.46088	2.61043	.78708	1.75399
.057	2.53690	.31685	.45685	2.62871	.78590	1.76747
.056	2.55010	.31461	.45278	2.64744	.78471	1.78134
.055	2.56356	.31235	.44866	2.66665	.78350	1.79562
.054	2.57730	.31006	.44449	2.68635	.78229	1.81034
.053	2.59132	.30774	.44028	2.70657	.78107	1.82550
.052	2.60564	.30540	.43603	2.72734	.77983	1.84114
.051	2.62027	.30303	.43172	2.74867	.77858	1.85728

TABLE VI. Isentropic Exponent = 1.67

p/p_t	M	T/T_t	GAMMA	ISOTHERMAL T_t/T_t*	RAYLEIGH T_t/T_t*	SHOCK p_t1/p_t2
.050	2.63522	.30063	.42736	2.77059	.77733	1.87395
.049	2.65051	.29820	.42296	2.79314	.77605	1.89116
.048	2.66615	.29574	.41850	2.81634	.77477	1.90895
.047	2.68215	.29326	.41398	2.84023	.77347	1.92736
.046	2.69855	.29074	.40941	2.86484	.77216	1.94642
.045	2.71534	.28819	.40479	2.89021	.77084	1.96616
.044	2.73256	.28560	.40010	2.91639	.76950	1.98662
.043	2.75023	.28298	.39536	2.94341	.76814	2.00785
.042	2.76836	.28032	.39055	2.97133	.76677	2.02990
.041	2.78698	.27762	.38567	3.00020	.76539	2.05281
.040	2.80612	.27488	.38073	3.03007	.76399	2.07664
.039	2.82581	.27211	.37572	3.06100	.76257	2.10145
.038	2.84607	.26929	.37064	3.09307	.76113	2.12731
.037	2.86694	.26642	.36548	3.12634	.75967	2.15430
.036	2.88846	.26351	.36025	3.16090	.75819	2.18248
.035	2.91066	.26055	.35493	3.19682	.75670	2.21196
.034	2.93359	.25753	.34953	3.23422	.75518	2.24282
.033	2.95730	.25447	.34405	3.27319	.75364	2.27517
.032	2.98184	.25134	.33847	3.31385	.75207	2.30915
.031	3.00726	.24816	.33280	3.35633	.75048	2.34487
.030	3.03362	.24492	.32704	3.40077	.74887	2.38249
.029	3.06101	.24161	.32117	3.44734	.74723	2.42217
.028	3.08949	.23823	.31519	3.49622	.74556	2.46412
.027	3.11915	.23478	.30910	3.54760	.74386	2.50853
.026	3.15009	.23125	.30288	3.60173	.74212	2.55566
.025	3.18242	.22764	.29655	3.65885	.74036	2.60578
.024	3.21626	.22395	.29008	3.71927	.73856	2.65922
.023	3.25175	.22016	.28347	3.78332	.73671	2.71633
.022	3.28905	.21626	.27671	3.85140	.73483	2.77756
.021	3.32835	.21226	.26980	3.92395	.73290	2.84341
.020	3.36985	.20815	.26271	4.00152	.73093	2.91445
.019	3.41380	.20391	.25545	4.08472	.72890	2.99141
.018	3.46050	.19954	.24799	4.17429	.72682	3.07510
.017	3.51028	.19501	.24032	4.27112	.72468	3.16656
.016	3.56356	.19033	.23242	4.37628	.72247	3.26702
.015	3.62083	.18546	.22428	4.49107	.72019	3.37800
.014	3.68268	.18040	.21587	4.61712	.71783	3.50143
.013	3.74986	.17511	.20717	4.75646	.71538	3.63976
.012	3.82332	.16958	.19814	4.91168	.71283	3.79613
.011	3.90425	.16376	.18873	5.08617	.71016	3.97472
.010	3.99420	.15762	.17892	5.28442	.70736	4.18118
.009	4.09528	.15109	.16863	5.51259	.70440	4.42334
.008	4.21038	.14412	.15779	5.77933	.70127	4.71249
.007	4.34364	.13660	.14630	6.09739	.69792	5.06555
.006	4.50126	.12841	.13403	6.48639	.69429	5.50925
.005	4.69313	.11935	.12079	6.97863	.69032	6.08885
.004	4.93638	.10913	.10630	7.63221	.68589	6.88853
.003	5.25443	.09724	.09007	8.56596	.68079	8.08778
.002	5.75650	.08264	.07122	0.07914	.67462	10.16436
.001	6.68719	.06258	.04754	3.31059	.66629	15.09960

TABLE VII. Isentropic Exponent = 1.00

M	p/p_t	T/T_t	GAMMA	ISOTHERMAL T_t/T_t^*	RAYLEIGH T_t/T_t^*	SHOCK p_{t1}/p_{t2}
.000	1.00000	1.	.00000	1.	.00000	1.
.005	.99999	1.	.00824	1.	.00010	1.
.010	.99995	1.	.01649	1.	.00040	1.
.015	.99989	1.	.02473	1.	.00090	1.
.020	.99980	1.	.03297	1.	.00160	1.
.025	.99969	1.	.04121	1.	.00250	1.
.030	.99955	1.	.04944	1.	.00359	1.
.035	.99939	1.	.05767	1.	.00489	1.
.040	.99920	1.	.06590	1.	.00638	1.
.045	.99899	1.	.07412	1.	.00807	1.
.050	.99875	1.	.08233	1.	.00995	1.
.055	.99849	1.	.09054	1.	.01203	1.
.060	.99820	1.	.09875	1.	.01430	1.
.065	.99789	1.	.10694	1.	.01676	1.
.070	.99755	1.	.11513	1.	.01941	1.
.075	.99719	1.	.12331	1.	.02225	1.
.080	.99681	1.	.13148	1.	.02528	1.
.085	.99639	1.	.13964	1.	.02849	1.
.090	.99596	1.	.14779	1.	.03188	1.
.095	.99550	1.	.15592	1.	.03546	1.
.100	.99501	1.	.16405	1.	.03921	1.
.105	.99450	1.	.17216	1.	.04314	1.
.110	.99397	1.	.18027	1.	.04725	1.
.115	.99341	1.	.18835	1.	.05153	1.
.120	.99283	1.	.19643	1.	.05598	1.
.125	.99222	1.	.20449	1.	.06059	1.
.130	.99159	1.	.21253	1.	.06537	1.
.135	.99093	1.	.22056	1.	.07031	1.
.140	.99025	1.	.22857	1.	.07541	1.
.145	.98954	1.	.23656	1.	.08067	1.
.150	.98881	1.	.24454	1.	.08608	1.
.155	.98806	1.	.25250	1.	.09164	1.
.160	.98728	1.	.26044	1.	.09735	1.
.165	.98648	1.	.26836	1.	.10320	1.
.170	.98565	1.	.27626	1.	.10920	1.
.175	.98480	1.	.28414	1.	.11533	1.
.180	.98393	1.	.29200	1.	.12159	1.
.185	.98303	1.	.29984	1.	.12799	1.
.190	.98211	1.	.30765	1.	.13451	1.
.195	.98117	1.	.31545	1.	.14116	1.
.200	.98020	1.	.32321	1.	.14793	1.
.205	.97921	1.	.33096	1.	.15481	1.
.210	.97819	1.	.33868	1.	.16181	1.
.215	.97715	1.	.34638	1.	.16892	1.
.220	.97609	1.	.35405	1.	.17614	1.
.225	.97501	1.	.36169	1.	.18345	1.
.230	.97390	1.	.36931	1.	.19087	1.
.235	.97277	1.	.37690	1.	.19838	1.
.240	.97161	1.	.38446	1.	.20599	1.
.245	.97043	1.	.39199	1.	.21368	1.

TABLE VII. Isentropic Exponent = 1.00

M	p/p_t	T/T_t	GAMMA	ISOTHERMAL T_t/T_t^*	RAYLEIGH T_t/T_t^*	SHOCK p_{t1}/p_{t2}
.250	.96923	1.	.39950	1.	.22145	1.
.255	.96801	1.	.40697	1.	.22931	1.
.260	.96676	1.	.41442	1.	.23724	1.
.265	.96550	1.	.42184	1.	.24525	1.
.270	.96421	1.	.42922	1.	.25332	1.
.275	.96289	1.	.43657	1.	.26146	1.
.280	.96156	1.	.44390	1.	.26966	1.
.285	.96020	1.	.45118	1.	.27792	1.
.290	.95882	1.	.45844	1.	.28623	1.
.295	.95742	1.	.46566	1.	.29459	1.
.300	.95600	1.	.47285	1.	.30300	1.
.305	.95455	1.	.48001	1.	.31146	1.
.310	.95309	1.	.48713	1.	.31995	1.
.315	.95160	1.	.49421	1.	.32848	1.
.320	.95009	1.	.50126	1.	.33704	1.
.325	.94856	1.	.50827	1.	.34563	1.
.330	.94701	1.	.51524	1.	.35424	1.
.335	.94543	1.	.52218	1.	.36288	1.
.340	.94384	1.	.52908	1.	.37154	1.
.345	.94222	1.	.53595	1.	.38021	1.
.350	.94059	1.	.54277	1.	.38889	1.
.355	.93893	1.	.54955	1.	.39758	1.
.360	.93725	1.	.55630	1.	.40627	1.
.365	.93556	1.	.56300	1.	.41497	1.
.370	.93384	1.	.56967	1.	.42366	1.
.375	.93210	1.	.57629	1.	.43235	1.
.380	.93034	1.	.58287	1.	.44103	1.
.385	.92857	1.	.58942	1.	.44970	1.
.390	.92677	1.	.59591	1.	.45836	1.
.395	.92495	1.	.60237	1.	.46700	1.
.400	.92312	1.	.60878	1.	.47562	1.
.405	.92126	1.	.61516	1.	.48422	1.
.410	.91939	1.	.62148	1.	.49280	1.
.415	.91749	1.	.62776	1.	.50134	1.
.420	.91558	1.	.63400	1.	.50986	1.
.425	.91365	1.	.64020	1.	.51834	1.
.430	.91169	1.	.64635	1.	.52679	1.
.435	.90973	1.	.65245	1.	.53519	1.
.440	.90774	1.	.65851	1.	.54356	1.
.445	.90573	1.	.66452	1.	.55188	1.
.450	.90371	1.	.67048	1.	.56016	1.
.455	.90166	1.	.67640	1.	.56839	1.
.460	.89960	1.	.68227	1.	.57658	1.
.465	.89753	1.	.68809	1.	.58471	1.
.470	.89543	1.	.69387	1.	.59278	1.
.475	.89332	1.	.69960	1.	.60080	1.
.480	.89119	1.	.70527	1.	.60877	1.
.485	.88904	1.	.71090	1.	.61667	1.
.490	.88688	1.	.71648	1.	.62451	1.
.495	.88469	1.	.72201	1.	.63229	1.

TABLE VII. Isentropic Exponent = 1.00

M	p/p_t	T/T_t	GAMMA	ISOTHERMAL T_t/T_t^*	RAYLEIGH T_t/T_t^*	SHOCK p_{t1}/p_{t2}
.500	.88250	1.	.72750	1.	.64000	1.
.505	.88028	1.	.73293	1.	.64765	1.
.510	.87805	1.	.73831	1.	.65522	1.
.515	.87580	1.	.74364	1.	.66273	1.
.520	.87354	1.	.74892	1.	.67017	1.
.525	.87126	1.	.75415	1.	.67754	1.
.530	.86897	1.	.75932	1.	.68483	1.
.535	.86666	1.	.76445	1.	.69204	1.
.540	.86433	1.	.76952	1.	.69918	1.
.545	.86199	1.	.77454	1.	.70625	1.
.550	.85963	1.	.77951	1.	.71323	1.
.555	.85726	1.	.78443	1.	.72013	1.
.560	.85488	1.	.78929	1.	.72696	1.
.565	.85247	1.	.79410	1.	.73370	1.
.570	.85006	1.	.79886	1.	.74036	1.
.575	.84763	1.	.80356	1.	.74694	1.
.580	.84518	1.	.80821	1.	.75343	1.
.585	.84273	1.	.81281	1.	.75984	1.
.590	.84025	1.	.81735	1.	.76616	1.
.595	.83777	1.	.82184	1.	.77240	1.
.600	.83527	1.	.82628	1.	.77855	1.
.605	.83276	1.	.83066	1.	.78461	1.
.610	.83023	1.	.83498	1.	.79058	1.
.615	.82769	1.	.83925	1.	.79647	1.
.620	.82514	1.	.84347	1.	.80227	1.
.625	.82258	1.	.84763	1.	.80798	1.
.630	.82000	1.	.85173	1.	.81360	1.
.635	.81741	1.	.85578	1.	.81913	1.
.640	.81481	1.	.85977	1.	.82457	1.
.645	.81220	1.	.86371	1.	.82992	1.
.650	.80957	1.	.86759	1.	.83518	1.
.655	.80693	1.	.87142	1.	.84036	1.
.660	.80429	1.	.87519	1.	.84544	1.
.665	.80163	1.	.87890	1.	.85043	1.
.670	.79896	1.	.88256	1.	.85533	1.
.675	.79627	1.	.88616	1.	.86014	1.
.680	.79358	1.	.88971	1.	.86486	1.
.685	.79088	1.	.89320	1.	.86949	1.
.690	.78816	1.	.89663	1.	.87403	1.
.695	.78544	1.	.90000	1.	.87848	1.
.700	.78270	1.	.90332	1.	.88284	1.
.705	.77996	1.	.90659	1.	.88712	1.
.710	.77721	1.	.90979	1.	.89130	1.
.715	.77444	1.	.91294	1.	.89539	1.
.720	.77167	1.	.91603	1.	.89940	1.
.725	.76889	1.	.91907	1.	.90332	1.
.730	.76609	1.	.92205	1.	.90715	1.
.735	.76329	1.	.92497	1.	.91089	1.
.740	.76048	1.	.92783	1.	.91455	1.
.745	.75767	1.	.93064	1.	.91812	1.

TABLE VII. Isentropic Exponent = 1.00

M	p/p_t	T/T_t	GAMMA	ISOTHERMAL T_t/T_t^*	RAYLEIGH T_t/T_t^*	SHOCK p_{t1}/p_{t2}
.750	.75484	1.	.93339	1.	.92160	1.
.755	.75200	1.	.93608	1.	.92500	1.
.760	.74916	1.	.93872	1.	.92831	1.
.765	.74631	1.	.94130	1.	.93154	1.
.770	.74345	1.	.94382	1.	.93468	1.
.775	.74059	1.	.94629	1.	.93774	1.
.780	.73771	1.	.94870	1.	.94072	1.
.785	.73483	1.	.95105	1.	.94362	1.
.790	.73195	1.	.95335	1.	.94643	1.
.795	.72905	1.	.95559	1.	.94916	1.
.800	.72615	1.	.95777	1.	.95181	1.
.805	.72324	1.	.95990	1.	.95439	1.
.810	.72033	1.	.96197	1.	.95688	1.
.815	.71741	1.	.96398	1.	.95929	1.
.820	.71448	1.	.96594	1.	.96163	1.
.825	.71155	1.	.96784	1.	.96389	1.
.830	.70861	1.	.96969	1.	.96607	1.
.835	.70567	1.	.97148	1.	.96818	1.
.840	.70272	1.	.97321	1.	.97021	1.
.845	.69976	1.	.97489	1.	.97216	1.
.850	.69680	1.	.97651	1.	.97405	1.
.855	.69384	1.	.97808	1.	.97586	1.
.860	.69087	1.	.97959	1.	.97759	1.
.865	.68790	1.	.98104	1.	.97926	1.
.870	.68492	1.	.98244	1.	.98085	1.
.875	.68194	1.	.98379	1.	.98238	1.
.880	.67896	1.	.98508	1.	.98383	1.
.885	.67597	1.	.98631	1.	.98522	1.
.890	.67297	1.	.98749	1.	.98654	1.
.895	.66998	1.	.98862	1.	.98779	1.
.900	.66698	1.	.98969	1.	.98898	1.
.905	.66397	1.	.99071	1.	.99010	1.
.910	.66097	1.	.99167	1.	.99116	1.
.915	.65796	1.	.99258	1.	.99215	1.
.920	.65495	1.	.99344	1.	.99308	1.
.925	.65193	1.	.99424	1.	.99395	1.
.930	.64892	1.	.99499	1.	.99475	1.
.935	.64590	1.	.99569	1.	.99550	1.
.940	.64288	1.	.99633	1.	.99618	1.
.945	.63986	1.	.99692	1.	.99681	1.
.950	.63683	1.	.99746	1.	.99737	1.
.955	.63381	1.	.99795	1.	.99788	1.
.960	.63078	1.	.99838	1.	.99834	1.
.965	.62775	1.	.99876	1.	.99873	1.
.970	.62472	1.	.99909	1.	.99907	1.
.975	.62169	1.	.99937	1.	.99936	1.
.980	.61866	1.	.99960	1.	.99959	1.
.985	.61563	1.	.99977	1.	.99977	1.
.990	.61260	1.	.99990	1.	.99990	1.
.995	.60956	1.	.99997	1.	.99997	1.

TABLE VII. Isentropic Exponent = 1.00

M	p/p_t	T/T_t	GAMMA	ISOTHERMAL T_t/T_t^*	RAYLEIGH T_t/T_t^*	SHOCK p_{t1}/p_{t2}
1.000	.60653	1.	1.00000	1.	1.00000	1.00000
1.005	.60350	1.	.99998	1.	.99998	1.00000
1.010	.60047	1.	.99990	1.	.99990	1.00000
1.015	.59743	1.	.99978	1.	.99990	1.00000
1.020	.59440	1.	.99960	1.	.99961	1.00001
1.025	.59137	1.	.99938	1.	.99939	1.00002
1.030	.58834	1.	.99911	1.	.99913	1.00003
1.035	.58531	1.	.99879	1.	.99882	1.00005
1.040	.58228	1.	.99842	1.	.99846	1.00008
1.045	.57926	1.	.99801	1.	.99807	1.00011
1.050	.57623	1.	.99754	1.	.99762	1.00015
1.055	.57320	1.	.99703	1.	.99714	1.00020
1.060	.57018	1.	.99648	1.	.99661	1.00026
1.065	.56716	1.	.99587	1.	.99604	1.00033
1.070	.56414	1.	.99522	1.	.99544	1.00041
1.075	.56112	1.	.99452	1.	.99479	1.00050
1.080	.55811	1.	.99378	1.	.99410	1.00061
1.085	.55510	1.	.99299	1.	.99337	1.00073
1.090	.55209	1.	.99216	1.	.99261	1.00085
1.095	.54908	1.	.99128	1.	.99181	1.00100
1.100	.54607	1.	.99036	1.	.99097	1.00116
1.105	.54307	1.	.98939	1.	.99010	1.00133
1.110	.54007	1.	.98838	1.	.98919	1.00152
1.115	.53708	1.	.98732	1.	.98824	1.00173
1.120	.53409	1.	.98622	1.	.98727	1.00195
1.125	.53110	1.	.98508	1.	.98625	1.00219
1.130	.52811	1.	.98390	1.	.98521	1.00244
1.135	.52513	1.	.98267	1.	.98413	1.00272
1.140	.52215	1.	.98140	1.	.98303	1.00301
1.145	.51918	1.	.98009	1.	.98189	1.00333
1.150	.51621	1.	.97874	1.	.98072	1.00366
1.155	.51324	1.	.97735	1.	.97952	1.00401
1.160	.51028	1.	.97591	1.	.97829	1.00439
1.165	.50732	1.	.97444	1.	.97703	1.00478
1.170	.50437	1.	.97293	1.	.97575	1.00520
1.175	.50142	1.	.97137	1.	.97444	1.00564
1.180	.49848	1.	.96978	1.	.97310	1.00610
1.185	.49554	1.	.96815	1.	.97173	1.00658
1.190	.49260	1.	.96648	1.	.97034	1.00709
1.195	.48968	1.	.96477	1.	.96892	1.00762
1.200	.48675	1.	.96302	1.	.96748	1.00817
1.205	.48383	1.	.96124	1.	.96602	1.00874
1.210	.48092	1.	.95942	1.	.96453	1.00935
1.215	.47802	1.	.95756	1.	.96301	1.00997
1.220	.47511	1.	.95566	1.	.96148	1.01062
1.225	.47222	1.	.95373	1.	.95992	1.01130
1.230	.46933	1.	.95177	1.	.95834	1.01200
1.235	.46645	1.	.94976	1.	.95674	1.01273
1.240	.46357	1.	.94773	1.	.95512	1.01349
1.245	.46070	1.	.94566	1.	.95348	1.01427

TABLE VII.   Isentropic Exponent = 1.00

M	p/p_t	T/T_t	GAMMA	ISOTHERMAL T_t/T_t^*	RAYLEIGH T_t/T_t^*	SHOCK p_{t1}/p_{t2}
1.250	.45783	1.	.94355	1.	.95181	1.01508
1.255	.45498	1.	.94141	1.	.95013	1.01591
1.260	.45212	1.	.93924	1.	.94843	1.01677
1.265	.44928	1.	.93703	1.	.94671	1.01767
1.270	.44644	1.	.93479	1.	.94498	1.01859
1.275	.44361	1.	.93252	1.	.94322	1.01953
1.280	.44078	1.	.93021	1.	.94145	1.02051
1.285	.43797	1.	.92788	1.	.93967	1.02152
1.290	.43516	1.	.92551	1.	.93786	1.02255
1.295	.43235	1.	.92311	1.	.93604	1.02362
1.300	.42956	1.	.92069	1.	.93420	1.02471
1.305	.42677	1.	.91823	1.	.93235	1.02584
1.310	.42399	1.	.91574	1.	.93049	1.02700
1.315	.42121	1.	.91322	1.	.92861	1.02818
1.320	.41845	1.	.91068	1.	.92672	1.02940
1.325	.41569	1.	.90810	1.	.92481	1.03065
1.330	.41294	1.	.90550	1.	.92289	1.03193
1.335	.41020	1.	.90287	1.	.92095	1.03324
1.340	.40747	1.	.90021	1.	.91901	1.03459
1.345	.40474	1.	.89752	1.	.91705	1.03596
1.350	.40202	1.	.89481	1.	.91508	1.03737
1.355	.39931	1.	.89207	1.	.91310	1.03882
1.360	.39661	1.	.88930	1.	.91111	1.04029
1.365	.39392	1.	.88651	1.	.90911	1.04180
1.370	.39123	1.	.88370	1.	.90709	1.04334
1.375	.38856	1.	.88086	1.	.90507	1.04492
1.380	.38589	1.	.87799	1.	.90304	1.04653
1.385	.38323	1.	.87510	1.	.90099	1.04818
1.390	.38058	1.	.87219	1.	.89894	1.04986
1.395	.37794	1.	.86926	1.	.89688	1.05158
1.400	.37531	1.	.86630	1.	.89481	1.05333
1.405	.37269	1.	.86332	1.	.89274	1.05511
1.410	.37008	1.	.86031	1.	.89065	1.05694
1.415	.36747	1.	.85729	1.	.88856	1.05879
1.420	.36488	1.	.85424	1.	.88646	1.06069
1.425	.36229	1.	.85117	1.	.88435	1.06262
1.430	.35971	1.	.84808	1.	.88224	1.06459
1.435	.35715	1.	.84498	1.	.88012	1.06660
1.440	.35459	1.	.84185	1.	.87799	1.06864
1.445	.35204	1.	.83870	1.	.87586	1.07072
1.450	.34950	1.	.83553	1.	.87372	1.07284
1.455	.34697	1.	.83235	1.	.87158	1.07500
1.460	.34445	1.	.82914	1.	.86943	1.07720
1.465	.34194	1.	.82592	1.	.86727	1.07943
1.470	.33944	1.	.82268	1.	.86511	1.08171
1.475	.33695	1.	.81942	1.	.86295	1.08402
1.480	.33447	1.	.81615	1.	.86078	1.08638
1.485	.33200	1.	.81286	1.	.85861	1.08877
1.490	.32954	1.	.80955	1.	.85643	1.09120
1.495	.32709	1.	.80623	1.	.85425	1.09368

TABLE VII. Isentropic Exponent = 1.00

M	p/p_t	T/T_t	GAMMA	ISOTHERMAL T_t/T_t^*	RAYLEIGH T_t/T_t^*	SHOCK p_{t1}/p_{t2}
1.500	.32465	1.	.80289	1.	.85207	1.09620
1.505	.32222	1.	.79954	1.	.84988	1.09876
1.510	.31980	1.	.79617	1.	.84770	1.10136
1.515	.31739	1.	.79279	1.	.84550	1.10400
1.520	.31499	1.	.78939	1.	.84331	1.10668
1.525	.31261	1.	.78598	1.	.84111	1.10941
1.530	.31023	1.	.78256	1.	.83891	1.11218
1.535	.30786	1.	.77913	1.	.83671	1.11499
1.540	.30550	1.	.77568	1.	.83451	1.11785
1.545	.30315	1.	.77222	1.	.83230	1.12075
1.550	.30082	1.	.76875	1.	.83009	1.12370
1.555	.29849	1.	.76526	1.	.82789	1.12669
1.560	.29618	1.	.76177	1.	.82568	1.12972
1.565	.29387	1.	.75826	1.	.82347	1.13280
1.570	.29158	1.	.75475	1.	.82125	1.13593
1.575	.28929	1.	.75122	1.	.81904	1.13910
1.580	.28702	1.	.74768	1.	.81683	1.14232
1.585	.28476	1.	.74414	1.	.81462	1.14559
1.590	.28251	1.	.74058	1.	.81240	1.14890
1.595	.28027	1.	.73702	1.	.81019	1.15226
1.600	.27804	1.	.73345	1.	.80798	1.15567
1.605	.27582	1.	.72987	1.	.80577	1.15913
1.610	.27361	1.	.72628	1.	.80355	1.16263
1.615	.27141	1.	.72269	1.	.80134	1.16619
1.620	.26923	1.	.71909	1.	.79913	1.16979
1.625	.26705	1.	.71548	1.	.79692	1.17345
1.630	.26489	1.	.71186	1.	.79471	1.17716
1.635	.26273	1.	.70824	1.	.79250	1.18091
1.640	.26059	1.	.70462	1.	.79029	1.18472
1.645	.25846	1.	.70098	1.	.78809	1.18858
1.650	.25634	1.	.69735	1.	.78588	1.19249
1.655	.25423	1.	.69370	1.	.78368	1.19646
1.660	.25213	1.	.69006	1.	.78148	1.20047
1.665	.25005	1.	.68641	1.	.77928	1.20454
1.670	.24797	1.	.68275	1.	.77708	1.20867
1.675	.24590	1.	.67909	1.	.77489	1.21285
1.680	.24385	1.	.67543	1.	.77269	1.21708
1.685	.24181	1.	.67176	1.	.77050	1.22137
1.690	.23978	1.	.66810	1.	.76831	1.22572
1.695	.23776	1.	.66443	1.	.76612	1.23012
1.700	.23575	1.	.66076	1.	.76394	1.23458
1.705	.23375	1.	.65708	1.	.76176	1.23910
1.710	.23176	1.	.65341	1.	.75958	1.24367
1.715	.22978	1.	.64973	1.	.75740	1.24831
1.720	.22782	1.	.64605	1.	.75523	1.25300
1.725	.22587	1.	.64237	1.	.75306	1.25775
1.730	.22392	1.	.63869	1.	.75089	1.26256
1.735	.22199	1.	.63502	1.	.74872	1.26744
1.740	.22007	1.	.63134	1.	.74656	1.27237
1.745	.21816	1.	.62766	1.	.74440	1.27737

TABLE VII. Isentropic Exponent = 1.00

M	p/p_t	T/T_t	GAMMA	ISOTHERMAL T_t/T_t^*	RAYLEIGH T_t/T_t^*	SHOCK p_{t1}/p_{t2}
1.750	.21627	1.	.62398	1.	.74225	1.28243
1.755	.21438	1.	.62031	1.	.74010	1.28755
1.760	.21250	1.	.61663	1.	.73795	1.29273
1.765	.21064	1.	.61296	1.	.73580	1.29798
1.770	.20879	1.	.60928	1.	.73366	1.30329
1.775	.20694	1.	.60561	1.	.73153	1.30867
1.780	.20511	1.	.60195	1.	.72939	1.31412
1.785	.20329	1.	.59828	1.	.72726	1.31963
1.790	.20148	1.	.59462	1.	.72514	1.32521
1.795	.19969	1.	.59096	1.	.72302	1.33085
1.800	.19790	1.	.58730	1.	.72090	1.33657
1.805	.19612	1.	.58365	1.	.71878	1.34235
1.810	.19436	1.	.58000	1.	.71667	1.34821
1.815	.19261	1.	.57636	1.	.71457	1.35413
1.820	.19086	1.	.57272	1.	.71247	1.36013
1.825	.18913	1.	.56908	1.	.71037	1.36620
1.830	.18741	1.	.56545	1.	.70828	1.37234
1.835	.18570	1.	.56182	1.	.70619	1.37855
1.840	.18400	1.	.55820	1.	.70411	1.38484
1.845	.18232	1.	.55459	1.	.70203	1.39120
1.850	.18064	1.	.55098	1.	.69995	1.39764
1.855	.17897	1.	.54737	1.	.69788	1.40416
1.860	.17732	1.	.54377	1.	.69582	1.41075
1.865	.17568	1.	.54018	1.	.69375	1.41742
1.870	.17404	1.	.53660	1.	.69170	1.42417
1.875	.17242	1.	.53302	1.	.68965	1.43100
1.880	.17081	1.	.52944	1.	.68760	1.43791
1.885	.16921	1.	.52588	1.	.68556	1.44490
1.890	.16762	1.	.52232	1.	.68352	1.45197
1.895	.16604	1.	.51877	1.	.68149	1.45913
1.900	.16447	1.	.51523	1.	.67946	1.46637
1.905	.16292	1.	.51169	1.	.67744	1.47369
1.910	.16137	1.	.50817	1.	.67542	1.48110
1.915	.15984	1.	.50465	1.	.67341	1.48860
1.920	.15831	1.	.50114	1.	.67140	1.49618
1.925	.15680	1.	.49764	1.	.66940	1.50385
1.930	.15529	1.	.49414	1.	.66741	1.51161
1.935	.15380	1.	.49066	1.	.66541	1.51947
1.940	.15232	1.	.48718	1.	.66343	1.52741
1.945	.15084	1.	.48372	1.	.66145	1.53544
1.950	.14938	1.	.48026	1.	.65947	1.54357
1.955	.14793	1.	.47682	1.	.65750	1.55179
1.960	.14649	1.	.47338	1.	.65553	1.56011
1.965	.14506	1.	.46995	1.	.65357	1.56852
1.970	.14364	1.	.46654	1.	.65162	1.57703
1.975	.14223	1.	.46313	1.	.64967	1.58564
1.980	.14083	1.	.45974	1.	.64772	1.59435
1.985	.13944	1.	.45635	1.	.64578	1.60316
1.990	.13806	1.	.45298	1.	.64385	1.61207
1.995	.13669	1.	.44961	1.	.64192	1.62109

TABLE VII. Isentropic Exponent = 1.00

M	p/p_t	T/T_t	GAMMA	ISOTHERMAL T_t/T_t^*	RAYLEIGH T_t/T_t^*	SHOCK p_{t1}/p_{t2}
2.000	.13534	1.	.44626	1.	.64000	1.63020
2.010	.13265	1.	.43959	1.	.63617	1.64876
2.020	.13000	1.	.43296	1.	.63236	1.66774
2.030	.12740	1.	.42638	1.	.62858	1.68716
2.040	.12483	1.	.41985	1.	.62482	1.70703
2.050	.12230	1.	.41337	1.	.62107	1.72736
2.060	.11982	1.	.40694	1.	.61736	1.74816
2.070	.11737	1.	.40056	1.	.61366	1.76944
2.080	.11496	1.	.39423	1.	.60998	1.79122
2.090	.11258	1.	.38795	1.	.60633	1.81349
2.100	.11025	1.	.38172	1.	.60270	1.83629
2.110	.10795	1.	.37555	1.	.59910	1.85961
2.120	.10569	1.	.36943	1.	.59551	1.88347
2.130	.10347	1.	.36337	1.	.59195	1.90789
2.140	.10129	1.	.35737	1.	.58841	1.93287
2.150	.09914	1.	.35142	1.	.58490	1.95844
2.160	.09702	1.	.34552	1.	.58140	1.98460
2.170	.09495	1.	.33969	1.	.57793	2.01136
2.180	.09290	1.	.33391	1.	.57448	2.03876
2.190	.09090	1.	.32819	1.	.57105	2.06679
2.200	.08892	1.	.32254	1.	.56765	2.09547
2.210	.08698	1.	.31694	1.	.56427	2.12483
2.220	.08508	1.	.31140	1.	.56091	2.15487
2.230	.08320	1.	.30591	1.	.55757	2.18562
2.240	.08137	1.	.30050	1.	.55425	2.21710
2.250	.07956	1.	.29514	1.	.55096	2.24931
2.260	.07779	1.	.28984	1.	.54769	2.28228
2.270	.07604	1.	.28460	1.	.54444	2.31604
2.280	.07433	1.	.27942	1.	.54122	2.35059
2.290	.07265	1.	.27431	1.	.53801	2.38596
2.300	.07101	1.	.26926	1.	.53483	2.42217
2.310	.06939	1.	.26427	1.	.53167	2.45924
2.320	.06780	1.	.25934	1.	.52853	2.49720
2.330	.06624	1.	.25447	1.	.52541	2.53606
2.340	.06471	1.	.24966	1.	.52231	2.57586
2.350	.06321	1.	.24492	1.	.51924	2.61661
2.360	.06174	1.	.24023	1.	.51619	2.65834
2.370	.06030	1.	.23561	1.	.51315	2.70108
2.380	.05888	1.	.23106	1.	.51014	2.74485
2.390	.05750	1.	.22656	1.	.50715	2.78969
2.400	.05613	1.	.22212	1.	.50418	2.83561
2.410	.05480	1.	.21775	1.	.50124	2.88265
2.420	.05349	1.	.21343	1.	.49831	2.93084
2.430	.05221	1.	.20918	1.	.49540	2.98020
2.440	.05096	1.	.20499	1.	.49252	3.03078
2.450	.04972	1.	.20086	1.	.48965	3.08260
2.460	.04852	1.	.19679	1.	.48680	3.13570
2.470	.04734	1.	.19278	1.	.48398	3.19012
2.480	.04618	1.	.18882	1.	.48117	3.24588
2.490	.04505	1.	.18493	1.	.47839	3.30303

TABLE VII. Isentropic Exponent = 1.00

M	p/p_t	T/T_t	GAMMA	ISOTHERMAL T_t/T_t^*	RAYLEIGH T_t/T_t^*	SHOCK p_{t1}/p_{t2}
2.500	.04394	1.	.18110	1.	.47562	3.36160
2.510	.04285	1.	.17733	1.	.47288	3.42164
2.520	.04179	1.	.17361	1.	.47015	3.48318
2.530	.04074	1.	.16995	1.	.46745	3.54627
2.540	.03972	1.	.16635	1.	.46476	3.61095
2.550	.03873	1.	.16281	1.	.46209	3.67726
2.560	.03775	1.	.15933	1.	.45944	3.74526
2.570	.03679	1.	.15590	1.	.45681	3.81498
2.580	.03586	1.	.15253	1.	.45420	3.88648
2.590	.03494	1.	.14921	1.	.45161	3.95981
2.600	.03405	1.	.14595	1.	.44904	4.03502
2.610	.03317	1.	.14274	1.	.44648	4.11216
2.620	.03232	1.	.13959	1.	.44395	4.19130
2.630	.03148	1.	.13650	1.	.44143	4.27248
2.640	.03066	1.	.13345	1.	.43893	4.35577
2.650	.02986	1.	.13046	1.	.43645	4.44122
2.660	.02908	1.	.12752	1.	.43398	4.52891
2.670	.02831	1.	.12463	1.	.43154	4.61889
2.680	.02757	1.	.12180	1.	.42911	4.71124
2.690	.02683	1.	.11901	1.	.42670	4.80602
2.700	.02612	1.	.11628	1.	.42431	4.90330
2.710	.02542	1.	.11360	1.	.42193	5.00316
2.720	.02474	1.	.11096	1.	.41957	5.10568
2.730	.02408	1.	.10838	1.	.41723	5.21093
2.740	.02343	1.	.10584	1.	.41490	5.31899
2.750	.02279	1.	.10335	1.	.41260	5.42996
2.760	.02217	1.	.10091	1.	.41030	5.54391
2.770	.02157	1.	.09851	1.	.40803	5.66094
2.780	.02098	1.	.09616	1.	.40577	5.78114
2.790	.02040	1.	09385	1.	.40353	5.90460
2.800	.01984	1.	.09159	1.	.40130	6.03143
2.810	.01929	1.	.08938	1.	.39909	6.16173
2.820	.01876	1.	.08721	1.	.39690	6.29560
2.830	.01823	1.	.08508	1.	.39472	6.43315
2.840	.01772	1.	.08299	1.	.39256	6.57450
2.850	.01723	1.	.08095	1.	.39041	6.71976
2.860	.01674	1.	.07895	1.	.38828	6.86906
2.870	.01627	1.	.07699	1.	.38616	7.02252
2.880	.01581	1.	.07506	1.	.38406	7.18026
2.890	.01536	1.	.07318	1.	.38198	7.34243
2.900	.01492	1.	.07134	1.	.37991	7.50916
2.910	.01449	1.	.06954	1.	.37785	7.68059
2.920	.01408	1.	.06777	1.	.37581	7.85688
2.930	.01367	1.	.06604	1.	.37378	8.03817
2.940	.01328	1.	.06435	1.	.37177	8.22462
2.950	.01289	1.	.06270	1.	.36977	8.41641
2.960	.01252	1.	.06108	1.	.36779	8.61368
2.970	.01215	1.	.05949	1.	.36582	8.81664
2.980	.01179	1.	.05794	1.	.36387	9.02544
2.990	.01145	1.	.05643	1.	.36193	9.24029

TABLE VII. Isentropic Exponent = 1.00

M	p/p_t	T/T_t	GAMMA	ISOTHERMAL T_t/T_t^*	RAYLEIGH T_t/T_t^*	SHOCK p_{t1}/p_{t2}
3.000	.01111	1.	.05495	1.	.36000	9.46140
3.010	.01078	1.	.05350	1.	.35809	9.68892
3.020	.01046	1.	.05208	1.	.35619	9.92309
3.030	.01015	1.	.05070	1.	.35430	10.16412
3.040	.00984	1.	.04934	1.	.35243	10.41224
3.050	.00955	1.	.04802	1.	.35057	10.66767
3.060	.00926	1.	.04673	1.	.34872	10.93066
3.070	.00898	1.	.04547	1.	.34689	11.20146
3.080	.00871	1.	.04423	1.	.34507	11.48031
3.090	.00845	1.	.04303	1.	.34326	11.76750
3.100	.00819	1.	.04185	1.	.34147	12.06328
3.110	.00794	1.	.04070	1.	.33969	12.36796
3.120	.00769	1.	.03958	1.	.33792	12.68182
3.130	.00746	1.	.03849	1.	.33616	13.00517
3.140	.00723	1.	.03742	1.	.33442	13.33833
3.150	.00700	1.	.03638	1.	.33269	13.68163
3.160	.00679	1.	.03536	1.	.33097	14.03541
3.170	.00658	1.	.03436	1.	.32926	14.40002
3.180	.00637	1.	.03340	1.	.32757	14.77583
3.190	.00617	1.	.03245	1.	.32588	15.16322
3.200	.00598	1.	.03153	1.	.32421	15.56259
3.210	.00579	1.	.03063	1.	.32255	15.97433
3.220	.00560	1.	.02975	1.	.32090	16.39888
3.230	.00543	1.	.02890	1.	.31927	16.83668
3.240	.00525	1.	.02807	1.	.31764	17.28818
3.250	.00509	1.	.02725	1.	.31603	17.75385
3.260	.00492	1.	.02646	1.	.31442	18.23418
3.270	.00477	1.	.02569	1.	.31283	18.72968
3.280	.00461	1.	.02494	1.	.31125	19.24088
3.290	.00446	1.	.02421	1.	.30968	19.76833
3.300	.00432	1.	.02349	1.	.30812	20.31258
3.310	.00418	1.	.02280	1.	.30658	20.87425
3.320	.00404	1.	.02212	1.	.30504	21.45392
3.330	.00391	1.	.02146	1.	.30351	22.05224
3.340	.00378	1.	.02082	1.	.30200	22.66988
3.350	.00366	1.	.02020	1.	.30049	23.30750
3.360	.00354	1.	.01959	1.	.29899	23.96582
3.370	.00342	1.	.01900	1.	.29751	24.64558
3.380	.00331	1.	.01842	1.	.29603	25.34754
3.390	.00320	1.	.01786	1.	.29457	26.07250
3.400	.00309	1.	.01731	1.	.29312	26.82128
3.410	.00299	1.	.01678	1.	.29167	27.59474
3.420	.00289	1.	.01627	1.	.29024	28.39376
3.430	.00279	1.	.01577	1.	.28881	29.21929
3.440	.00269	1.	.01528	1.	.28740	30.07226
3.450	.00260	1.	.01480	1.	.28599	30.95369
3.460	.00251	1.	.01434	1.	.28459	31.86461
3.470	.00243	1.	.01389	1.	.28321	32.80610
3.480	.00235	1.	.01346	1.	.28183	33.77927
3.490	.00227	1.	.01303	1.	.28046	34.78529

TABLE VII.   Isentropic Exponent = 1.00

M	p/p_t	T/T_t	GAMMA	ISOTHERMAL T_t/T_t^*	RAYLEIGH T_t/T_t^*	SHOCK p_{tf}/p_{t2}
3.500	.00219	1.	.01262	1.	.27910	35.82537
3.510	.00211	1.	.01222	1.	.27775	36.90077
3.520	.00204	1.	.01183	1.	.27641	38.01279
3.530	.00197	1.	.01146	1.	.27508	39.16280
3.540	.00190	1.	.01109	1.	.27376	40.35220
3.550	.00183	1.	.01073	1.	.27245	41.58247
3.560	.00177	1.	.01039	1.	.27114	42.85512
3.570	.00171	1.	.01005	1.	.26984	44.17177
3.580	.00165	1.	.00973	1.	.26856	45.53404
3.590	.00159	1.	.00941	1.	.26728	46.94367
3.600	.00153	1.	.00910	1.	.26601	48.40245
3.610	.00148	1.	.00881	1.	.26475	49.91222
3.620	.00143	1.	.00852	1.	.26349	51.47494
3.630	.00138	1.	.00824	1.	.26225	53.09261
3.640	.00133	1.	.00796	1.	.26101	54.76733
3.650	.00128	1.	.00770	1.	.25978	56.50128
3.660	.00123	1.	.00744	1.	.25856	58.29673
3.670	.00119	1.	.00720	1.	.25735	60.15606
3.680	.00115	1.	.00696	1.	.25614	62.08171
3.690	.00110	1.	.00672	1.	.25495	64.07625
3.700	.00106	1.	.00650	1.	.25376	66.14235
3.710	.00103	1.	.00628	1.	.25258	68.28279
3.720	.00099	1.	.00606	1.	.25140	70.50045
3.730	.00095	1.	.00586	1.	.25024	72.79836
3.740	.00092	1.	.00566	1.	.24908	75.17965
3.750	.00088	1.	.00546	1.	.24793	77.64759
3.760	.00085	1.	.00528	1.	.24679	80.20558
3.770	.00082	1.	.00510	1.	.24565	82.85717
3.780	.00079	1.	.00492	1.	.24452	85.60607
3.790	.00076	1.	.00475	1.	.24340	88.45612
3.800	.00073	1.	.00458	1.	.24229	91.41132
3.810	.00070	1.	.00443	1.	.24118	94.47588
3.820	.00068	1.	.00427	1.	.24008	97.65415
3.830	.00065	1.	.00412	1.	.23899	100.95068
3.840	.00063	1.	.00398	1.	.23791	104.37020
3.850	.00060	1.	.00384	1.	.23683	107.91766
3.860	.00058	1.	.00370	1.	.23576	111.59821
3.870	.00056	1.	.00357	1.	.23469	115.41721
3.880	.00054	1.	.00344	1.	.23363	119.38028
3.890	.00052	1.	.00332	1.	.23258	123.49325
3.900	.00050	1.	.00320	1.	.23154	127.76222
3.910	.00048	1.	.00309	1.	.23050	132.19355
3.920	.00046	1.	.00298	1.	.22947	136.79387
3.930	.00044	1.	.00287	1.	.22845	141.57010
3.940	.00043	1.	.00277	1.	.22743	146.52945
3.950	.00041	1.	.00267	1.	.22642	151.67947
3.960	.00039	1.	.00257	1.	.22541	157.02803
3.970	.00038	1.	.00247	1.	.22441	162.58331
3.980	.00036	1.	.00238	1.	.22342	168.35389
3.990	.00035	1.	.00230	1.	.22243	174.34872

TABLE VII. Isentropic Exponent = 1.00

M	p/p_t	T/T_t	GAMMA	ISOTHERMAL T_t/T_t^*	RAYLEIGH T_t/T_t^*	SHOCK p_{t1}/p_{t2}
4.000	.00034	1.	.00221	1.	.22145	180.57772
4.010	.00032	1.	.00213	1.	.22048	187.04948
4.020	.00031	1.	.00205	1.	.21951	193.77472
4.030	.00030	1.	.00198	1.	.21855	200.76411
4.040	.00029	1.	.00190	1.	.21759	208.02872
4.050	.00027	1.	.00183	1.	.21664	215.58018
4.060	.00026	1.	.00176	1.	.21570	223.43055
4.070	.00025	1.	.00170	1.	.21476	231.59252
4.080	.00024	1.	.00163	1.	.21383	240.07929
4.090	.00023	1.	.00157	1.	.21290	248.90465
4.100	.00022	1.	.00151	1.	.21198	258.08305
4.110	.00021	1.	.00146	1.	.21107	267.62959
4.120	.00021	1.	.00140	1.	.21016	277.55999
4.130	.00020	1.	.00135	1.	.21016	287.89073
4.140	.00019	1.	.00130	1.	.20836	298.63907
4.150	.00018	1.	.00125	1.	.20746	309.82294
4.160	.00017	1.	.00120	1.	.20658	321.46119
4.170	.00017	1.	.00115	1.	.20569	333.57354
4.180	.00016	1.	.00111	1.	.20482	346.18050
4.190	.00015	1.	.00106	1.	.20395	359.30362
4.200	.00015	1.	.00102	1.	.20308	372.96535
4.210	.00014	1.	.00098	1.	.20222	387.18933
4.220	.00014	1.	.00094	1.	.20136	402.00008
4.230	.00013	1.	.00091	1.	.20051	417.42345
4.240	.00012	1.	.00087	1.	.19967	433.48629
4.250	.00012	1.	.00084	1.	.19883	450.21690
4.260	.00011	1.	.00081	1.	.19799	467.64473
4.270	.00011	1.	.00077	1.	.19716	485.80069
4.280	.00011	1.	.00074	1.	.19634	504.71706
4.290	.00010	1.	.00071	1.	.19552	524.42774
4.300	.00010	1.	.00068	1.	.19470	544.96805
4.310	.00009	1.	.00066	1.	.19389	566.37518
4.320	.00009	1.	.00063	1.	.19309	588.68787
4.330	.00008	1.	.00061	1.	.19229	611.94681
4.340	.00008	1.	.00058	1.	.19149	636.19463
4.350	.00008	1.	.00056	1.	.19070	661.47585
4.360	.00007	1.	.00054	1.	.18991	687.83708
4.370	.00007	1.	.00051	1.	.18913	715.32741
4.380	.00007	1.	.00049	1.	.18835	743.99793
4.390	.00007	1.	.00047	1.	.18758	773.90243
4.400	.00006	1.	.00045	1.	.18681	805.09717
4.410	.00006	1.	.00043	1.	.18605	837.64101
4.420	.00006	1.	.00042	1.	.18529	871.59577
4.430	.00005	1.	.00040	1.	.18454	907.02622
4.440	.00005	1.	.00038	1.	.18379	944.00016
4.450	.00005	1.	.00037	1.	.18304	982.58877
4.460	.00005	1.	.00035	1.	.18230	1022.86680
4.470	.00005	1.	.00034	1.	.18156	1064.91193
4.480	.00004	1.	.00032	1.	.18083	1108.80669
4.490	.00004	1.	.00031	1.	.18010	1154.63695

TABLE VIII. Isentropic Exponent = 1.10

M	p/p_t	T/T_t	GAMMA	ISOTHERMAL T_t/T_t^*	RAYLEIGH T_t/T_t^*	SHOCK p_{t1}/p_{t2}
.000	1.00000	1.00000	.00000	.95652	.00000	1.
.005	.99999	.00000	.00830	.95652	.00010	1.
.010	.99995	.00000	.01668	.95653	.00042	1.
.015	.99988	.99999	.02500	.95653	.00094	1.
.020	.99978	.99998	.03336	.95654	.00168	1.
.025	.99966	.99997	.04171	.95655	.00262	1.
.030	.99951	.99996	.05004	.95656	.00377	1.
.035	.99933	.99994	.05837	.95658	.00513	1.
.040	.99912	.99992	.06670	.95660	.00670	1.
.045	.99889	.99990	.07502	.95662	.00847	1.
.050	.99863	.99988	.08334	.95664	.01044	1.
.055	.99834	.99985	.09165	.95667	.01262	1.
.060	.99802	.99982	.09995	.95669	.01500	1.
.065	.99768	.99979	.10825	.95672	.01758	1.
.070	.99731	.99976	.11653	.95676	.02036	1.
.075	.99691	.99972	.12481	.95679	.02334	1.
.080	.99649	.99968	.13308	.95683	.02651	1.
.085	.99603	.99964	.14134	.95687	.02988	1.
.090	.99556	.99960	.14958	.95691	.03343	1.
.095	.99505	.99955	.15781	.95695	.03718	1.
.100	.99452	.99950	.16604	.95700	.04111	1.
.105	.99396	.99945	.17424	.95705	.04523	1.
.110	.99337	.99940	.18244	.95710	.04952	1.
.115	.99276	.99934	.19062	.95715	.05400	1.
.120	.99211	.99928	.19879	.95721	.05865	1.
.125	.99145	.99922	.20693	.95727	.06347	1.
.130	.99075	.99916	.21507	.95733	.06847	1.
.135	.99003	.99909	.22318	.95739	.07363	1.
.140	.98928	.99902	.23128	.95746	.07896	1.
.145	.98851	.99895	.23937	.95753	.08445	1.
.150	.98771	.99888	.24743	.95760	.09009	1.
.155	.98688	.99880	.25547	.95767	.09589	1.
.160	.98603	.99872	.26350	.95775	.10184	1.
.165	.98515	.99864	.27150	.95782	.10794	1.
.170	.98424	.99856	.27948	.95790	.11418	1.
.175	.98331	.99847	.28744	.95799	.12056	1.
.180	.98235	.99838	.29538	.95807	.12708	1.
.185	.98137	.99829	.30329	.95816	.13373	1.
.190	.98036	.99820	.31118	.95825	.14051	1.
.195	.97932	.99810	.31905	.95834	.14742	1.
.200	.97826	.99800	.32689	.95843	.15444	1.
.205	.97718	.99790	.33471	.95853	.16159	1.
.210	.97606	.99780	.34250	.95863	.16885	1.
.215	.97493	.99769	.35027	.95873	.17622	1.
.220	.97376	.99759	.35800	.95884	.18369	1.
.225	.97257	.99748	.36571	.95894	.19127	1.
.230	.97136	.99736	.37339	.95905	.19894	1.
.235	.97012	.99725	.38105	.95916	.20671	1.
.240	.96886	.99713	.38867	.95928	.21457	1.
.245	.96757	.99701	.39627	.95939	.22251	1.

TABLE VIII. Isentropic Exponent = 1.10

M	p/p_t	T/T_t	GAMMA	ISOTHERMAL T_t/T_t^*	RAYLEIGH T_t/T_t^*	SHOCK p_{t1}/p_{t2}
.250	.96626	.99688	.40383	.95951	.23053	1.
.255	.96492	.99676	.41136	.95963	.23863	1.
.260	.96356	.99663	.41886	.95975	.24681	1.
.265	.96218	.99650	.42633	.95988	.25505	1.
.270	.96077	.99637	.43377	.96001	.26336	1.
.275	.95933	.99623	.44117	.96014	.27174	1.
.280	.95788	.99610	.44854	.96027	.28016	1.
.285	.95640	.99596	.45588	.96041	.28865	1.
.290	.95489	.99581	.46318	.96054	.29718	1.
.295	.95336	.99567	.47045	.96068	.30575	1.
.300	.95181	.99552	.47768	.96083	.31437	1.
.305	.95024	.99537	.48487	.96097	.32303	1.
.310	.94864	.99522	.49203	.96112	.33172	1.
.315	.94702	.99506	.49915	.96127	.34044	1.
.320	.94537	.99491	.50623	.96142	.34919	1.
.325	.94371	.99475	.51328	.96157	.35796	1.
.330	.94202	.99458	.52028	.96173	.36674	1.
.335	.94030	.99442	.52725	.96189	.37555	1.
.340	.93857	.99425	.53418	.96205	.38436	1.
.345	.93681	.99408	.54106	.96221	.39318	1.
.350	.93504	.99391	.54791	.96238	.40201	1.
.355	.93324	.99374	.55472	.96255	.41084	1.
.360	.93142	.99356	.56148	.96272	.41966	1.
.365	.92957	.99338	.56820	.96289	.42848	1.
.370	.92771	.99320	.57489	.96307	.43729	1.
.375	.92582	.99302	.58152	.96325	.44609	1.
.380	.92392	.99283	.58812	.96343	.45488	1.
.385	.92199	.99264	.59467	.96361	.46364	1.
.390	.92004	.99245	.60118	.96380	.47238	1.
.395	.91807	.99226	.60764	.96398	.48110	1.
.400	.91608	.99206	.61406	.96417	.48980	1.
.405	.91407	.99187	.62044	.96437	.49846	1.
.410	.91204	.99167	.62676	.96456	.50709	1.
.415	.90999	.99146	.63305	.96476	.51568	1.
.420	.90792	.99126	.63928	.96496	.52423	1.
.425	.90584	.99105	.64547	.96516	.53274	1.
.430	.90373	.99084	.65162	.96536	.54122	1.
.435	.90160	.99063	.65771	.96557	.54964	1.
.440	.89945	.99041	.66376	.96578	.55801	1.
.445	.89729	.99020	.66976	.96599	.56634	1.
.450	.89510	.98998	.67571	.96621	.57461	1.
.455	.89290	.98975	.68161	.96642	.58283	1.
.460	.89068	.98953	.68747	.96664	.59099	1.
.465	.88844	.98930	.69327	.96686	.59909	1.
.470	.88619	.98908	.69903	.96709	.60713	1.
.475	.88391	.98884	.70473	.96731	.61510	1.
.480	.88162	.98861	.71039	.96754	.62302	1.
.485	.87931	.98838	.71599	.96777	.63086	1.
.490	.87698	.98814	.72155	.96800	.63864	1.
.495	.87464	.98790	.72705	.96824	.64634	1.

TABLE VIII. Isentropic Exponent = 1.10

M	p/p$_t$	T/T$_t$	GAMMA	ISOTHERMAL T$_t$/T$_t^*$	RAYLEIGH T$_t$/T$_t^*$	SHOCK p$_{t1}$/p$_{t2}$
.500	.87228	.98765	.73250	.96848	.65398	1.
.505	.86990	.98741	.73790	.96872	.66154	1.
.510	.86751	.98716	.74325	.96896	.66903	1.
.515	.86510	.98691	.74855	.96921	.67644	1.
.520	.86267	.98666	.75379	.96945	.68378	1.
.525	.86023	.98641	.75898	.96970	.69103	1.
.530	.85777	.98615	.76412	.96996	.69821	1.
.535	.85530	.98589	.76921	.97021	.70531	1.
.540	.85281	.98563	.77424	.97047	.71232	1.
.545	.85030	.98537	.77922	.97073	.71925	1.
.550	.84778	.98510	.78414	.97099	.72610	1.
.555	.84525	.98483	.78901	.97125	.73286	1.
.560	.84270	.98456	.79383	.97152	.73954	1.
.565	.84014	.98429	.79859	.97179	.74614	1.
.570	.83756	.98401	.80330	.97206	.75264	1.
.575	.83497	.98374	.80796	.97233	.75906	1.
.580	.83237	.98346	.81255	.97261	.76539	1.
.585	.82975	.98318	.81710	.97289	.77163	1.
.590	.82712	.98289	.82159	.97317	.77778	1.
.595	.82447	.98261	.82602	.97345	.78385	1.
.600	.82182	.98232	.83040	.97374	.78982	1.
.605	.81915	.98203	.83472	.97403	.79570	1.
.610	.81646	.98173	.83899	.97432	.80150	1.
.615	.81377	.98144	.84320	.97461	.80720	1.
.620	.81106	.98114	.84736	.97491	.81281	1.
.625	.80834	.98084	.85146	.97520	.81833	1.
.630	.80561	.98054	.85550	.97550	.82376	1.
.635	.80287	.98024	.85949	.97581	.82909	1.
.640	.80011	.97993	.86342	.97611	.83434	1.
.645	.79735	.97962	.86729	.97642	.83949	1.
.650	.79457	.97931	.87111	.97673	.84455	1.
.655	.79178	.97900	.87487	.97704	.84953	1.
.660	.78898	.97868	.87857	.97735	.85441	1.
.665	.78618	.97837	.88222	.97767	.85919	1.
.670	.78336	.97805	.88581	.97799	.86389	1.
.675	.78053	.97773	.88935	.97831	.86850	1.
.680	.77769	.97740	.89283	.97864	.87302	1.
.685	.77484	.97708	.89625	.97896	.87744	1.
.690	.77199	.97675	.89961	.97929	.88178	1.
.695	.76912	.97642	.90292	.97962	.88603	1.
.700	.76625	.97609	.90617	.97996	.89018	1.
.705	.76336	.97575	.90936	.98029	.89425	1.
.710	.76047	.97541	.91250	.98063	.89823	1.
.715	.75757	.97508	.91558	.98097	.90213	1.
.720	.75466	.97473	.91860	.98131	.90593	1.
.725	.75174	.97439	.92157	.98166	.90965	1.
.730	.74882	.97405	.92448	.98201	.91328	1.
.735	.74589	.97370	.92733	.98236	.91683	1.
.740	.74295	.97335	.93013	.98271	.92029	1.
.745	.74000	.97300	.93287	.98307	.92366	1.

TABLE VIII. Isentropic Exponent = 1.10

M	p/p_t	T/T_t	GAMMA	ISOTHERMAL T_t/T_t^*	RAYLEIGH T_t/T_t	SHOCK p_{t1}/p_{t2}
.750	.73705	.97264	.93555	.98342	.92695	1.
.755	.73409	.97229	.93818	.98378	.93016	1.
.760	.73112	.97193	.94074	.98415	.93329	1.
.765	.72815	.97157	.94326	.98451	.93633	1.
.770	.72517	.97121	.94571	.98488	.93929	1.
.775	.72218	.97084	.94811	.98525	.94217	1.
.780	.71919	.97048	.95046	.98562	.94496	1.
.785	.71619	.97011	.95274	.98599	.94768	1.
.790	.71319	.96974	.95498	.98637	.95032	1.
.795	.71018	.96937	.95715	.98675	.95288	1.
.800	.70717	.96899	.95927	.98713	.95537	1.
.805	.70415	.96862	.96133	.98751	.95777	1.
.810	.70113	.96824	.96334	.98790	.96010	1.
.815	.69810	.96786	.96529	.98829	.96236	1.
.820	.69507	.96747	.96719	.98868	.96454	1.
.825	.69204	.96709	.96903	.98907	.96664	1.
.830	.68900	.96670	.97082	.98947	.96868	1.
.835	.68596	.96631	.97255	.98987	.97064	1.
.840	.68291	.96592	.97423	.99027	.97253	1.
.845	.67986	.96553	.97585	.99067	.97434	1.
.850	.67681	.96513	.97741	.99108	.97609	1.
.855	.67375	.96474	.97893	.99148	.97777	1.
.860	.67070	.96434	.98039	.99189	.97938	1.
.865	.66764	.96394	.98179	.99231	.98093	1.
.870	.65457	.96354	.98314	.99272	.98240	1.
.875	.66151	.96313	.98444	.99314	.98381	1.
.880	.65844	.96272	.98568	.99356	.98516	1.
.885	.65537	.96231	.98687	.99398	.98644	1.
.890	.65230	.96190	.98801	.99440	.98766	1.
.895	.64923	.96149	.98909	.99483	.98881	1.
.900	.64616	.96108	.99012	.99526	.98990	1.
.905	.64308	.96066	.99110	.99569	.99094	1.
.910	.64001	.96024	.99202	.99613	.99191	1.
.915	.63693	.95982	.99290	.99656	.99282	1.
.920	.63385	.95940	.99372	.99700	.99367	1.
.925	.63078	.95897	.99449	.99744	.99447	1.
.930	.62770	.95855	.99521	.99789	.99521	1.
.935	.62462	.95812	.99588	.99833	.99589	1.
.940	.62154	.95769	.99649	.99878	.99652	1.
.945	.61847	.95726	.99706	.99923	.99709	1.
.950	.61539	.95682	.99757	.99968	.99761	1.
.955	.61231	.95639	.99804	1.00014	.99807	1.
.960	.60924	.95595	.99845	1.00060	.99848	1.
.965	.60616	.95551	.99882	1.00106	.99885	1.
.970	.60309	.95507	.99913	1.00152	.99916	1.
.975	.60002	.95463	.99940	1.00199	.99942	1.
.980	.59695	.95418	.99962	1.00245	.99963	1.
.985	.59388	.95373	.99978	1.00292	.99979	1.
.990	.59081	.95328	.99990	1.00340	.99991	1.
.995	.58774	.95283	.99998	1.00387	.99998	1.

TABLE VIII. Isentropic Exponent = 1.10

M	p/p_t	T/T_t	GAMMA	ISOTHERMAL T_t/T_t^*	RAYLEIGH T_t/T_t^*	SHOCK p_{t1}/p_{t2}
1.000	.58468	.95238	1.00000	1.00435	1.00000	1.00000
1.005	.58162	.95193	.99998	1.00483	.99998	1.00000
1.010	.57856	.95147	.99991	1.00531	.99991	1.00000
1.015	.57550	.95101	.99979	1.00579	.99980	1.00000
1.020	.57245	.95055	.99962	1.00628	.99965	1.00001
1.025	.56940	.95009	.99941	1.00677	.99945	1.00002
1.030	.56635	.94963	.99915	1.00726	.99921	1.00003
1.035	.56330	.94916	.99885	1.00775	.99893	1.00005
1.040	.56026	.94869	.99850	1.00825	.99861	1.00008
1.045	.55722	.94823	.99811	1.00875	.99825	1.00011
1.050	.55419	.94776	.99767	1.00925	.99785	1.00015
1.055	.55116	.94728	.99718	1.00975	.99742	1.00020
1.060	.54813	.94681	.99666	1.01026	.99694	1.00026
1.065	.54511	.94633	.99608	1.01077	.99643	1.00033
1.070	.54209	.94585	.99547	1.01128	.99589	1.00041
1.075	.53907	.94538	.99481	1.01179	.99530	1.00050
1.080	.53606	.94489	.99411	1.01231	.99469	1.00060
1.085	.53305	.94441	.99336	1.01282	.99404	1.00071
1.090	.53005	.94393	.99257	1.01334	.99335	1.00084
1.095	.52705	.94344	.99174	1.01387	.99263	1.00098
1.100	.52406	.94295	.99087	1.01439	.99188	1.00114
1.105	.52107	.94246	.98996	1.01492	.99110	1.00131
1.110	.51809	.94197	.98901	1.01545	.99029	1.00149
1.115	.51512	.94148	.98801	1.01598	.98945	1.00169
1.120	.51214	.94098	.98698	1.01651	.98857	1.00191
1.125	.50918	.94048	.98590	1.01705	.98767	1.00215
1.130	.50622	.93999	.98479	1.01759	.98674	1.00240
1.135	.50326	.93949	.98364	1.01813	.98578	1.00266
1.140	.50032	.93898	.98244	1.01868	.98479	1.00295
1.145	.49737	.93848	.98121	1.01922	.98378	1.00326
1.150	.49444	.93798	.97994	1.01977	.98274	1.00358
1.155	.49151	.93747	.97863	1.02032	.98167	1.00392
1.160	.48858	.93696	.97729	1.02088	.98058	1.00429
1.165	.48567	.93645	.97591	1.02143	.97947	1.00467
1.170	.48275	.93594	.97449	1.02199	.97833	1.00507
1.175	.47985	.93543	.97303	1.02255	.97716	1.00550
1.180	.47695	.93491	.97154	1.02311	.97598	1.00594
1.185	.47406	.93439	.97001	1.02368	.97477	1.00641
1.190	.47118	.93388	.96845	1.02425	.97353	1.00689
1.195	.46830	.93336	.96685	1.02482	.97228	1.00740
1.200	.46543	.93284	.96521	1.02539	.97101	1.00794
1.205	.46257	.93231	.96355	1.02597	.96971	1.00849
1.210	.45972	.93179	.96184	1.02654	.96839	1.00907
1.215	.45687	.93126	.96011	1.02712	.96706	1.00967
1.220	.45403	.93073	.95834	1.02771	.96570	1.01030
1.225	.45120	.93021	.95654	1.02829	.96433	1.01095
1.230	.44838	.92967	.95470	1.02888	.96294	1.01162
1.235	.44556	.92914	.95284	1.02947	.96153	1.01232
1.240	.44275	.92861	.95094	1.03006	.96010	1.01304
1.245	.43995	.92807	.94901	1.03065	.95866	1.01379

TABLE VIII.   Isentropic Exponent = 1.10

M	p/p_t	T/T_t	GAMMA	ISOTHERMAL T_t/T_t^*	RAYLEIGH T_t/T_t^*	SHOCK p_{t1}/p_{t2}
1.250	.43716	.92754	.94705	1.03125	.95719	1.01456
1.255	.43438	.92700	.94506	1.03185	.95572	1.01536
1.260	.43160	.92646	.94303	1.03245	.95422	1.01618
1.265	.42883	.92592	.94098	1.03305	.95271	1.01703
1.270	.42608	.92537	.93890	1.03366	.95119	1.01791
1.275	.42333	.92483	.93679	1.03427	.94965	1.01881
1.280	.42058	.92428	.93465	1.03488	.94810	1.01974
1.285	.41785	.92374	.93248	1.03549	.94653	1.02069
1.290	.41513	.92319	.93028	1.03611	.94495	1.02167
1.295	.41241	.92264	.92806	1.03673	.94336	1.02268
1.300	.40971	.92208	.92581	1.03735	.94175	1.02372
1.305	.40701	.92153	.92353	1.03797	.94014	1.02479
1.310	.40432	.92098	.92122	1.03860	.93851	1.02588
1.315	.40165	.92042	.91889	1.03922	.93686	1.02700
1.320	.39898	.91986	.91653	1.03985	.93521	1.02815
1.325	.39632	.91930	.91415	1.04049	.93355	1.02933
1.330	.39367	.91874	.91174	1.04112	.93187	1.03053
1.335	.39103	.91818	.90931	1.04176	.93019	1.03177
1.340	.38839	.91762	.90685	1.04240	.92849	1.03303
1.345	.38577	.91705	.90437	1.04304	.92678	1.03433
1.350	.38316	.91649	.90186	1.04368	.92507	1.03565
1.355	.38056	.91592	.89933	1.04433	.92335	1.03700
1.360	.37797	.91535	.89678	1.04498	.92161	1.03839
1.365	.37538	.91478	.89420	1.04563	.91987	1.03980
1.370	.37281	.91421	.89161	1.04629	.91812	1.04124
1.375	.37025	.91363	.88899	1.04694	.91637	1.04272
1.380	.36769	.91306	.88634	1.04760	.91460	1.04422
1.385	.36515	.91248	.88368	1.04826	.91283	1.04576
1.390	.36262	.91191	.88100	1.04893	.91105	1.04732
1.395	.36009	.91133	.87829	1.04959	.90927	1.04892
1.400	.35758	.91075	.87557	1.05026	.90747	1.05055
1.405	.35508	.91017	.87283	1.05093	.90567	1.05221
1.410	.35259	.90958	.87006	1.05160	.90387	1.05390
1.415	.35010	.90900	.86728	1.05228	.90206	1.05563
1.420	.34763	.90841	.86448	1.05296	.90024	1.05739
1.425	.34517	.90783	.86166	1.05364	.89842	1.05917
1.430	.34272	.90724	.85882	1.05432	.89659	1.06100
1.435	.34028	.90665	.85597	1.05501	.89476	1.06285
1.440	.33785	.90606	.85309	1.05569	.89293	1.06474
1.445	.33543	.90547	.85020	1.05638	.89109	1.06666
1.450	.33302	.90488	.84730	1.05708	.88924	1.06861
1.455	.33063	.90428	.84437	1.05777	.88739	1.07060
1.460	.32824	.90369	.84144	1.05847	.88554	1.07262
1.465	.32586	.90309	.83848	1.05917	.88368	1.07467
1.470	.32350	.90249	.83551	1.05987	.88182	1.07676
1.475	.32114	.90189	.83253	1.06057	.87996	1.07888
1.480	.31880	.90129	.82953	1.06128	.87810	1.08104
1.485	.31646	.90069	.82651	1.06199	.87623	1.08323
1.490	.31414	.90009	.82349	1.06270	.87436	1.08546
1.495	.31183	.89948	.82044	1.06341	.87248	1.08772

TABLE VIII. Isentropic Exponent = 1.10

M	p/p_t	T/T_t	GAMMA	ISOTHERMAL T_t/T_t^*	RAYLEIGH T_t/T_t^*	SHOCK p_{t1}/p_{t2}
1.500	.30953	.89888	.81739	1.06413	.87061	1.09002
1.505	.30724	.89827	.81432	1.06485	.86873	1.09235
1.510	.30496	.89766	.81124	1.06557	.86685	1.09472
1.515	.30269	.89705	.80815	1.06629	.86497	1.09713
1.520	.30043	.89644	.80504	1.06702	.86309	1.09957
1.525	.29819	.89583	.80193	1.06775	.86120	1.10204
1.530	.29595	.89522	.79880	1.06848	.85932	1.10455
1.535	.29373	.89461	.79566	1.06921	.85743	1.10710
1.540	.29152	.89399	.79251	1.06995	.85554	1.10969
1.545	.28931	.89337	.78935	1.07068	.85365	1.11231
1.550	.28712	.89276	.78618	1.07142	.85177	1.11497
1.555	.28494	.89214	.78300	1.07217	.84988	1.11767
1.560	.28278	.89152	.77981	1.07291	.84799	1.12041
1.565	.28062	.89090	.77661	1.07366	.84610	1.12318
1.570	.27847	.89028	.77340	1.07441	.84421	1.12600
1.575	.27634	.88965	.77019	1.07516	.84232	1.12885
1.580	.27421	.88903	.76696	1.07591	.84043	1.13174
1.585	.27210	.88841	.76373	1.07667	.83854	1.13466
1.590	.27000	.88778	.76049	1.07743	.83665	1.13763
1.595	.26791	.88715	.75724	1.07819	.83477	1.14064
1.600	.26583	.88652	.75398	1.07896	.83288	1.14369
1.605	.26376	.88590	.75072	1.07972	.83099	1.14677
1.610	.26170	.88527	.74745	1.08049	.82911	1.14990
1.615	.25966	.88463	.74418	1.08126	.82722	1.15307
1.620	.25762	.88400	.74090	1.08204	.82534	1.15627
1.625	.25560	.88337	.73761	1.08281	.82346	1.15952
1.630	.25359	.88273	.73432	1.08359	.82158	1.16281
1.635	.25158	.88210	.73102	1.08437	.81970	1.16614
1.640	.24959	.88146	.72772	1.08515	.81782	1.16951
1.645	.24762	.88082	.72441	1.08594	.81595	1.17293
1.650	.24565	.88018	.72110	1.08673	.81407	1.17638
1.655	.24369	.87955	.71779	1.08752	.81220	1.17988
1.660	.24175	.87890	.71447	1.08831	.81033	1.18342
1.665	.23981	.87826	.71115	1.08911	.80847	1.18701
1.670	.23789	.87762	.70782	1.08990	.80660	1.19064
1.675	.23598	.87698	.70449	1.09070	.80474	1.19431
1.680	.23408	.87633	.70116	1.09151	.80288	1.19802
1.685	.23219	.87569	.69783	1.09231	.80102	1.20178
1.690	.23031	.87504	.69449	1.09312	.79916	1.20559
1.695	.22844	.87439	.69115	1.09393	.79731	1.20943
1.700	.22658	.87374	.68781	1.09474	.79546	1.21333
1.705	.22474	.87309	.68447	1.09555	.79361	1.21726
1.710	.22290	.87244	.68113	1.09637	.79177	1.22125
1.715	.22108	.87179	.67779	1.09719	.78993	1.22528
1.720	.21927	.87114	.67444	1.09801	.78809	1.22935
1.725	.21747	.87049	.67110	1.09883	.78625	1.23348
1.730	.21567	.86983	.66775	1.09966	.78442	1.23765
1.735	.21389	.86918	.66441	1.10049	.78259	1.24186
1.740	.21213	.86852	.66106	1.10132	.78076	1.24612
1.745	.21037	.86787	.65772	1.10215	.77894	1.25044

TABLE VIII.   Isentropic Exponent=1.10

M	p/p_t	T/T_t	GAMMA	ISOTHERMAL T_t/T_t^*	RAYLEIGH T_t/T_t^*	SHOCK p_{t1}/p_{t2}
1.750	.20862	.86721	.65437	1.10299	.77712	1.25480
1.755	.20689	.86655	.65103	1.10383	.77530	1.25920
1.760	.20516	.86589	.64768	1.10467	.77349	1.26366
1.765	.20345	.86523	.64434	1.10551	.77168	1.26817
1.770	.20174	.86457	.64100	1.10636	.76987	1.27272
1.775	.20005	.86391	.63766	1.10720	.76807	1.27733
1.780	.19837	.86324	.63433	1.10805	.76627	1.28198
1.785	.19670	.86258	.63099	1.10891	.76448	1.28669
1.790	.19504	.86192	.62766	1.10976	.76269	1.29144
1.795	.19339	.86125	.62433	1.11062	.76090	1.29625
1.800	.19175	.86059	.62101	1.11148	.75912	1.30111
1.805	.19012	.85992	.61768	1.11234	.75734	1.30602
1.810	.18850	.85925	.61436	1.11320	.75556	1.31099
1.815	.18690	.85858	.61104	1.11407	.75379	1.31600
1.820	.18530	.85791	.60773	1.11494	.75202	1.32107
1.825	.18371	.85724	.60442	1.11581	.75026	1.32620
1.830	.18214	.85657	.60111	1.11669	.74850	1.33137
1.835	.18057	.85590	.59781	1.11756	.74675	1.33660
1.840	.17902	.85523	.59451	1.11844	.74500	1.33660
1.845	.17747	.85455	.59122	1.11932	.74325	1.34723
1.850	.17594	.85388	.58793	1.12021	.74151	1.35263
1.855	.17442	.85321	.58465	1.12109	.73977	1.35808
1.860	.17290	.85253	.58137	1.12198	.73804	1.36359
1.865	.17140	.85185	.57809	1.12287	.73631	1.36915
1.870	.16991	.85118	.57482	1.12376	.73458	1.37478
1.875	.16843	.85050	.57156	1.12466	.73286	1.38046
1.880	.16695	.84982	.56830	1.12556	.73115	1.38620
1.885	.16549	.84914	.56505	1.12646	.72943	1.39199
1.890	.16404	.84846	.56180	1.12736	.72773	1.39785
1.895	.16260	.84778	.55856	1.12827	.72603	1.40377
1.900	.16117	.84710	.55533	1.12917	.72433	1.40974
1.905	.15974	.84642	.55210	1.13008	.72263	1.41578
1.910	.15833	.84573	.54888	1.13100	.72095	1.42187
1.915	.15693	.84505	.54566	1.13191	.71926	1.42803
1.920	.15554	.84437	.54246	1.13283	.71758	1.43425
1.925	.15416	.84368	.53925	1.13375	.71591	1.44053
1.930	.15279	.84300	.53606	1.13467	.71424	1.44688
1.935	.15142	.84231	.53287	1.13559	.71257	1.45329
1.940	.15007	.84162	.52970	1.13652	.71091	1.45976
1.945	.14873	.84094	.52652	1.13745	.70926	1.46630
1.950	.14739	.84025	.52336	1.13838	.70760	1.47290
1.955	.14607	.83956	.52020	1.13931	.70596	1.47956
1.960	.14476	.83887	.51706	1.14025	.70432	1.48630
1.965	.14345	.83818	.51392	1.14119	.70268	1.49310
1.970	.14216	.83749	.51078	1.14213	.70105	1.49996
1.975	.14087	.83680	.50766	1.14307	.69942	1.50689
1.980	.13960	.83611	.50454	1.14402	.69780	1.51390
1.985	.13833	.83541	.50144	1.14497	.69618	1.52097
1.990	.13707	.83472	.49834	1.14592	.69457	1.52810
1.995	.13583	.83403	.49525	1.14687	.69296	1.53531

TABLE VIII. Isentropic Exponent = 1.10

M	p/p_t	T/T_t	GAMMA	ISOTHERMAL T_t/T_t^*	RAYLEIGH T_t/T_t^*	SHOCK p_{t1}/p_{t2}
2.000	.13459	.83333	.49217	1.14783	.69136	1.54259
2.010	.13214	.83194	.48604	1.14974	.68817	1.55736
2.020	.12973	.83055	.47994	1.15167	.68500	1.57242
2.030	.12735	.82916	.47388	1.15361	.68184	1.58777
2.040	.12501	.82776	.46786	1.15555	.67871	1.60342
2.050	.12271	.82636	.46187	1.15751	.67560	1.61937
2.060	.12044	.82496	.45593	1.15948	.67251	1.63563
2.070	.11820	.82356	.45003	1.16145	.66944	1.65221
2.080	.11600	.82215	.44417	1.16344	.66638	1.66910
2.090	.11384	.82075	.43835	1.16543	.66335	1.68631
2.100	.11171	.81934	.43257	1.16743	.66034	1.70386
2.110	.10961	.81793	.42684	1.16945	.65734	1.72174
2.120	.10755	.81651	.42115	1.17147	.65437	1.73996
2.130	.10551	.81510	.41550	1.17350	.65142	1.75852
2.140	.10352	.81368	.40990	1.17555	.64848	1.77744
2.150	.10155	.81227	.40435	1.17760	.64557	1.79672
2.160	.09961	.81085	.39884	1.17966	.64268	1.81637
2.170	.09771	.80942	.39337	1.18173	.63980	1.83639
2.180	.09584	.80800	.38795	1.18381	.63695	1.85678
2.190	.09400	.80658	.38258	1.18590	.63411	1.87756
2.200	.09219	.80515	.37725	1.18800	.63130	1.89874
2.210	.09040	.80373	.37198	1.19011	.62850	1.92031
2.220	.08865	.80230	.36675	1.19223	.62572	1.94229
2.230	.08693	.80087	.36156	1.19436	.62297	1.96468
2.240	.08524	.79944	.35643	1.19649	.62023	1.98749
2.250	.08357	.79801	.35134	1.19864	.61751	2.01073
2.260	.08194	.79657	.34630	1.20080	.61481	2.03441
2.270	.08033	.79514	.34131	1.20296	.61213	2.05854
2.280	.07875	.79370	.33637	1.20514	.60947	2.08311
2.290	.07719	.79226	.33148	1.20733	.60683	2.10815
2.300	.07566	.79083	.32664	1.20952	.60420	2.13366
2.310	.07416	.78939	.32185	1.21173	.60160	2.15964
2.320	.07269	.78795	.31710	1.21394	.59901	2.18611
2.330	.07124	.78651	.31241	1.21616	.59644	2.21308
2.340	.06982	.78506	.30776	1.21840	.59390	2.24055
2.350	.06842	.78362	.30316	1.22064	.59137	2.26854
2.360	.06705	.78218	.29862	1.22289	.58885	2.29706
2.370	.06570	.78073	.29412	1.22516	.58636	2.32610
2.380	.06437	.77929	.28967	1.22743	.58388	2.35569
2.390	.06307	.77784	.28527	1.22971	.58143	2.38584
2.400	.06179	.77640	.28092	1.23200	.57899	2.41655
2.410	.06054	.77495	.27662	1.23430	.57657	2.44783
2.420	.05930	.77350	.27237	1.23661	.57416	2.47970
2.430	.05809	.77205	.26816	1.23893	.57178	2.51217
2.440	.05691	.77061	.26401	1.24126	.56941	2.54525
2.450	.05574	.76916	.25990	1.24360	.56706	2.57895
2.460	.05459	.76771	.25584	1.24595	.56472	2.61328
2.470	.05347	.76626	.25184	1.24830	.56240	2.64825
2.480	.05237	.76481	.24787	1.25067	.56010	2.68388
2.490	.05129	.76336	.24396	1.25305	.55782	2.72019

TABLE VIII. Isentropic Exponent = 1.10

M	p/p_t	T/T_t	GAMMA	ISOTHERMAL T_t/T^*	RAYLEIGH T_t/T^*	SHOCK p_{t1}/p_{t2}
2.500	.05022	.76190	.24010	1.25543	.55556	2.75717
2.510	.04918	.76045	.23628	1.25783	.55331	2.79484
2.520	.04816	.75900	.23251	1.26024	.55107	2.83323
2.530	.04715	.75755	.22878	1.26265	.54886	2.87234
2.540	.04617	.75610	.22511	1.26508	.54666	2.91218
2.550	.04520	.75465	.22148	1.26751	.54448	2.95278
2.560	.04426	.75319	.21790	1.26995	.54231	2.99414
2.570	.04333	.75174	.21436	1.27241	.54016	3.03628
2.580	.04241	.75029	.21087	1.27487	.53802	3.07921
2.590	.04152	.74884	.20742	1.27734	.53590	3.12296
2.600	.04064	.74738	.20402	1.27983	.53380	3.16753
2.610	.03978	.74593	.20066	1.28232	.53171	3.21294
2.620	.03894	.74448	.19735	1.28482	.52964	3.25922
2.630	.03811	.74303	.19409	1.28733	.52758	3.30637
2.640	.03730	.74158	.19086	1.28985	.52554	3.35441
2.650	.03650	.74012	.18769	1.29238	.52352	3.40336
2.660	.03572	.73867	.18455	1.29492	.52151	3.45324
2.670	.03496	.73722	.18146	1.29747	.51951	3.50407
2.680	.03421	.73577	.17841	1.30003	.51753	3.55586
2.690	.03348	.73432	.17540	1.30260	.51556	3.60864
2.700	.03276	.73287	.17244	1.30517	.51361	3.66242
2.710	.03205	.73142	.16951	1.30776	.51167	3.71723
2.720	.03136	.72997	.16663	1.31036	.50975	3.77308
2.730	.03068	.72852	.16379	1.31296	.50784	3.82999
2.740	.03002	.72707	.16099	1.31558	.50595	3.88799
2.750	.02936	.72562	.15823	1.31821	.50407	3.94710
2.760	.02873	.72418	.15551	1.32084	.50220	4.00734
2.770	.02810	.72273	.15283	1.32349	.50035	4.06873
2.780	.02749	.72128	.15019	1.32614	.49851	4.13129
2.790	.02689	.71984	.14759	1.32880	.49668	4.19505
2.800	.02630	.71839	.14502	1.33148	.49487	4.26004
2.810	.02572	.71695	.14250	1.33416	.49307	4.32627
2.820	.02516	.71550	.14001	1.33685	.49129	4.39378
2.830	.02461	.71406	.13756	1.33956	.48952	4.46258
2.840	.02407	.71262	.13514	1.34227	.48776	4.53271
2.850	.02354	.71117	.13276	1.34499	.48601	4.60418
2.860	.02302	.70973	.13042	1.34772	.48428	4.67704
2.870	.02251	.70829	.12812	1.35046	.48256	4.75129
2.880	.02201	.70685	.12585	1.35321	.48086	4.82698
2.890	.02152	.70542	.12361	1.35597	.47916	4.90414
2.900	.02104	.70398	.12141	1.35874	.47748	4.98278
2.910	.02058	.70254	.11924	1.36152	.47581	5.06295
2.920	.02012	.70111	.11711	1.36431	.47415	5.14467
2.930	.01967	.69967	.11501	1.36710	.47251	5.22797
2.940	.01923	.69824	.11294	1.36991	.47087	5.31289
2.950	.01880	.69680	.11091	1.37273	.46925	5.39946
2.960	.01838	.69537	.10891	1.37555	.46764	5.48771
2.970	.01797	.69394	.10694	1.37839	.46605	5.57768
2.980	.01757	.69251	.10500	1.38124	.46446	5.66940
2.990	.01717	.69108	.10309	1.38409	.46289	5.76291

TABLE VIII. Isentropic Exponent = 1.10

M	p/p_t	T/T_t	GAMMA	ISOTHERMAL T_t/T_t^*	RAYLEIGH T_t/T_t^*	SHOCK p_{t1}/p_{t2}
3.000	.01679	.68966	.10122	1.38696	.46132	5.85824
3.010	.01641	.68823	.09937	1.38983	.45977	5.95543
3.020	.01604	.68680	.09755	1.39271	.45823	6.05452
3.030	.01568	.68538	.09577	1.39561	.45671	6.15554
3.040	.01532	.68396	.09401	1.39851	.45519	6.25854
3.050	.01498	.68254	.09228	1.40142	.45368	6.36357
3.060	.01464	.68112	.09058	1.40435	.45219	6.47064
3.070	.01430	.67970	.08891	1.40728	.45070	6.57981
3.080	.01398	.67828	.08726	1.41022	.44923	6.69114
3.090	.01366	.67686	.08564	1.41317	.44777	6.80464
3.100	.01335	.67545	.08405	1.41613	.44631	6.92038
3.110	.01305	.67403	.08249	1.41910	.44487	7.03840
3.120	.01275	.67262	.08095	1.42208	.44344	7.15874
3.130	.01246	.67121	.07944	1.42507	.44202	7.28145
3.140	.01217	.66980	.07796	1.42807	.44061	7.40658
3.150	.01189	.66839	.07650	1.43108	.43921	7.53419
3.160	.01162	.66699	.07506	1.43409	.43782	7.66431
3.170	.01136	.66558	.07365	1.43712	.43643	7.79701
3.180	.01110	.66418	.07226	1.44016	.43506	7.93233
3.190	.01084	.66278	.07090	1.44320	.43370	8.07034
3.200	.01059	.66138	.06956	1.44626	.43235	8.21108
3.210	.01035	.65998	.06824	1.44933	.43101	8.35461
3.220	.01011	.65858	.06695	1.45240	.42968	8.50099
3.230	.00988	.65718	.06568	1.45549	.42835	8.65028
3.240	.00965	.65579	.06443	1.45858	.42704	8.80254
3.250	.00942	.65440	.06320	1.46168	.42574	8.95783
3.260	.00921	.65301	.06199	1.46480	.42444	9.11620
3.270	.00899	.65162	.06081	1.46792	.42316	9.27774
3.280	.00878	.65023	.05964	1.47105	.42188	9.44249
3.290	.00858	.64884	.05850	1.47420	.42061	9.61055
3.300	.00838	.64746	.05738	1.47735	.41936	9.78194
3.310	.00819	.64608	.05627	1.48051	.41811	9.95677
3.320	.00800	.64470	.05519	1.48368	.41687	10.13510
3.330	.00781	.64332	.05412	1.48686	.41563	10.31699
3.340	.00763	.64194	.05308	1.49005	.41441	10.50252
3.350	.00745	.64056	.05205	1.49325	.41320	10.69178
3.360	.00728	.63919	.05104	1.49646	.41199	10.88483
3.370	.00711	.63782	.05005	1.49968	.41079	11.08177
3.380	.00694	.63645	.04908	1.50291	.40960	11.28264
3.390	.00678	.63508	.04813	1.50614	.40842	11.48756
3.400	.00662	.63371	.04719	1.50939	.40725	11.69661
3.410	.00646	.63235	.04627	1.51265	.40608	11.90986
3.420	.00631	.63099	.04537	1.51591	.40493	12.12740
3.430	.00616	.62963	.04448	1.51919	.40378	12.34933
3.440	.00602	.62827	.04361	1.52248	.40264	12.57574
3.450	.00588	.62691	.04275	1.52577	.40150	12.80672
3.460	.00574	.62556	.04191	1.52908	.40038	13.04235
3.470	.00561	.62420	.04109	1.53239	.39926	13.28275
3.480	.00547	.62285	.04028	1.53571	.39815	13.52802
3.490	.00534	.62150	.03949	1.53905	.39705	13.77824

TABLE VIII.   Isentropic Exponent = 1.10

M	p/p_t	T/T_t	GAMMA	ISOTHERMAL T_t/T_t^*	RAYLEIGH T_t/T_t^*	SHOCK p_{t1}/p_{t2}
3.500	.00522	.62016	.03871	1.54239	.39596	14.03354
3.510	.00509	.61881	.03794	1.54574	.39487	14.29400
3.520	.00497	.61747	.03719	1.54911	.39379	14.55974
3.530	.00486	.61613	.03646	1.55248	.39272	14.83087
3.540	.00474	.61479	.03574	1.55586	.39165	15.10751
3.550	.00463	.61345	.03503	1.55925	.39060	15.38976
3.560	.00452	.61212	.03433	1.56265	.38955	15.67775
3.570	.00441	.61078	.03365	1.56606	.38850	15.97159
3.580	.00431	.60945	.03298	1.56948	.38747	16.27141
3.590	.00421	.60812	.03232	1.57291	.38644	16.57733
3.600	.00411	.60680	.03168	1.57635	.38542	16.88949
3.610	.00401	.60547	.03104	1.57980	.38440	17.20800
3.620	.00391	.60415	.03042	1.58325	.38339	17.53301
3.630	.00382	.60283	.02981	1.58672	.38239	17.86464
3.640	.00373	.60151	.02922	1.59020	.38140	18.20305
3.650	.00364	.60020	.02863	1.59368	.38041	18.54836
3.660	.00355	.59888	.02806	1.59718	.37943	18.90074
3.670	.00347	.59757	.02749	1.60069	.37845	19.26031
3.680	.00339	.59626	.02694	1.60420	.37748	19.62724
3.690	.00331	.59495	.02640	1.60773	.37652	20.00167
3.700	.00323	.59365	.02587	1.61126	.37557	20.38379
3.710	.00315	.59235	.02535	1.61480	.37462	20.77371
3.720	.00307	.59104	.02483	1.61836	.37367	21.17163
3.730	.00300	.58975	.02433	1.62192	.37274	21.57771
3.740	.00293	.58845	.02384	1.62549	.37181	21.99212
3.750	.00286	.58716	.02336	1.62908	.37088	22.41503
3.760	.00279	.58586	.02288	1.63267	.36997	22.84663
3.770	.00272	.58457	.02242	1.63627	.36905	23.28709
3.780	.00266	.58329	.02197	1.63988	.36815	23.73661
3.790	.00260	.58200	.02152	1.64350	.36725	24.19537
3.800	.00253	.58072	.02108	1.64713	.36635	24.66355
3.810	.00247	.57944	.02065	1.65077	.36546	25.14138
3.820	.00241	.57816	.02023	1.65442	.36458	25.62905
3.830	.00235	.57689	.01982	1.65808	.36370	26.12676
3.840	.00230	.57561	.01942	1.66175	.36283	26.63471
3.850	.00224	.57434	.01902	1.66542	.36197	27.15316
3.860	.00219	.57307	.01863	1.66911	.36111	27.68227
3.870	.00214	.57181	.01825	1.67281	.36025	28.22231
3.880	.00209	.57054	.01788	1.67651	.35941	28.77349
3.890	.00204	.56928	.01751	1.68023	.35856	29.33606
3.900	.00199	.56802	.01715	1.68396	.35772	29.91021
3.910	.00194	.56676	.01680	1.68769	.35689	30.49626
3.920	.00189	.56551	.01646	1.69144	.35606	31.09440
3.930	.00185	.56426	.01612	1.69519	.35524	31.70492
3.940	.00180	.56301	.01579	1.69895	.35443	32.32805
3.950	.00176	.56176	.01547	1.70273	.35362	32.96410
3.960	.00172	.56051	.01515	1.70651	.35281	33.61330
3.970	.00167	.55927	.01484	1.71030	.35201	34.27594
3.980	.00163	.55803	.01453	1.71411	.35121	34.95230
3.990	.00159	.55679	.01423	1.71792	.35042	35.64266

TABLE VIII. Isentropic Exponent = 1.10

M	p/p_t	T/T_t	GAMMA	ISOTHERMAL T_t/T_t^*	RAYLEIGH T_t/T_t^*	SHOCK p_{t1}/p_{t2}
4.000	.00156	.55556	.01394	1.72174	.34964	36.34742
4.010	.00152	.55432	.01365	1.72557	.34886	37.06672
4.020	.00148	.55309	.01337	1.72941	.34808	37.80091
4.030	.00145	.55186	.01309	1.73326	.34731	38.55034
4.040	.00141	.55064	.01282	1.73712	.34654	39.31531
4.050	.00138	.54941	.01256	1.74099	.34578	40.09615
4.060	.00134	.54819	.01230	1.74487	.34502	40.89323
4.070	.00131	.54697	.01204	1.74876	.34427	41.70686
4.080	.00128	.54576	.01179	1.75265	.34353	42.53735
4.090	.00125	.54454	.01155	1.75656	.34278	43.38514
4.100	.00122	.54333	.01131	1.76048	.34205	44.25051
4.110	.00119	.54212	.01108	1.76440	.34131	45.13389
4.120	.00116	.54091	.01085	1.76834	.34058	46.03564
4.130	.00113	.53971	.01062	1.77229	.33986	46.95616
4.140	.00110	.53851	.01040	1.77624	.33914	47.89577
4.150	.00108	.53731	.01018	1.78021	.33843	48.85498
4.160	.00105	.53611	.00997	1.78418	.33772	49.83414
4.170	.00103	.53492	.00976	1.78816	.33701	50.83368
4.180	.00100	.53373	.00956	1.79216	.33631	51.85403
4.190	.00098	.53254	.00936	1.79616	.33561	52.89562
4.200	.00095	.53135	.00917	1.80017	.33492	53.95892
4.210	.00093	.53017	.00898	1.80420	.33423	55.04438
4.220	.00091	.52898	.00879	1.80823	.33354	56.15242
4.230	.00089	.52780	.00861	1.81227	.33286	57.28359
4.240	.00086	.52663	.00843	1.81632	.33219	58.43833
4.250	.00084	.52545	.00825	1.82038	.33151	59.61711
4.260	.00082	.52428	.00808	1.82445	.33085	60.82051
4.270	.00080	.52311	.00791	1.82853	.33018	62.04897
4.280	.00078	.52194	.00774	1.83262	.32952	63.30311
4.290	.00076	.52078	.00758	1.83672	.32887	64.58339
4.300	.00075	.51962	.00742	1.84083	.32821	65.89037
4.310	.00073	.51846	.00727	1.84494	.32757	67.22459
4.320	.00071	.51730	.00712	1.84907	.32692	68.58670
4.330	.00069	.51614	.00697	1.85321	.32628	69.97723
4.340	.00068	.51499	.00682	1.85735	.32564	71.39680
4.350	.00066	.51384	.00668	1.86151	.32501	72.84597
4.360	.00064	.51269	.00654	1.86568	.32438	74.32544
4.370	.00063	.51155	.00640	1.86985	.32376	75.83583
4.380	.00061	.51041	.00627	1.87404	.32314	77.37772
4.390	.00060	.50927	.00614	1.87823	.32252	78.95184
4.400	.00058	.50813	.00601	1.88243	.32190	80.55882
4.410	.00057	.50700	.00588	1.88665	.32129	82.19940
4.420	.00056	.50586	.00576	1.89087	.32069	83.87425
4.430	.00054	.50473	.00564	1.89510	.32008	85.58409
4.440	.00053	.50361	.00552	1.89935	.31948	87.32963
4.450	.00052	.50248	.00540	1.90360	.31889	89.11167
4.460	.00050	.50136	.00529	1.90786	.31829	90.93099
4.470	.00049	.50024	.00518	1.91213	.31771	92.78831
4.480	.00048	.49912	.00507	1.91641	.31712	94.68447
4.490	.00047	.49801	.00496	1.92070	.31654	96.62021

TABLE VIII. Isentropic Exponent = 1.10

M	p/p_t	T/T_t	GAMMA	ISOTHERMAL T_t/T_t^*	RAYLEIGH T_t/T_t^*	SHOCK p_{t1}/p_{t2}
4.500	.00046	.49689	.00486	1.92500	.31596	98.59644
4.510	.00044	.49578	.00476	1.92931	.31538	100.61405
4.520	.00043	.49468	.00466	1.93363	.31481	102.67373
4.530	.00042	.49357	.00456	1.93796	.31424	104.77652
4.540	.00041	.49247	.00446	1.94229	.31368	106.92330
4.550	.00040	.49137	.00437	1.94664	.31312	109.11495
4.560	.00039	.49027	.00428	1.95100	.31256	111.35242
4.570	.00038	.48918	.00419	1.95536	.31200	113.63667
4.580	.00037	.48809	.00410	1.95974	.31145	115.96866
4.590	.00037	.48700	.00401	1.96413	.31090	118.34949
4.600	.00036	.48591	.00393	1.96852	.31035	120.78001
4.610	.00035	.48482	.00384	1.97293	.30981	123.26142
4.620	.00034	.48374	.00376	1.97734	.30927	125.79464
4.630	.00033	.48266	.00368	1.98176	.30873	128.38088
4.640	.00032	.48158	.00361	1.98620	.30820	131.02116
4.650	.00032	.48051	.00353	1.99064	.30767	133.71663
4.660	.00031	.47944	.00346	1.99509	.30714	136.46850
4.670	.00030	.47837	.00338	1.99956	.30662	139.27787
4.680	.00029	.47730	.00331	2.00403	.30610	142.14598
4.690	.00029	.47623	.00324	2.00851	.30558	145.07413
4.700	.00028	.47517	.00317	2.01300	.30506	148.06337
4.710	.00027	.47411	.00311	2.01750	.30455	151.11514
4.720	.00027	.47305	.00304	2.02201	.30404	154.23066
4.730	.00026	.47200	.00298	2.02653	.30353	157.41142
4.740	.00025	.47095	.00291	2.03106	.30303	160.65855
4.750	.00025	.46990	.00285	2.03560	.30253	163.97356
4.760	.00024	.46885	.00279	2.04015	.30203	167.35785
4.770	.00023	.46780	.00273	2.04470	.30153	170.81289
4.780	.00023	.46676	.00268	2.04927	.30104	174.34012
4.790	.00022	.46572	.00262	2.05385	.30055	177.94096
4.800	.00022	.46468	.00256	2.05843	.30006	181.61707
4.810	.00021	.46365	.00251	2.06303	.29958	185.36993
4.820	.00021	.46262	.00246	2.06764	.29910	189.20129
4.830	.00020	.46159	.00240	2.07225	.29862	193.11261
4.840	.00020	.46056	.00235	2.07688	.29814	197.10554
4.850	.00019	.45953	.00230	2.08151	.29767	201.18201
4.860	.00019	.45851	.00226	2.08615	.29719	205.34356
4.870	.00018	.45749	.00221	2.09081	.29672	209.59174
4.880	.00018	.45647	.00216	2.09547	.29626	213.92888
4.890	.00017	.45546	.00212	2.10014	.29579	218.35654
4.900	.00017	.45444	.00207	2.10483	.29533	222.87633
4.910	.00017	.45343	.00203	2.10952	.29488	227.49059
4.920	.00016	.45242	.00198	2.11422	.29442	232.20115
4.930	.00016	.45142	.00194	2.11893	.29397	237.00988
4.940	.00015	.45041	.00190	2.12365	.29351	241.91906
4.950	.00015	.44941	.00186	2.12838	.29307	246.93057
4.960	.00015	.44841	.00182	2.13312	.29262	252.04644
4.970	.00014	.44742	.00178	2.13787	.29218	257.26916
4.980	.00014	.44642	.00175	2.14263	.29173	262.60053
4.990	.00014	.44543	.00171	2.14740	.29130	268.04316

TABLE IX. Isentropic Exponent = 1.20

M	p/p$_t$	T/T$_t$	GAMMA	ISOTHERMAL T$_t$/T$_t$*	RAYLEIGH T$_t$/T$_t$*	SHOCK p$_{t1}$/p$_{t2}$
.000	1.00000	1.00000	.00000	.92308	.00000	1.
.005	.99999	.00000	.00840	.92308	.00011	1.
.010	.99994	.99999	.01689	.92309	.00044	1.
.015	.99987	.99998	.02533	.92310	.00099	1.
.020	.99976	.99996	.03377	.92311	.00176	1.
.025	.99963	.99994	.04221	.92313	.00275	1.
.030	.99946	.99991	.05064	.92316	.00395	1.
.035	.99927	.99988	.05907	.92319	.00537	1.
.040	.99904	.99984	.06750	.92322	.00701	1.
.045	.99879	.99980	.07592	.92326	.00887	1.
.050	.99850	.99975	.08434	.92331	.01094	1.
.055	.99819	.99970	.09274	.92336	.01322	1.
.060	.99784	.99964	.10115	.92341	.01571	1.
.065	.99747	.99958	.10954	.92347	.01841	1.
.070	.99707	.99951	.11792	.92353	.02132	1.
.075	.99663	.99944	.12629	.92360	.02443	1.
.080	.99617	.99936	.13465	.92367	.02775	1.
.085	.99568	.99928	.14301	.92374	.03127	1.
.090	.99515	.99919	.15134	.92382	.03499	1.
.095	.99460	.99910	.15967	.92391	.03890	1.
.100	.99402	.99900	.16798	.92400	.04301	1.
.105	.99341	.99890	.17628	.92409	.04730	1.
.110	.99277	.99879	.18457	.92419	.05179	1.
.115	.99210	.99868	.19284	.92430	.05646	1.
.120	.99140	.99856	.20110	.92441	.06131	1.
.125	.99068	.99844	.20933	.92452	.06635	1.
.130	.98992	.99831	.21755	.92464	.07155	1.
.135	.98913	.99818	.22576	.92476	.07693	1.
.140	.98832	.99804	.23394	.92489	.08248	1.
.145	.98748	.99790	.24211	.92502	.08820	1.
.150	.98661	.99776	.25025	.92515	.09407	1.
.155	.98571	.99760	.25838	.92529	.10011	1.
.160	.98478	.99745	.26648	.92544	.10630	1.
.165	.98382	.99728	.27457	.92559	.11264	1.
.170	.98283	.99712	.28263	.92574	.11912	1.
.175	.98182	.99695	.29066	.92590	.12575	1.
.180	.98078	.99677	.29868	.92607	.13252	1.
.185	.97971	.99659	.30667	.92624	.13942	1.
.190	.97861	.99640	.31463	.92641	.14645	1.
.195	.97749	.99621	.32257	.92659	.15361	1.
.200	.97633	.99602	.33049	.92677	.16089	1.
.205	.97515	.99582	.33837	.92696	.16829	1.
.210	.97394	.99561	.34623	.92715	.17580	1.
.215	.97271	.99540	.35406	.92734	.18342	1.
.220	.97145	.99518	.36187	.92754	.19114	1.
.225	.97016	.99496	.36964	.92775	.19897	1.
.230	.96884	.99474	.37739	.92796	.20689	1.
.235	.96750	.99451	.38510	.92817	.21490	1.
.240	.96613	.99427	.39278	.92839	.22301	1.
.245	.95473	.99403	.40043	.92862	.23119	1.

TABLE IX. Isentropic Exponent = 1.20

M	p/p_t	T/T_t	GAMMA	ISOTHERMAL T_t/T_t^*	RAYLEIGH T_t/T_t^*	SHOCK p_{t1}/p_{t2}
.250	.96331	.99379	.40805	.92885	.23945	1.
.255	.96186	.99354	.41564	.92908	.24779	1.
.260	.96038	.99329	.42319	.92932	.25620	1.
.265	.95888	.99303	.43071	.92956	.26467	1.
.270	.95735	.99276	.43820	.92981	.27321	1.
.275	.95580	.99249	.44565	.93006	.28180	1.
.280	.95422	.99222	.45307	.93031	.29044	1.
.285	.95262	.99194	.46045	.93057	.29914	1.
.290	.95099	.99166	.46779	.93084	.30787	1.
.295	.94934	.99137	.47510	.93111	.31665	1.
.300	.94766	.99108	.48237	.93138	.32547	1.
.305	.94596	.99078	.48960	.93166	.33431	1.
.310	.94423	.99048	.49679	.93195	.34319	1.
.315	.94248	.99017	.50395	.93224	.35208	1.
.320	.94070	.98986	.51106	.93253	.36100	1.
.325	.93890	.98955	.51814	.93283	.36994	1.
.330	.93708	.98923	.52517	.93313	.37888	1.
.335	.93523	.98890	.53217	.93344	.38784	1.
.340	.93336	.98857	.53912	.93375	.39680	1.
.345	.93147	.98824	.54603	.93406	.40576	1.
.350	.92955	.98790	.55290	.93438	.41471	1.
.355	.92761	.98755	.55972	.93471	.42367	1.
.360	.92565	.98721	.56651	.93504	.43261	1.
.365	.92366	.98685	.57325	.93537	.44154	1.
.370	.92166	.98649	.57994	.93571	.45045	1.
.375	.91963	.98613	.58659	.93606	.45934	1.
.380	.91758	.98577	.59320	.93641	.46821	1.
.385	.91550	.98539	.59976	.93676	.47706	1.
.390	.91341	.98502	.60627	.93712	.48587	1.
.395	.91129	.98464	.61274	.93748	.49465	1.
.400	.90915	.98425	.61916	.93785	.50340	1.
.405	.90700	.98386	.62554	.93822	.51211	1.
.410	.90482	.98347	.63187	.93859	.52078	1.
.415	.90262	.98307	.63815	.93897	.52940	1.
.420	.90040	.98267	.64438	.93936	.53798	1.
.425	.89816	.98226	.65056	.93975	.54652	1.
.430	.89590	.98185	.65670	.94014	.55500	1.
.435	.89362	.98143	.66278	.94054	.56342	1.
.440	.89132	.98101	.66882	.94095	.57179	1.
.445	.88900	.98058	.67481	.94136	.58011	1.
.450	.88667	.98015	.68075	.94177	.58836	1.
.455	.88431	.97972	.68663	.94219	.59655	1.
.460	.88194	.97928	.69247	.94261	.60468	1.
.465	.87954	.97884	.69826	.94304	.61274	1.
.470	.87713	.97839	.70399	.94347	.62073	1.
.475	.87470	.97794	.70967	.94390	.62865	1.
.480	.87226	.97748	.71531	.94434	.63650	1.
.485	.86979	.97702	.72088	.94479	.64428	1.
.490	.86731	.97655	.72641	.94524	.65198	1.
.495	.85481	.97608	.73189	.94569	.65961	1.

TABLE IX. Isentropic Exponent = 1.20

M	p/p_t	T/T_t	GAMMA	ISOTHERMAL T_t/T_t^*	RAYLEIGH T_t/T_t^*	SHOCK p_{t1}/p_{t2}
.500	.86230	.97561	.73731	.94615	.66116	1.
.505	.85976	.97513	.74268	.94662	.67463	1.
.510	.85722	.97465	.74799	.94709	.68202	1.
.515	.85465	.97416	.75325	.94756	.68933	1.
.520	.85207	.97367	.75846	.94804	.69656	1.
.525	.84947	.97318	.76361	.94852	.70370	1.
.530	.84686	.97268	.76871	.94901	.71076	1.
.535	.84424	.97217	.77376	.94950	.71773	1.
.540	.84159	.97167	.77875	.94999	.72461	1.
.545	.83894	.97115	.78368	.95049	.73141	1.
.550	.83627	.97064	.78857	.95100	.73812	1.
.555	.83358	.97012	.79339	.95151	.74475	1.
.560	.83088	.96959	.79816	.95202	.75128	1.
.565	.82817	.96907	.80288	.95254	.75772	1.
.570	.82544	.96853	.80754	.95307	.76407	1.
.575	.82270	.96800	.81214	.95360	.77033	1.
.580	.81994	.96745	.81669	.95413	.77650	1.
.585	.81718	.96691	.82118	.95467	.78258	1.
.590	.81440	.96636	.82562	.95521	.78856	1.
.595	.81161	.96581	.83000	.95576	.79446	1.
.600	.80880	.96525	.83432	.95631	.80026	1.
.605	.80598	.96469	.83859	.95686	.80596	1.
.610	.80316	.96412	.84280	.95742	.81158	1.
.615	.80032	.96356	.84695	.95799	.81710	1.
.620	.79746	.96298	.85105	.95856	.82253	1.
.625	.79460	.96241	.85509	.95913	.82786	1.
.630	.79173	.96183	.85907	.95971	.83311	1.
.635	.78885	.96124	.86300	.96030	.83826	1.
.640	.78595	.96065	.86687	.96089	.84331	1.
.645	.78305	.96006	.87068	.96148	.84828	1.
.650	.78013	.95946	.87444	.96208	.85315	1.
.655	.77721	.95886	.87813	.96268	.85793	1.
.660	.77428	.95826	.88178	.96329	.86262	1.
.665	.77133	.95765	.88536	.96390	.86721	1.
.670	.76838	.95704	.88889	.96451	.87172	1.
.675	.76542	.95642	.89236	.96513	.87613	1.
.680	.76245	.95580	.89577	.96576	.88046	1.
.685	.75947	.95518	.89913	.96639	.88469	1.
.690	.75649	.95455	.90242	.96702	.88883	1.
.695	.75349	.95392	.90567	.96766	.89289	1.
.700	.75049	.95329	.90885	.96831	.89686	1.
.705	.74749	.95265	.91198	.96896	.90074	1.
.710	.74447	.95201	.91505	.96961	.90453	1.
.715	.74145	.95136	.91806	.97027	.90823	1.
.720	.73842	.95071	.92102	.97093	.91185	1.
.725	.73538	.95006	.92392	.97160	.91538	1.
.730	.73234	.94941	.92676	.97227	.91883	1.
.735	.72929	.94875	.92955	.97294	.92219	1.
.740	.72624	.94808	.93228	.97362	.92547	1.
.745	.72318	.94742	.93495	.97431	.92866	1.

TABLE IX. Isentropic Exponent = 1.20

M	p/p_t	T/T_t	GAMMA	ISOTHERMAL $T_t/T_t{}^*$	RAYLEIGH $T_t/T_t{}^*$	SHOCK p_{t1}/p_{t2}
.750	.72011	.94675	.93757	.97500	.93178	1.
.755	.71704	.94607	.94013	.97569	.93481	1.
.760	.71397	.94539	.94264	.97639	.93776	1.
.765	.71089	.94471	.94508	.97710	.94063	1.
.770	.70780	.94403	.94748	.97781	.94342	1.
.775	.70471	.94334	.94981	.97852	.94613	1.
.780	.70162	.94265	.95209	.97924	.94877	1.
.785	.69852	.94195	.95432	.97996	.95132	1.
.790	.69542	.94126	.95649	.98069	.95380	1.
.795	.69232	.94055	.95860	.98142	.95621	1.
.800	.68921	.93985	.96066	.98215	.95854	1.
.805	.68610	.93914	.96267	.98289	.96079	1.
.810	.68298	.93843	.96462	.98364	.96298	1.
.815	.67987	.93771	.96651	.98439	.96509	1.
.820	.67675	.93700	.96835	.98514	.96713	1.
.825	.67363	.93627	.97013	.98590	.96910	1.
.830	.67051	.93555	.97187	.98667	.97099	1.
.835	.66738	.93482	.97354	.98744	.97282	1.
.840	.66425	.93409	.97517	.98821	.97459	1.
.845	.66113	.93336	.97674	.98899	.97628	1.
.850	.65800	.93262	.97825	.98977	.97791	1.
.855	.65487	.93188	.97971	.99056	.97947	1.
.860	.65174	.93113	.98112	.99135	.98097	1.
.865	.64860	.93039	.98248	.99214	.98240	1.
.870	.64547	.92964	.98378	.99294	.98377	1.
.875	.64234	.92888	.98504	.99375	.98508	1.
.880	.63921	.92813	.98623	.99456	.98633	1.
.885	.63607	.92737	.98738	.99537	.98751	1.
.890	.63294	.92660	.98848	.99619	.98864	1.
.895	.62981	.92584	.98952	.99702	.98971	1.
.900	.62668	.92507	.99051	.99785	.99072	1.
.905	.62355	.92430	.99146	.99868	.99167	1.
.910	.62042	.92352	.99235	.99952	.99257	1.
.915	.61729	.92275	.99319	1.00036	.99341	1.
.920	.61417	.92196	.99398	1.00121	.99419	1.
.925	.61104	.92118	.99472	1.00206	.99493	1.
.930	.60792	.92040	.99541	1.00291	.99560	1.
.935	.60480	.91961	.99605	1.00377	.99623	1.
.940	.60168	.91881	.99664	1.00464	.99681	1.
.945	.59856	.91802	.99718	1.00551	.99733	1.
.950	.59545	.91722	.99768	1.00638	.99781	1.
.955	.59234	.91642	.99812	1.00726	.99824	1.
.960	.58923	.91562	.99852	1.00815	.99861	1.
.965	.58612	.91481	.99887	1.00904	.99895	1.
.970	.58302	.91400	.99917	1.00993	.99923	1.
.975	.57992	.91319	.99943	1.01083	.99947	1.
.980	.57682	.91238	.99963	1.01173	.99966	1.
.985	.57373	.91156	.99979	1.01264	.99981	1.
.990	.57064	.91074	.99991	1.01355	.99992	1.
.995	.56756	.90992	.99998	1.01446	.99998	1.

TABLE IX. Isentropic Exponent = 1.20

M	p/p_t	T/T_t	GAMMA	ISOTHERMAL T_t/T_t^*	RAYLEIGH T_t/T_t^*	SHOCK p_{t1}/p_{t2}
1.000	.55447	.90909	1.00000	1.01538	1.00000	1.00000
1.005	.56140	.90826	.99998	1.01631	.99998	1.00000
1.010	.55832	.90743	.99991	1.01724	.99992	1.00000
1.015	.55526	.90660	.99980	1.01817	.99982	1.00000
1.020	.55219	.90576	.99964	1.01911	.99968	1.00001
1.025	.54913	.90493	.99944	1.02006	.99950	1.00002
1.030	.54608	.90409	.99919	1.02101	.99928	1.00003
1.035	.54303	.90324	.99890	1.02196	.99903	1.00005
1.040	.53999	.90240	.99857	1.02292	.99874	1.00008
1.045	.53695	.90155	.99820	1.02388	.99841	1.00011
1.050	.53392	.90070	.99778	1.02485	.99805	1.00015
1.055	.53089	.89985	.99732	1.02582	.99766	1.00020
1.060	.52787	.89899	.99682	1.02679	.99723	1.00026
1.065	.52486	.89813	.99628	1.02777	.99677	1.00032
1.070	.52185	.89727	.99569	1.02876	.99627	1.00040
1.075	.51884	.89641	.99507	1.02975	.99575	1.00049
1.080	.51585	.89554	.99440	1.03074	.99519	1.00059
1.085	.51286	.89468	.99369	1.03174	.99460	1.00070
1.090	.50987	.89381	.99295	1.03275	.99399	1.00083
1.095	.50690	.89293	.99216	1.03376	.99334	1.00097
1.100	.50393	.89206	.99134	1.03477	.99267	1.00112
1.105	.50096	.89118	.99047	1.03579	.99196	1.00128
1.110	.49801	.89031	.98957	1.03681	.99123	1.00147
1.115	.49506	.88942	.98863	1.03784	.99047	1.00166
1.120	.49211	.88854	.98765	1.03887	.98969	1.00187
1.125	.48918	.88766	.98664	1.03990	.98888	1.00210
1.130	.48625	.88677	.98559	1.04094	.98804	1.00234
1.135	.48333	.88588	.98450	1.04199	.98718	1.00261
1.140	.48042	.88499	.98337	1.04304	.98630	1.00288
1.145	.47752	.88409	.98221	1.04409	.98539	1.00318
1.150	.47462	.88320	.98101	1.04515	.98446	1.00349
1.155	.47173	.88230	.97978	1.04622	.98351	1.00383
1.160	.46885	.88140	.97851	1.04729	.98253	1.00418
1.165	.46598	.88050	.97721	1.04836	.98153	1.00455
1.170	.46312	.87959	.97588	1.04944	.98051	1.00494
1.175	.46026	.87869	.97450	1.05052	.97947	1.00535
1.180	.45741	.87778	.97310	1.05161	.97842	1.00578
1.185	.45457	.87687	.97166	1.05270	.97734	1.00623
1.190	.45175	.87596	.97020	1.05379	.97624	1.00670
1.195	.44892	.87504	.96869	1.05489	.97512	1.00719
1.200	.44611	.87413	.96716	1.05600	.97399	1.00771
1.205	.44331	.87321	.96559	1.05711	.97283	1.00824
1.210	.44051	.87229	.96400	1.05822	.97166	1.00880
1.215	.43773	.87137	.96237	1.05934	.97047	1.00937
1.220	.43495	.87044	.96071	1.06047	.96927	1.00997
1.225	.43219	.86952	.95902	1.06160	.96805	1.01060
1.230	.42943	.86859	.95730	1.06273	.96681	1.01124
1.235	.42668	.86766	.95555	1.06387	.96556	1.01191
1.240	.42394	.86673	.95378	1.06501	.96430	1.01260
1.245	.42121	.86580	.95197	1.06616	.96302	1.01332

TABLE IX. Isentropic Exponent = 1.20

M	p/p$_t$	T/T$_t$	GAMMA	ISOTHERMAL T$_t$/T$_t$*	RAYLEIGH T$_t$/T$_t$*	SHOCK p$_{ti}$/p$_{t2}$
1.250	.41849	.86486	.95013	1.06731	.96172	1.01406
1.255	.41578	.86393	.94827	1.06846	.96041	1.01482
1.260	.41308	.86299	.94638	1.06962	.95909	1.01561
1.265	.41039	.86205	.94446	1.07079	.95775	1.01642
1.270	.40771	.86111	.94252	1.07196	.95641	1.01725
1.275	.40504	.86017	.94055	1.07313	.95505	1.01811
1.280	.40238	.85922	.93855	1.07431	.95367	1.01899
1.285	.39973	.85828	.93652	1.07550	.95229	1.01990
1.290	.39709	.85733	.93448	1.07669	.95090	1.02084
1.295	.39446	.85638	.93240	1.07788	.94949	1.02180
1.300	.39184	.85543	.93030	1.07908	.94807	1.02278
1.305	.38923	.85448	.92818	1.08028	.94665	1.02379
1.310	.38664	.85353	.92603	1.08149	.94521	1.02483
1.315	.38405	.85257	.92386	1.08270	.94376	1.02589
1.320	.38147	.85161	.92166	1.08391	.94231	1.02697
1.325	.37890	.85066	.91944	1.08513	.94084	1.02809
1.330	.37634	.84970	.91720	1.08636	.93937	1.02923
1.335	.37380	.84874	.91494	1.08759	.93789	1.03039
1.340	.37126	.84777	.91265	1.08882	.93640	1.03159
1.345	.36874	.84681	.91034	1.09006	.93490	1.03281
1.350	.36622	.84584	.90801	1.09131	.93339	1.03405
1.355	.36372	.84488	.90566	1.09256	.93188	1.03533
1.360	.36123	.84391	.90329	1.09381	.93036	1.03663
1.365	.35874	.84294	.90090	1.09507	.92884	1.03796
1.370	.35627	.84197	.89849	1.09633	.92730	1.03931
1.375	.35381	.84100	.89606	1.09760	.92576	1.04070
1.380	.35136	.84003	.89361	1.09887	.92422	1.04211
1.385	.34892	.83905	.89114	1.10014	.92266	1.04354
1.390	.34650	.83808	.88865	1.10142	.92111	1.04501
1.395	.34408	.83710	.88615	1.10271	.91955	1.04651
1.400	.34167	.83612	.88362	1.10400	.91798	1.04803
1.405	.33928	.83514	.88108	1.10529	.91640	1.04958
1.410	.33690	.83416	.87852	1.10659	.91483	1.05116
1.415	.33453	.83318	.87594	1.10790	.91325	1.05277
1.420	.33217	.83220	.87335	1.10921	.91166	1.05440
1.425	.32982	.83121	.87074	1.11052	.91007	1.05607
1.430	.32748	.83023	.86812	1.11184	.90848	1.05776
1.435	.32515	.82924	.86548	1.11316	.90688	1.05949
1.440	.32283	.82825	.86282	1.11449	.90528	1.06124
1.445	.32053	.82727	.86015	1.11582	.90367	1.06302
1.450	.31824	.82628	.85746	1.11715	.90207	1.06483
1.455	.31595	.82529	.85476	1.11849	.90046	1.06668
1.460	.31368	.82429	.85205	1.11984	.89884	1.06855
1.465	.31142	.82330	.84932	1.12119	.89723	1.07045
1.470	.30918	.82231	.84658	1.12254	.89561	1.07238
1.475	.30694	.82131	.84382	1.12390	.89399	1.07434
1.480	.30471	.82032	.84105	1.12527	.89237	1.07633
1.485	.30250	.81932	.83827	1.12664	.89075	1.07835
1.490	.30030	.81832	.83548	1.12801	.88912	1.08040
1.495	.29811	.81733	.83267	1.12939	.88749	1.08248

TABLE IX.   Isentropic Exponent = 1.20

M	p/p_t	T/T_t	GAMMA	ISOTHERMAL T_t/T_t^*	RAYLEIGH T_t/T_t^*	SHOCK p_{t1}/p_{t2}
1.500	.29593	.81633	.82986	1.13077	.88587	1.08460
1.505	.29376	.81533	.82703	1.13216	.88424	1.08674
1.510	.29160	.81433	.82419	1.13355	.88261	1.08892
1.515	.28946	.81332	.82134	1.13494	.88097	1.09112
1.520	.28732	.81232	.81848	1.13634	.87934	1.09336
1.525	.28520	.81132	.81561	1.13775	.87771	1.09563
1.530	.28309	.81031	.81273	1.13916	.87608	1.09793
1.535	.28099	.80931	.80983	1.14057	.87444	1.10026
1.540	.27890	.80830	.80693	1.14199	.87281	1.10262
1.545	.27682	.80730	.80403	1.14342	.87118	1.10502
1.550	.27475	.80629	.80111	1.14485	.86954	1.10744
1.555	.27270	.80528	.79818	1.14628	.86791	1.10990
1.560	.27066	.80427	.79525	1.14772	.86627	1.11239
1.565	.26862	.80326	.79230	1.14916	.86464	1.11492
1.570	.26660	.80225	.78935	1.15061	.86301	1.11747
1.575	.26460	.80124	.78639	1.15206	.86138	1.12006
1.580	.26260	.80023	.78343	1.15351	.85975	1.12268
1.585	.26061	.79922	.78046	1.15497	.85812	1.12534
1.590	.25864	.79821	.77748	1.15644	.85649	1.12803
1.595	.25667	.79719	.77449	1.15791	.85486	1.13075
1.600	.25472	.79618	.77150	1.15938	.85323	1.13350
1.605	.25278	.79516	.76850	1.16086	.85161	1.13629
1.610	.25085	.79415	.76550	1.16235	.84998	1.13911
1.615	.24893	.79313	.76249	1.16384	.84836	1.14197
1.620	.24702	.79212	.75948	1.16533	.84674	1.14486
1.625	.24513	.79110	.75646	1.16683	.84512	1.14778
1.630	.24324	.79008	.75344	1.16833	.84350	1.15074
1.635	.24137	.78907	.75041	1.16984	.84188	1.15373
1.640	.23950	.78805	.74738	1.17135	.84027	1.15676
1.645	.23765	.78703	.74434	1.17286	.83865	1.15982
1.650	.23581	.78601	.74130	1.17438	.83704	1.16292
1.655	.23398	.78499	.73826	1.17591	.83543	1.16605
1.660	.23216	.78397	.73521	1.17744	.83383	1.16921
1.665	.23036	.78295	.73216	1.17897	.83222	1.17242
1.670	.22856	.78193	.72911	1.18051	.83062	1.17565
1.675	.22678	.78091	.72606	1.18206	.82902	1.17893
1.680	.22500	.77989	.72300	1.18361	.82743	1.18224
1.685	.22324	.77886	.71994	1.18516	.82583	1.18558
1.690	.22149	.77784	.71688	1.18672	.82424	1.18896
1.695	.21974	.77682	.71382	1.18828	.82265	1.19238
1.700	.21801	.77580	.71075	1.18985	.82106	1.19583
1.705	.21629	.77477	.70769	1.19142	.81948	1.19933
1.710	.21458	.77375	.70462	1.19299	.81790	1.20285
1.715	.21289	.77272	.70155	1.19457	.81632	1.20642
1.720	.21120	.77170	.69848	1.19616	.81475	1.21002
1.725	.20952	.77068	.69541	1.19775	.81318	1.21366
1.730	.20786	.76965	.69234	1.19934	.81161	1.21734
1.735	.20620	.76863	.68928	1.20094	.81004	1.22105
1.740	.20456	.76760	.68621	1.20255	.80848	1.22481
1.745	.20292	.76658	.68314	1.20416	.80692	1.22860

TABLE IX. Isentropic Exponent = 1.20

M	p/p_t	T/T_t	GAMMA	ISOTHERMAL T_t/T_t^*	RAYLEIGH T_t/T_t^*	SHOCK p_{ti}/p_{t2}
1.750	.20130	.76555	.68007	1.20577	.80536	1.23243
1.755	.19969	.76452	.67700	1.20739	.80381	1.23630
1.760	.19808	.76350	.67394	1.20901	.80226	1.24020
1.765	.19649	.76247	.67087	1.21064	.80072	1.24415
1.770	.19491	.76145	.66780	1.21227	.79917	1.24814
1.775	.19334	.76042	.66474	1.21390	.79763	1.25216
1.780	.19178	.75939	.66168	1.21554	.79610	1.25622
1.785	.19023	.75837	.65862	1.21719	.79457	1.26033
1.790	.18869	.75734	.65556	1.21884	.79304	1.26447
1.795	.18716	.75631	.65251	1.22049	.79151	1.26866
1.800	.18564	.75529	.64945	1.22215	.78999	1.27288
1.805	.18413	.75426	.64640	1.22382	.78848	1.27715
1.810	.18263	.75323	.64335	1.22549	.78696	1.28145
1.815	.18114	.75221	.64031	1.22716	.78545	1.28580
1.820	.17966	.75118	.63727	1.22884	.78395	1.29019
1.825	.17820	.75015	.63423	1.23052	.78244	1.29462
1.830	.17674	.74913	.63119	1.23221	.78095	1.29909
1.835	.17529	.74810	.62816	1.23390	.77945	1.30360
1.840	.17385	.74707	.62513	1.23559	.77796	1.30816
1.845	.17242	.74604	.62210	1.23729	.77648	1.31276
1.850	.17100	.74502	.61908	1.23900	.77499	1.31740
1.855	.16959	.74399	.61606	1.24071	.77351	1.32208
1.860	.16819	.74296	.61305	1.24242	.77204	1.32681
1.865	.16680	.74194	.61004	1.24414	.77057	1.33158
1.870	.16542	.74091	.60704	1.24587	.76910	1.33639
1.875	.15405	.73988	.60404	1.24760	.76764	1.34125
1.880	.16269	.73886	.60104	1.24933	.76618	1.34615
1.885	.16134	.73783	.59805	1.25107	.76473	1.35109
1.890	.16000	.73681	.59506	1.25281	.76328	1.35608
1.895	.15867	.73578	.59208	1.25456	.76183	1.36112
1.900	.15734	.73475	.58911	1.25631	.76039	1.36620
1.905	.15603	.73373	.58614	1.25806	.75896	1.37132
1.910	.15473	.73270	.58317	1.25982	.75752	1.37650
1.915	.15343	.73168	.58021	1.26159	.75609	1.38171
1.920	.15215	.73065	.57726	1.26336	.75467	1.38698
1.925	.15087	.72963	.57431	1.26513	.75325	1.39228
1.930	.14960	.72860	.57137	1.26691	.75183	1.39764
1.935	.14835	.72758	.56843	1.26870	.75042	1.40304
1.940	.14710	.72655	.56550	1.27049	.74901	1.40849
1.945	.14586	.72553	.56258	1.27228	.74761	1.41399
1.950	.14463	.72451	.55966	1.27408	.74621	1.41953
1.955	.14341	.72348	.55675	1.27588	.74482	1.42513
1.960	.14219	.72246	.55385	1.27769	.74343	1.43077
1.965	.14099	.72144	.55095	1.27950	.74204	1.43646
1.970	.13980	.72041	.54806	1.28131	.74066	1.44220
1.975	.13861	.71939	.54518	1.28313	.73928	1.44798
1.980	.13743	.71837	.54230	1.28496	.73791	1.45382
1.985	.13626	.71735	.53943	1.28679	.73654	1.45971
1.990	.13510	.71633	.53657	1.28862	.73518	1.46565
1.995	.13395	.71531	.53371	1.29046	.73382	1.47163

TABLE IX. Isentropic Exponent = 1.20

M	p/p_t	T/T_t	GAMMA	ISOTHERMAL T_t/T_t^*	RAYLEIGH T_t/T_t^*	SHOCK p_{ti}/p_{t2}
2.000	.13281	.71429	.53087	1.29231	.73246	1.47767
2.010	.13055	.71225	.52519	1.29601	.72976	1.48990
2.020	.12832	.71021	.51955	1.29973	.72708	1.50234
2.030	.12613	.70817	.51394	1.30347	.72442	1.51498
2.040	.12397	.70613	.50836	1.30722	.72178	1.52784
2.050	.12185	.70410	.50281	1.31100	.71915	1.54091
2.060	.11975	.70207	.49730	1.31479	.71654	1.55419
2.070	.11769	.70004	.49182	1.31861	.71395	1.56770
2.080	.11566	.69801	.48637	1.32244	.71138	1.58142
2.090	.11366	.69599	.48096	1.32629	.70882	1.59538
2.100	.11169	.69396	.47558	1.33015	.70628	1.60956
2.110	.10975	.69194	.47024	1.33404	.70376	1.62397
2.120	.10784	.68992	.46494	1.33794	.70126	1.63861
2.130	.10597	.68790	.45967	1.34187	.69877	1.65349
2.140	.10412	.68589	.45444	1.34581	.69630	1.66861
2.150	.10230	.68388	.44924	1.34977	.69385	1.68397
2.160	.10051	.68187	.44408	1.35375	.69142	1.69958
2.170	.09875	.67986	.43896	1.35774	.68900	1.71544
2.180	.09701	.67786	.43388	1.36176	.68660	1.73155
2.190	.09531	.67585	.42884	1.36579	.68422	1.74792
2.200	.09363	.67385	.42383	1.36985	.68186	1.76454
2.210	.09197	.67186	.41887	1.37392	.67951	1.78143
2.220	.09035	.66986	.41394	1.37801	.67718	1.79858
2.230	.08875	.66787	.40906	1.38211	.67487	1.81600
2.240	.08718	.66589	.40421	1.38624	.67257	1.83369
2.250	.08563	.66390	.39940	1.39038	.67029	1.85166
2.260	.08411	.66192	.39463	1.39455	.66803	1.86991
2.270	.08261	.65994	.38991	1.39873	.66578	1.88845
2.280	.08114	.65796	.38522	1.40293	.66355	1.90727
2.290	.07969	.65599	.38057	1.40715	.66134	1.92638
2.300	.07826	.65402	.37596	1.41138	.65914	1.94579
2.310	.07686	.65206	.37140	1.41564	.65696	1.96550
2.320	.07548	.65009	.36687	1.41991	.65480	1.98551
2.330	.07413	.64813	.36239	1.42421	.65265	2.00583
2.340	.07280	.64618	.35794	1.42852	.65052	2.02646
2.350	.07149	.64423	.35354	1.43285	.64840	2.04741
2.360	.07020	.64228	.34918	1.43719	.64630	2.06867
2.370	.06893	.64033	.34486	1.44156	.64422	2.09026
2.380	.06769	.63839	.34057	1.44594	.64215	2.11218
2.390	.06647	.63645	.33633	1.45035	.64009	2.13444
2.400	.06526	.63452	.33213	1.45477	.63806	2.15703
2.410	.06408	.63259	.32797	1.45921	.63603	2.17996
2.420	.06292	.63066	.32385	1.46367	.63403	2.20324
2.430	.06178	.62874	.31978	1.46814	.63204	2.22688
2.440	.06065	.62682	.31574	1.47264	.63006	2.25087
2.450	.05955	.62490	.31174	1.47715	.62810	2.27522
2.460	.05846	.62299	.30778	1.48169	.62615	2.29994
2.470	.05740	.62108	.30387	1.48624	.62422	2.32503
2.480	.05635	.61918	.29999	1.49081	.62230	2.35050
2.490	.05532	.61728	.29615	1.49539	.62040	2.37635

TABLE IX. Isentropic Exponent = 1.20

M	p/p_t	T/T_t	GAMMA	ISOTHERMAL T_t/T_t^*	RAYLEIGH T_t/T_t^*	SHOCK p_{t1}/p_{t2}
2.500	.05431	.61538	.29235	1.50000	.61851	2.40258
2.510	.05332	.61349	.28859	1.50462	.61664	2.42922
2.520	.05234	.61161	.28488	1.50927	.61478	2.45624
2.530	.05138	.60972	.28120	1.51393	.61293	2.48368
2.540	.05044	.60784	.27756	1.51861	.61110	2.51152
2.550	.04951	.60597	.27395	1.52331	.60929	2.53978
2.560	.04860	.60410	.27039	1.52802	.60748	2.56846
2.570	.04771	.60223	.26687	1.53276	.60569	2.59756
2.580	.04683	.60037	.26338	1.53751	.60392	2.62710
2.590	.04597	.59851	.25993	1.54229	.60216	2.65708
2.600	.04512	.59666	.25652	1.54708	.60041	2.68750
2.610	.04429	.59481	.25315	1.55189	.59867	2.71837
2.620	.04347	.59297	.24982	1.55671	.59695	2.74971
2.630	.04267	.59113	.24652	1.56156	.59524	2.78150
2.640	.04188	.58929	.24326	1.56642	.59355	2.81377
2.650	.04110	.58746	.24004	1.57131	.59186	2.84651
2.660	.04034	.58563	.23685	1.57621	.59019	2.87973
2.670	.03959	.58381	.23370	1.58113	.58854	2.91345
2.680	.03886	.58199	.23059	1.58607	.58689	2.94766
2.690	.03814	.58018	.22751	1.59102	.58526	2.98238
2.700	.03743	.57837	.22447	1.59600	.58364	3.01761
2.710	.03674	.57657	.22146	1.60099	.58203	3.05336
2.720	.03605	.57477	.21849	1.60601	.58044	3.08963
2.730	.03538	.57297	.21555	1.61104	.57885	3.12644
2.740	.03472	.57118	.21265	1.61609	.57728	3.16379
2.750	.03408	.56940	.20978	1.62115	.57572	3.20168
2.760	.03344	.56761	.20695	1.62624	.57418	3.24013
2.770	.03282	.56584	.20415	1.63134	.57264	3.27915
2.780	.03221	.56407	.20138	1.63647	.57112	3.31873
2.790	.03161	.56230	.19865	1.64161	.56961	3.35890
2.800	.03102	.56054	.19595	1.64677	.56811	3.39965
2.810	.03044	.55878	.19328	1.65195	.56662	3.44100
2.820	.02987	.55703	.19065	1.65714	.56514	3.48296
2.830	.02931	.55528	.18805	1.66236	.56367	3.52552
2.840	.02877	.55354	.18548	1.66759	.56222	3.56871
2.850	.02823	.55180	.18294	1.67285	.56077	3.61253
2.860	.02770	.55007	.18043	1.67812	.55934	3.65699
2.870	.02718	.54834	.17796	1.68341	.55792	3.70210
2.880	.02667	.54662	.17551	1.68871	.55650	3.74786
2.890	.02617	.54490	.17310	1.69404	.55510	3.79429
2.900	.02568	.54318	.17071	1.69938	.55371	3.84139
2.910	.02520	.54147	.16836	1.70475	.55233	3.88918
2.920	.02473	.53977	.16603	1.71013	.55096	3.93767
2.930	.02427	.53807	.16374	1.71553	.54960	3.98686
2.940	.02381	.53638	.16147	1.72095	.54825	4.03676
2.950	.02337	.53469	.15923	1.72638	.54691	4.08739
2.960	.02293	.53300	.15702	1.73184	.54558	4.13876
2.970	.02250	.53132	.15484	1.73731	.54426	4.19086
2.980	.02208	.52965	.15269	1.74281	.54295	4.24373
2.990	.02166	.52798	.15057	1.74832	.54165	4.29736

TABLE IX.   Isentropic Exponent = 1.20

M	p/p_t	T/T_t	GAMMA	ISOTHERMAL T_t/T_t^*	RAYLEIGH T_t/T_t^*	SHOCK p_{ti}/p_{t2}
3.000	.02126	.52632	.14847	1.75385	.54036	4.35177
3.010	.02086	.52466	.14640	1.75939	.53908	4.40696
3.020	.02047	.52300	.14436	1.76496	.53781	4.46296
3.030	.02008	.52135	.14234	1.77054	.53655	4.51976
3.040	.01970	.51971	.14035	1.77615	.53530	4.57738
3.050	.01933	.51807	.13838	1.78177	.53405	4.63584
3.060	.01897	.51643	.13645	1.78741	.53282	4.69514
3.070	.01861	.51480	.13453	1.79307	.53159	4.75529
3.080	.01826	.51318	.13264	1.79874	.53038	4.81632
3.090	.01792	.51156	.13078	1.80444	.52917	4.87822
3.100	.01758	.50994	.12894	1.81015	.52797	4.94102
3.110	.01725	.50833	.12713	1.81589	.52679	5.00472
3.120	.01693	.50673	.12534	1.82164	.52561	5.06934
3.130	.01661	.50513	.12357	1.82741	.52443	5.13489
3.140	.01630	.50353	.12183	1.83319	.52327	5.20138
3.150	.01599	.50195	.12011	1.83900	.52212	5.26883
3.160	.01569	.50036	.11841	1.84482	.52097	5.33725
3.170	.01540	.49878	.11674	1.85067	.51983	5.40665
3.180	.01511	.49721	.11509	1.85653	.51870	5.47705
3.190	.01482	.49564	.11346	1.86241	.51758	5.54847
3.200	.01455	.49407	.11185	1.86831	.51647	5.62090
3.210	.01427	.49251	.11027	1.87422	.51536	5.69437
3.220	.01400	.49096	.10871	1.88016	.51427	5.76890
3.230	.01374	.48941	.10716	1.88611	.51318	5.84450
3.240	.01348	.48786	.10564	1.89209	.51210	5.92117
3.250	.01323	.48632	.10414	1.89808	.51102	5.99895
3.260	.01298	.48479	.10266	1.90409	.50996	6.07784
3.270	.01274	.48326	.10120	1.91011	.50890	6.15786
3.280	.01250	.48173	.09976	1.91616	.50785	6.23902
3.290	.01226	.48021	.09834	1.92222	.50681	6.32135
3.300	.01203	.47870	.09694	1.92831	.50577	6.40484
3.310	.01181	.47719	.09556	1.93441	.50474	6.48954
3.320	.01159	.47568	.09420	1.94053	.50372	6.57544
3.330	.01137	.47418	.09286	1.94667	.50271	6.66256
3.340	.01115	.47269	.09153	1.95282	.50170	6.75093
3.350	.01095	.47120	.09022	1.95900	.50070	6.84056
3.360	.01074	.46971	.08894	1.96519	.49971	6.93146
3.370	.01054	.46823	.08767	1.97141	.49873	7.02366
3.380	.01034	.46676	.08641	1.97764	.49775	7.11717
3.390	.01015	.46529	.08518	1.98389	.49678	7.21202
3.400	.00996	.46382	.08396	1.99015	.49582	7.30821
3.410	.00977	.46236	.08276	1.99644	.49486	7.40576
3.420	.00959	.46091	.08157	2.00274	.49391	7.50471
3.430	.00941	.45946	.08041	2.00907	.49296	7.60506
3.440	.00923	.45801	.07926	2.01541	.49203	7.70683
3.450	.00906	.45657	.07812	2.02177	.49110	7.81005
3.460	.00889	.45513	.07700	2.02815	.49017	7.91473
3.470	.00872	.45370	.07590	2.03454	.48926	8.02089
3.480	.03856	.45228	.07481	2.04096	.48834	8.12856
3.490	.00840	.45085	.07374	2.04739	.48744	8.23775

TABLE IX. Isentropic Exponent = 1.20

M	p/p$_t$	T/T$_t$	GAMMA	ISOTHERMAL T$_t$/T$_t^*$	RAYLEIGH T$_t$/T$_t^*$	SHOCK p$_{t1}$/p$_{t2}$
3.500	.00824	.44944	.07268	2.05385	.48654	8.34849
3.510	.00809	.44803	.07164	2.06032	.48565	8.46079
3.520	.00794	.44662	.07061	2.06681	.48476	8.57469
3.530	.00779	.44522	.06960	2.07331	.48388	8.69019
3.540	.00764	.44382	.06860	2.07984	.48301	8.80732
3.550	.00750	.44243	.06761	2.08638	.48214	8.92611
3.560	.00736	.44104	.06664	2.09295	.48128	9.04657
3.570	.00722	.43966	.06568	2.09953	.48042	9.16873
3.580	.00709	.43828	.06474	2.10613	.47957	9.29261
3.590	.00696	.43691	.06381	2.11275	.47873	9.41824
3.600	.00683	.43554	.06290	2.11938	.47789	9.54564
3.610	.00670	.43418	.06199	2.12604	.47706	9.67482
3.620	.00657	.43282	.06110	2.13271	.47623	9.80583
3.630	.00645	.43146	.06022	2.13941	.47541	9.93868
3.640	.00633	.43012	.05936	2.14612	.47459	10.07339
3.650	.00621	.42877	.05850	2.15285	.47378	10.21000
3.660	.00610	.42743	.05766	2.15959	.47298	10.34852
3.670	.00598	.42610	.05684	2.16636	.47218	10.48899
3.680	.00587	.42477	.05602	2.17314	.47139	10.63143
3.690	.00576	.42344	.05521	2.17995	.47060	10.77586
3.700	.00566	.42212	.05442	2.18677	.46981	10.92231
3.710	.00555	.42080	.05364	2.19361	.46904	11.07082
3.720	.00545	.41949	.05287	2.20047	.46826	11.22140
3.730	.00535	.41818	.05211	2.20734	.46750	11.37409
3.740	.00525	.41688	.05136	2.21424	.46673	11.52891
3.750	.00515	.41558	.05062	2.22115	.46598	11.68589
3.760	.00506	.41429	.04989	2.22809	.46522	11.84507
3.770	.00496	.41300	.04917	2.23504	.46448	12.00646
3.780	.00487	.41172	.04847	2.24201	.46373	12.17011
3.790	.00478	.41044	.04777	2.24899	.46300	12.33603
3.800	.00469	.40917	.04709	2.25600	.46226	12.50427
3.810	.00461	.40790	.04641	2.26302	.46154	12.67485
3.820	.00452	.40663	.04574	2.27007	.46081	12.84780
3.830	.00444	.40537	.04509	2.27713	.46010	13.02315
3.840	.00436	.40411	.04444	2.28421	.45938	13.20094
3.850	.00427	.40286	.04380	2.29131	.45867	13.38120
3.860	.00420	.40161	.04317	2.29842	.45797	13.56396
3.870	.00412	.40037	.04255	2.30556	.45727	13.74926
3.880	.00404	.39913	.04194	2.31271	.45658	13.93712
3.890	.00397	.39790	.04134	2.31989	.45589	14.12759
3.900	.00390	.39667	.04074	2.32708	.45520	14.32069
3.910	.00382	.39544	.04016	2.33429	.45452	14.51647
3.920	.00375	.39422	.03958	2.34151	.45384	14.71495
3.930	.00368	.39301	.03902	2.34876	.45317	14.91618
3.940	.00362	.39179	.03846	2.35602	.45250	15.12018
3.950	.00355	.39059	.03791	2.36331	.45184	15.32700
3.960	.00349	.38938	.03736	2.37061	.45118	15.53667
3.970	.00342	.38819	.03683	2.37793	.45052	15.74924
3.980	.00336	.38699	.03630	2.38527	.44987	15.96472
3.990	.00330	.38580	.03578	2.39262	.44923	16.18318

TABLE IX. Isentropic Exponent = 1.20

M	p/p_t	T/T_t	GAMMA	ISOTHERMAL T_t/T_t^*	RAYLEIGH T_t/T_t^*	SHOCK p_{ti}/p_{t2}
4.000	.00324	.38462	.03527	2.40000	.44858	16.40467
4.010	.00318	.38343	.03476	2.40739	.44794	16.62917
4.020	.00312	.38226	.03426	2.41481	.44731	16.85677
4.030	.00306	.38108	.03377	2.42224	.44668	17.08748
4.040	.00301	.37992	.03329	2.42969	.44605	17.32137
4.050	.00295	.37875	.03281	2.43715	.44543	17.55846
4.060	.00290	.37759	.03235	2.44464	.44481	17.79880
4.070	.00285	.37644	.03188	2.45214	.44420	18.04244
4.080	.00279	.37529	.03143	2.45967	.44359	18.28941
4.090	.00274	.37414	.03098	2.46721	.44298	18.53975
4.100	.00269	.37300	.03054	2.47477	.44238	18.79352
4.110	.00264	.37186	.03010	2.48235	.44178	19.05075
4.120	.00260	.37072	.02967	2.48994	.44118	19.31150
4.130	.00255	.36959	.02925	2.49756	.44059	19.57581
4.140	.00250	.36847	.02883	2.50519	.44000	19.84372
4.150	.00246	.36734	.02842	2.51285	.43942	20.11527
4.160	.00241	.36623	.02801	2.52052	.43884	20.39053
4.170	.00237	.36511	.02761	2.52821	.43826	20.66952
4.180	.00233	.36400	.02722	2.53591	.43769	20.95232
4.190	.00228	.36290	.02683	2.54364	.43712	21.23896
4.200	.00224	.36179	.02645	2.55138	.43655	21.52948
4.210	.00220	.36070	.02608	2.55915	.43599	21.82395
4.220	.00216	.35960	.02570	2.56693	.43543	22.12241
4.230	.00212	.35851	.02534	2.57473	.43487	22.42492
4.240	.00209	.35743	.02498	2.58255	.43432	22.73152
4.250	.00205	.35635	.02462	2.59038	.43377	23.04227
4.260	.00201	.35527	.02427	2.59824	.43323	23.35722
4.270	.00197	.35420	.02393	2.60611	.43268	23.67643
4.280	.00194	.35313	.02359	2.61401	.43215	23.99994
4.290	.00190	.35206	.02326	2.62192	.43161	24.32783
4.300	.00187	.35100	.02293	2.62985	.43108	24.66013
4.310	.00184	.34994	.02260	2.63779	.43055	24.99691
4.320	.00180	.34889	.02228	2.64576	.43002	25.33823
4.330	.00177	.34784	.02196	2.65374	.42950	25.68414
4.340	.00174	.34679	.02165	2.66175	.42898	26.03470
4.350	.00171	.34575	.02135	2.66977	.42846	26.38996
4.360	.00168	.34471	.02105	2.67781	.42795	26.75000
4.370	.00165	.34368	.02075	2.68587	.42744	27.11486
4.380	.00162	.34265	.02046	2.69394	.42693	27.48463
4.390	.00159	.34162	.02017	2.70204	.42643	27.85934
4.400	.00156	.34060	.01988	2.71015	.42593	28.23906
4.410	.00153	.33958	.01960	2.71829	.42543	28.62387
4.420	.00151	.33857	.01932	2.72644	.42493	29.01383
4.430	.00148	.33755	.01905	2.73461	.42444	29.40898
4.440	.00145	.33655	.01878	2.74279	.42395	29.80941
4.450	.00143	.33554	.01852	2.75100	.42347	30.21518
4.460	.00140	.33454	.01826	2.75922	.42298	30.62636
4.470	.00138	.33355	.01800	2.76747	.42250	31.04302
4.480	.00135	.33255	.01775	2.77573	.42202	31.46521
4.490	.00133	.33156	.01750	2.78401	.42155	31.89302

TABLE IX. Isentropic Exponent = 1.20

M	p/p_t	T/T_t	GAMMA	ISOTHERMAL T_t/T_t^*	RAYLEIGH T_t/T_t^*	SHOCK p_{t1}/p_{t2}
4.500	.00131	.33058	.01725	2.79231	.42108	32.32653
4.510	.00128	.32960	.01701	2.80062	.42061	32.76578
4.520	.00126	.32862	.01677	2.80896	.42014	33.21086
4.530	.00124	.32764	.01654	2.81731	.41968	33.66184
4.540	.00122	.32667	.01631	2.82569	.41922	34.11880
4.550	.00119	.32571	.01608	2.83408	.41876	34.58179
4.560	.00117	.32474	.01585	2.84249	.41830	35.05092
4.570	.00115	.32378	.01563	2.85091	.41785	35.52625
4.580	.00113	.32283	.01541	2.85936	.41740	36.00785
4.590	.00111	.32187	.01520	2.86782	.41695	36.49581
4.600	.00109	.32092	.01498	2.87631	.41651	36.99019
4.610	.00107	.31998	.01477	2.88481	.41606	37.49110
4.620	.00105	.31904	.01457	2.89333	.41562	37.99860
4.630	.00104	.31810	.01437	2.90187	.41519	38.51277
4.640	.00102	.31716	.01417	2.91042	.41475	39.03370
4.650	.00100	.31623	.01397	2.91900	.41432	39.56148
4.660	.00098	.31530	.01377	2.92759	.41389	40.09619
4.670	.00097	.31438	.01358	2.93621	.41346	40.63790
4.680	.00095	.31346	.01339	2.94484	.41304	41.18670
4.690	.00093	.31254	.01321	2.95349	.41261	41.74270
4.700	.00092	.31162	.01302	2.96215	.41219	42.30597
4.710	.00090	.31071	.01284	2.97084	.41177	42.87658
4.720	.00088	.30980	.01266	2.97954	.41136	43.45467
4.730	.00087	.30890	.01249	2.98827	.41095	44.04031
4.740	.00085	.30800	.01232	2.99701	.41054	44.63356
4.750	.00084	.30710	.01215	3.00577	.41013	45.23456
4.760	.00082	.30621	.01198	3.01455	.40972	45.84338
4.770	.00081	.30532	.01181	3.02334	.40932	46.46012
4.780	.00080	.30443	.01165	3.03216	.40892	47.08487
4.790	.00078	.30354	.01149	3.04099	.40852	47.71774
4.800	.00077	.30266	.01133	3.04985	.40812	48.35882
4.810	.00076	.30179	.01117	3.05872	.40772	49.00821
4.820	.00074	.30091	.01102	3.06761	.40733	49.66603
4.830	.00073	.30004	.01087	3.07651	.40694	50.33235
4.840	.00072	.29917	.01072	3.08544	.40655	51.00728
4.850	.00070	.29831	.01057	3.09438	.40617	51.69094
4.860	.00069	.29745	.01042	3.10335	.40578	52.38344
4.870	.00068	.29659	.01028	3.11233	.40540	53.08488
4.880	.00067	.29573	.01014	3.12133	.40502	53.79535
4.890	.00066	.29488	.01000	3.13035	.40464	54.51497
4.900	.00065	.29403	.00986	3.13938	.40427	55.24386
4.910	.00064	.29319	.00973	3.14844	.40389	55.98211
4.920	.00062	.29234	.00959	3.15751	.40352	56.72985
4.930	.00061	.29150	.00946	3.16661	.40315	57.48720
4.940	.00060	.29067	.00933	3.17572	.40279	58.25427
4.950	.00059	.28983	.00921	3.18485	.40242	59.03115
4.960	.00058	.28900	.00908	3.19399	.40206	59.81800
4.970	.00057	.28818	.00896	3.20316	.40170	60.61491
4.980	.00056	.28735	.00883	3.21234	.40134	61.42201
4.990	.00055	.28653	.00871	3.22155	.40098	62.23942

TABLE X.   Isentropic Exponent = 1.30

M	p/p_t	T/T_t	GAMMA	ISOTHERMAL T_t/T_t^*	RAYLEIGH T_t/T_t^*	SHOCK p_{ti}/p_{t2}
.000	1.00000	1.00000	.00000	.89655	.00000	1.
.005	.99998	.00000	.00853	.89656	.00011	1.
.010	.99994	.99999	.01707	.89657	.00046	1.
.015	.99985	.99997	.02562	.89658	.00103	1.
.020	.99974	.99994	.03416	.89661	.00184	1.
.025	.99959	.99991	.04270	.89664	.00287	1.
.030	.99942	.99987	.05123	.89667	.00413	1.
.035	.99920	.99982	.05976	.89672	.00562	1.
.040	.99896	.99976	.06829	.89677	.00733	1.
.045	.99868	.99970	.07680	.89682	.00927	1.
.050	.99838	.99963	.08531	.89689	.01143	1.
.055	.99804	.99955	.09382	.89696	.01381	1.
.060	.99766	.99946	.10231	.89704	.01641	1.
.065	.99726	.99937	.11080	.89712	.01923	1.
.070	.99682	.99927	.11927	.89721	.02227	1.
.075	.99635	.99916	.12774	.89731	.02552	1.
.080	.99585	.99904	.13620	.89741	.02898	1.
.085	.99532	.99892	.14464	.89752	.03265	1.
.090	.99475	.99879	.15307	.89764	.03653	1.
.095	.99415	.99865	.16149	.89777	.04061	1.
.100	.99353	.99850	.16989	.89790	.04489	1.
.105	.99287	.99835	.17828	.89803	.04937	1.
.110	.99217	.99819	.18666	.89818	.05405	1.
.115	.99145	.99802	.19502	.89833	.05891	1.
.120	.99069	.99784	.20336	.89849	.06397	1.
.125	.98991	.99766	.21168	.89865	.06920	1.
.130	.98909	.99747	.21999	.89882	.07462	1.
.135	.98824	.99727	.22828	.89900	.08022	1.
.140	.98736	.99707	.23655	.89919	.08599	1.
.145	.98645	.99686	.24479	.89938	.09193	1.
.150	.98551	.99664	.25302	.89958	.09803	1.
.155	.98453	.99641	.26123	.89978	.10430	1.
.160	.98353	.99617	.26941	.89999	.11072	1.
.165	.98249	.99593	.27757	.90021	.11730	1.
.170	.98143	.99568	.28571	.90044	.12402	1.
.175	.98034	.99543	.29382	.90067	.13089	1.
.180	.97921	.99516	.30191	.90091	.13790	1.
.185	.97806	.99489	.30997	.90115	.14505	1.
.190	.97687	.99461	.31801	.90141	.15233	1.
.195	.97566	.99433	.32602	.90167	.15973	1.
.200	.97441	.99404	.33400	.90193	.16726	1.
.205	.97314	.99374	.34195	.90220	.17490	1.
.210	.97183	.99343	.34988	.90248	.18266	1.
.215	.97050	.99311	.35777	.90277	.19052	1.
.220	.96914	.99279	.36564	.90306	.19849	1.
.225	.96775	.99246	.37348	.90336	.20656	1.
.230	.96633	.99213	.38128	.90367	.21472	1.
.235	.96488	.99178	.38905	.90398	.22297	1.
.240	.96341	.99143	.39679	.90430	.23131	1.
.245	.96190	.99108	.40450	.90462	.23973	1.

TABLE X. Isentropic Exponent = 1.30

M	p/p_t	T/T_t	GAMMA	ISOTHERMAL T_t/T_t^*	RAYLEIGH T_t/T_t^*	SHOCK p_{ti}/p_{t2}
.250	.96037	.99071	.41217	.90496	.24822	1.
.255	.95881	.99034	.41981	.90530	.25678	1.
.260	.95722	.98996	.42742	.90564	.26541	1.
.265	.95561	.98958	.43499	.90600	.27411	1.
.270	.95397	.98918	.44252	.90636	.28285	1.
.275	.95230	.98878	.45002	.90672	.29166	1.
.280	.95060	.98838	.45748	.90710	.30050	1.
.285	.94888	.98796	.46490	.90748	.30940	1.
.290	.94713	.98754	.47228	.90786	.31833	1.
.295	.94535	.98711	.47963	.90826	.32730	1.
.300	.94355	.98668	.48694	.90866	.33629	1.
.305	.94172	.98624	.49420	.90906	.34532	1.
.310	.93986	.98579	.50143	.90948	.35436	1.
.315	.93799	.98533	.50862	.90990	.36342	1.
.320	.93608	.98487	.51576	.91032	.37250	1.
.325	.93415	.98440	.52286	.91076	.38159	1.
.330	.93220	.98393	.52992	.91120	.39068	1.
.335	.93022	.98344	.53694	.91164	.39977	1.
.340	.92821	.98296	.54392	.91210	.40886	1.
.345	.92618	.98246	.55085	.91256	.41794	1.
.350	.92413	.98196	.55774	.91303	.42702	1.
.355	.92205	.98145	.56458	.91350	.43608	1.
.360	.91995	.98093	.57138	.91398	.44512	1.
.365	.91783	.98041	.57813	.91447	.45415	1.
.370	.91568	.97988	.58484	.91496	.46315	1.
.375	.91351	.97934	.59150	.91546	.47212	1.
.380	.91132	.97880	.59811	.91597	.48106	1.
.385	.90911	.97825	.60468	.91649	.48997	1.
.390	.90687	.97769	.61120	.91701	.49885	1.
.395	.90461	.97713	.61767	.91753	.50768	1.
.400	.90233	.97656	.62410	.91807	.51647	1.
.405	.90003	.97599	.63047	.91861	.52521	1.
.410	.89771	.97541	.63680	.91916	.53391	1.
.415	.89536	.97482	.64307	.91971	.54256	1.
.420	.89300	.97422	.64930	.92027	.55115	1.
.425	.89061	.97362	.65548	.92084	.55968	1.
.430	.88821	.97301	.66161	.92142	.56816	1.
.435	.88578	.97240	.66768	.92200	.57658	1.
.440	.88334	.97178	.67371	.92259	.58494	1.
.445	.88087	.97115	.67968	.92318	.59323	1.
.450	.87839	.97052	.68560	.92378	.60145	1.
.455	.87589	.96988	.69147	.92439	.60960	1.
.460	.87336	.96924	.69729	.92501	.61769	1.
.465	.87082	.96859	.70306	.92563	.62570	1.
.470	.86827	.96793	.70877	.92626	.63363	1.
.475	.86569	.96726	.71443	.92689	.64149	1.
.480	.86310	.96659	.72003	.92754	.64928	1.
.485	.86048	.96592	.72559	.92819	.65698	1.
.490	.85785	.96524	.73109	.92884	.66460	1.
.495	.85521	.96455	.73653	.92950	.67214	1.

TABLE X. Isentropic Exponent = 1.30

M	p/p_t	T/T_t	GAMMA	ISOTHERMAL T_t/T_t^*	RAYLEIGH T_t/T_t^*	SHOCK p_{ti}/p_{t2}
.500	.85255	.96386	.74192	.93017	.67960	1.
.505	.84987	.96316	.74726	.93085	.68697	1.
.510	.84717	.96245	.75254	.93153	.69426	1.
.515	.84446	.96174	.75776	.93222	.70146	1.
.520	.84174	.96102	.76294	.93292	.70858	1.
.525	.83899	.96030	.76805	.93362	.71560	1.
.530	.83624	.95957	.77311	.93433	.72254	1.
.535	.83347	.95883	.77812	.93504	.72938	1.
.540	.83068	.95809	.78307	.93577	.73614	1.
.545	.82788	.95735	.78796	.93650	.74280	1.
.550	.82506	.95659	.79280	.93723	.74937	1.
.555	.82224	.95584	.79758	.93798	.75585	1.
.560	.81939	.95507	.80230	.93873	.76224	1.
.565	.81654	.95430	.80697	.93948	.76853	1.
.570	.81367	.95353	.81158	.94025	.77473	1.
.575	.81079	.95275	.81613	.94102	.78083	1.
.580	.80790	.95196	.82063	.94179	.78684	1.
.585	.80499	.95117	.82507	.94258	.79276	1.
.590	.80207	.95038	.82946	.94337	.79858	1.
.595	.79914	.94957	.83378	.94416	.80430	1.
.600	.79620	.94877	.83805	.94497	.80993	1.
.605	.79325	.94795	.84226	.94578	.81547	1.
.610	.79029	.94714	.84642	.94659	.82091	1.
.615	.78732	.94631	.85052	.94742	.82626	1.
.620	.78433	.94548	.85456	.94825	.83151	1.
.625	.78134	.94465	.85854	.94908	.83667	1.
.630	.77834	.94381	.86246	.94993	.84173	1.
.635	.77532	.94297	.86633	.95078	.84670	1.
.640	.77230	.94212	.87014	.95164	.85158	1.
.645	.76927	.94126	.87389	.95250	.85636	1.
.650	.76623	.94040	.87759	.95337	.86105	1.
.655	.76318	.93954	.88122	.95425	.86565	1.
.660	.76012	.93867	.88480	.95513	.87015	1.
.665	.75706	.93779	.88833	.95602	.87457	1.
.670	.75399	.93691	.89179	.95692	.87889	1.
.675	.75091	.93603	.89520	.95783	.88312	1.
.680	.74782	.93514	.89855	.95874	.88726	1.
.685	.74472	.93424	.90184	.95965	.89131	1.
.690	.74162	.93335	.90508	.96058	.89528	1.
.695	.73852	.93244	.90826	.96151	.89915	1.
.700	.73540	.93153	.91138	.96245	.90294	1.
.705	.73228	.93062	.91444	.96339	.90664	1.
.710	.72916	.92970	.91745	.96434	.91025	1.
.715	.72602	.92878	.92040	.96530	.91378	1.
.720	.72289	.92785	.92329	.96627	.91722	1.
.725	.71975	.92692	.92613	.96724	.92058	1.
.730	.71660	.92598	.92891	.96822	.92386	1.
.735	.71345	.92504	.93163	.96920	.92705	1.
.740	.71029	.92409	.93430	.97019	.93016	1.
.745	.70714	.92314	.93691	.97119	.93319	1.

TABLE X. Isentropic Exponent = 1.30

M	p/p_t	T/T_t	GAMMA	ISOTHERMAL $T_t/T_t{}^*$	RAYLEIGH $T_t/T_t{}^*$	SHOCK p_{ti}/p_{t2}
.750	.70397	.92219	.93947	.97220	.93614	1.
.755	.70081	.92123	.94197	.97321	.93901	1.
.760	.69764	.92027	.94441	.97423	.94180	1.
.765	.69446	.91930	.94680	.97525	.94451	1.
.770	.69129	.91833	.94913	.97629	.94715	1.
.775	.68811	.91735	.95140	.97733	.94970	1.
.780	.68493	.91637	.95363	.97837	.95219	1.
.785	.68174	.91539	.95579	.97942	.95460	1.
.790	.67856	.91440	.95790	.98048	.95693	1.
.795	.67537	.91341	.95996	.98155	.95919	1.
.800	.67218	.91241	.96196	.98262	.96139	1.
.805	.66899	.91141	.96391	.98370	.96350	1.
.810	.66580	.91040	.96580	.98479	.96555	1.
.815	.66261	.90939	.96764	.98588	.96753	1.
.820	.65942	.90838	.96943	.98698	.96944	1.
.825	.65623	.90736	.97116	.98808	.97129	1.
.830	.65303	.90634	.97284	.98920	.97306	1.
.835	.64984	.90532	.97447	.99032	.97478	1.
.840	.64665	.90429	.97604	.99144	.97642	1.
.845	.64345	.90326	.97756	.99258	.97800	1.
.850	.64026	.90222	.97903	.99372	.97952	1.
.855	.63707	.90118	.98044	.99486	.98098	1.
.860	.63388	.90014	.98181	.99602	.98238	1.
.865	.63069	.89909	.98312	.99718	.98371	1.
.870	.62750	.89804	.98438	.99834	.98499	1.
.875	.62432	.89699	.98559	.99951	.98620	1.
.880	.62113	.89593	.98675	1.00070	.98736	1.
.885	.61795	.89487	.98786	1.00188	.98846	1.
.890	.61477	.89380	.98891	1.00308	.98951	1.
.895	.61159	.89273	.98992	1.00428	.99050	1.
.900	.60842	.89166	.99088	1.00548	.99143	1.
.905	.60525	.89059	.99179	1.00670	.99232	1.
.910	.60208	.88951	.99264	1.00792	.99315	1.
.915	.59891	.88843	.99345	1.00914	.99392	1.
.920	.59575	.88734	.99421	1.01038	.99465	1.
.925	.59259	.88625	.99493	1.01162	.99533	1.
.930	.58943	.88516	.99559	1.01287	.99596	1.
.935	.58628	.88407	.99621	1.01412	.99653	1.
.940	.58313	.88297	.99678	1.01538	.99707	1.
.945	.57999	.88187	.99730	1.01665	.99755	1.
.950	.57685	.88077	.99777	1.01792	.99799	1.
.955	.57372	.87966	.99820	1.01920	.99838	1.
.960	.57058	.87855	.99858	1.02049	.99873	1.
.965	.56746	.87744	.99892	1.02179	.99903	1.
.970	.56434	.87632	.99921	1.02309	.99929	1.
.975	.56122	.87520	.99945	1.02439	.99951	1.
.980	.55811	.87408	.99965	1.02571	.99969	1.
.985	.55501	.87296	.99980	1.02703	.99983	1.
.990	.55191	.87183	.99991	1.02836	.99992	1.
.995	.54882	.87070	.99998	1.02969	.99998	1.

TABLE X.   Isentropic Exponent = 1.30

M	p/p_t	T/T_t	GAMMA	ISOTHERMAL T_t/T_t^*	RAYLEIGH T_t/T_t^*	SHOCK p_{tf}/p_{t2}
1.000	.54573	.86957	1.00000	1.03103	1.00000	1.00000
1.005	.54265	.86843	.99998	1.03238	.99998	1.00000
1.010	.53957	.86729	.99991	1.03374	.99993	1.00000
1.015	.53650	.86615	.99981	1.03510	.99983	1.00000
1.020	.53344	.86501	.99966	1.03647	.99971	1.00001
1.025	.53038	.86386	.99946	1.03784	.99954	1.00002
1.030	.52733	.86271	.99923	1.03922	.99934	1.00003
1.035	.52429	.86156	.99895	1.04061	.99911	1.00005
1.040	.52126	.86041	.99864	1.04201	.99885	1.00008
1.045	.51823	.85925	.99828	1.04341	.99855	1.00011
1.050	.51521	.85809	.99788	1.04482	.99823	1.00015
1.055	.51219	.85693	.99744	1.04623	.99787	1.00020
1.060	.50919	.85577	.99697	1.04766	.99748	1.00025
1.065	.50619	.85460	.99645	1.04909	.99706	1.00032
1.070	.50320	.85344	.99589	1.05052	.99661	1.00040
1.075	.50022	.85227	.99530	1.05196	.99613	1.00048
1.080	.49724	.85109	.99467	1.05341	.99563	1.00058
1.085	.49427	.84992	.99399	1.05487	.99509	1.00069
1.090	.49132	.84874	.99329	1.05633	.99454	1.00081
1.095	.48837	.84756	.99254	1.05780	.99395	1.00095
1.100	.48542	.84638	.99176	1.05928	.99334	1.00110
1.105	.48249	.84520	.99094	1.06076	.99270	1.00126
1.110	.47957	.84401	.99008	1.06225	.99204	1.00143
1.115	.47665	.84283	.98919	1.06374	.99136	1.00163
1.120	.47374	.84164	.98826	1.06525	.99065	1.00183
1.125	.47084	.84045	.98730	1.06676	.98992	1.00205
1.130	.46795	.83925	.98630	1.06827	.98916	1.00229
1.135	.46507	.83806	.98527	1.06980	.98839	1.00254
1.140	.46220	.83686	.98421	1.07133	.98759	1.00282
1.145	.45934	.83566	.98311	1.07286	.98677	1.00310
1.150	.45649	.83446	.98198	1.07441	.98593	1.00341
1.155	.45365	.83326	.98081	1.07596	.98508	1.00373
1.160	.45081	.83206	.97961	1.07751	.98420	1.00407
1.165	.44799	.83085	.97838	1.07908	.98330	1.00443
1.170	.44518	.82964	.97712	1.08065	.98239	1.00481
1.175	.44237	.82844	.97583	1.08222	.98145	1.00520
1.180	.43958	.82723	.97450	1.08381	.98050	1.00562
1.185	.43680	.82601	.97315	1.08540	.97953	1.00605
1.190	.43402	.82480	.97176	1.08699	.97855	1.00651
1.195	.43126	.82359	.97034	1.08860	.97755	1.00698
1.200	.42850	.82237	.96890	1.09021	.97653	1.00748
1.205	.42576	.82115	.96742	1.09182	.97550	1.00799
1.210	.42303	.81993	.96592	1.09345	.97445	1.00853
1.215	.42030	.81871	.96438	1.09508	.97338	1.00908
1.220	.41759	.81749	.96282	1.09672	.97231	1.00966
1.225	.41489	.81626	.96123	1.09836	.97122	1.01026
1.230	.41220	.81504	.95962	1.10001	.97011	1.01088
1.235	.40952	.81381	.95797	1.10167	.96899	1.01152
1.240	.40685	.81259	.95630	1.10333	.96786	1.01218
1.245	.40419	.81136	.95460	1.10500	.96672	1.01287

TABLE X. Isentropic Exponent = 1.30

M	p/p$_t$	T/T$_t$	GAMMA	ISOTHERMAL T$_t$/T$_t$*	RAYLEIGH T$_t$/T$_t$*	SHOCK p$_{ti}$/p$_{t2}$
1.250	.40154	.81013	.95288	1.10668	.96556	1.01357
1.255	.39890	.80890	.95113	1.10837	.96440	1.01430
1.260	.39628	.80766	.94935	1.11006	.96322	1.01506
1.265	.39366	.80643	.94755	1.11175	.96203	1.01583
1.270	.39106	.80520	.94573	1.11346	.96083	1.01663
1.275	.38846	.80396	.94388	1.11517	.95962	1.01745
1.280	.38588	.80272	.94201	1.11689	.95840	1.01829
1.285	.38331	.80149	.94011	1.11861	.95717	1.01916
1.290	.38075	.80025	.93819	1.12034	.95593	1.02004
1.295	.37820	.79901	.93624	1.12208	.95468	1.02096
1.300	.37566	.79777	.93428	1.12383	.95342	1.02189
1.305	.37313	.79652	.93229	1.12558	.95215	1.02285
1.310	.37062	.79528	.93028	1.12734	.95088	1.02384
1.315	.36811	.79404	.92824	1.12910	.94959	1.02484
1.320	.36562	.79280	.92619	1.13087	.94830	1.02588
1.325	.36314	.79155	.92411	1.13265	.94700	1.02693
1.330	.36067	.79030	.92202	1.13444	.94570	1.02801
1.335	.35821	.78906	.91990	1.13623	.94438	1.02912
1.340	.35576	.78781	.91776	1.13803	.94306	1.03024
1.345	.35333	.78656	.91561	1.13983	.94174	1.03140
1.350	.35090	.78531	.91343	1.14165	.94041	1.03258
1.355	.34849	.78407	.91124	1.14347	.93907	1.03378
1.360	.34609	.78282	.90902	1.14529	.93772	1.03501
1.365	.34370	.78157	.90679	1.14712	.93637	1.03626
1.370	.34133	.78031	.90454	1.14896	.93502	1.03754
1.375	.33896	.77906	.90227	1.15081	.93366	1.03884
1.380	.33660	.77781	.89999	1.15266	.93229	1.04017
1.385	.33426	.77656	.89769	1.15452	.93092	1.04152
1.390	.33193	.77531	.89537	1.15639	.92955	1.04290
1.395	.32961	.77405	.89303	1.15826	.92817	1.04430
1.400	.32730	.77280	.89068	1.16014	.92679	1.04573
1.405	.32501	.77154	.88831	1.16202	.92540	1.04718
1.410	.32272	.77029	.88593	1.16392	.92401	1.04866
1.415	.32045	.76903	.88353	1.16582	.92262	1.05017
1.420	.31819	.76778	.88112	1.16772	.92122	1.05170
1.425	.31594	.76652	.87869	1.16964	.91982	1.05326
1.430	.31370	.76527	.87625	1.17156	.91842	1.05485
1.435	.31148	.76401	.87379	1.17348	.91702	1.05646
1.440	.30927	.76275	.87132	1.17541	.91561	1.05809
1.445	.30706	.76150	.86883	1.17735	.91420	1.05975
1.450	.30487	.76024	.86634	1.17930	.91279	1.06144
1.455	.30269	.75898	.86383	1.18125	.91137	1.06316
1.460	.30053	.75773	.86130	1.18321	.90996	1.06490
1.465	.29837	.75647	.85877	1.18518	.90854	1.06667
1.470	.29623	.75521	.85622	1.18716	.90712	1.06846
1.475	.29410	.75395	.85366	1.18914	.90570	1.07029
1.480	.29198	.75269	.85109	1.19112	.90428	1.07213
1.485	.28987	.75144	.84851	1.19312	.90285	1.07401
1.490	.28777	.75018	.84592	1.19512	.90143	1.07591
1.495	.28569	.74892	.84331	1.19712	.90000	1.07784

TABLE X. Isentropic Exponent = 1.30

M	p/p_t	T/T_t	GAMMA	ISOTHERMAL T_t/T_t^*	RAYLEIGH T_t/T_t^*	SHOCK p_{ti}/p_{t2}
1.500	.28361	.74766	.84070	1.19914	.89858	1.07980
1.505	.28155	.74641	.83807	1.20116	.89715	1.08178
1.510	.27950	.74515	.83544	1.20319	.89572	1.08379
1.515	.27746	.74389	.83280	1.20522	.89429	1.08583
1.520	.27544	.74263	.83014	1.20726	.89287	1.08789
1.525	.27342	.74138	.82748	1.20931	.89144	1.08999
1.530	.27142	.74012	.82481	1.21136	.89001	1.09211
1.535	.26943	.73886	.82213	1.21342	.88858	1.09425
1.540	.26745	.73760	.81944	1.21549	.88716	1.09643
1.545	.26548	.73635	.81675	1.21757	.88573	1.09863
1.550	.26352	.73509	.81405	1.21965	.88430	1.10086
1.555	.26157	.73384	.81133	1.22173	.88288	1.10312
1.560	.25964	.73258	.80862	1.22383	.88145	1.10541
1.565	.25772	.73132	.80589	1.22593	.88003	1.10773
1.570	.25580	.73007	.80316	1.22804	.87860	1.11007
1.575	.25390	.72881	.80042	1.23015	.87718	1.11244
1.580	.25202	.72756	.79767	1.23227	.87576	1.11484
1.585	.25014	.72630	.79492	1.23440	.87434	1.11727
1.590	.24827	.72505	.79217	1.23654	.87292	1.11973
1.595	.24642	.72380	.78940	1.23868	.87150	1.12221
1.600	.24457	.72254	.78664	1.24083	.87008	1.12472
1.605	.24274	.72129	.78386	1.24298	.86866	1.12727
1.610	.24092	.72004	.78108	1.24514	.86725	1.12984
1.615	.23911	.71879	.77830	1.24731	.86584	1.13244
1.620	.23731	.71754	.77551	1.24949	.86443	1.13507
1.625	.23552	.71628	.77272	1.25167	.86302	1.13773
1.630	.23375	.71503	.76992	1.25386	.86161	1.14041
1.635	.23198	.71378	.76712	1.25605	.86021	1.14313
1.640	.23023	.71253	.76432	1.25826	.85880	1.14588
1.645	.22849	.71129	.76151	1.26047	.85740	1.14865
1.650	.22675	.71004	.75870	1.26268	.85600	1.15146
1.655	.22503	.70879	.75589	1.26490	.85461	1.15429
1.660	.22332	.70754	.75307	1.26713	.85321	1.15715
1.665	.22162	.70630	.75026	1.26937	.85182	1.16005
1.670	.21993	.70505	.74743	1.27161	.85043	1.16297
1.675	.21826	.70381	.74461	1.27386	.84904	1.16592
1.680	.21659	.70256	.74178	1.27612	.84766	1.16891
1.685	.21493	.70132	.73896	1.27838	.84628	1.17192
1.690	.21329	.70008	.73613	1.28065	.84490	1.17497
1.695	.21165	.69883	.73330	1.28292	.84352	1.17804
1.700	.21003	.69759	.73047	1.28521	.84215	1.18114
1.705	.20841	.69635	.72763	1.28750	.84077	1.18428
1.710	.20681	.69511	.72480	1.28979	.83941	1.18745
1.715	.20522	.69387	.72196	1.29210	.83804	1.19064
1.720	.20364	.69264	.71913	1.29441	.83668	1.19387
1.725	.20206	.69140	.71629	1.29672	.83532	1.19713
1.730	.20050	.69016	.71346	1.29905	.83396	1.20042
1.735	.19895	.68893	.71062	1.30137	.83260	1.20374
1.740	.19741	.68769	.70778	1.30371	.83125	1.20709
1.745	.19588	.68646	.70495	1.30605	.82991	1.21047

TABLE X. Isentropic Exponent = 1.30

M	p/p_t	T/T_t	GAMMA	ISOTHERMAL T_t/T_t^*	RAYLEIGH T_t/T_t^*	SHOCK p_{ti}/p_{t2}
1.750	.19436	.68522	.70211	1.30841	.82856	1.21389
1.755	.19285	.68399	.69928	1.31076	.82722	1.21733
1.760	.19135	.68276	.69644	1.31313	.82588	1.22081
1.765	.18986	.68153	.69361	1.31550	.82454	1.22432
1.770	.18838	.68030	.69078	1.31787	.82321	1.22786
1.775	.18691	.67907	.68795	1.32026	.82188	1.23144
1.780	.18545	.67785	.68512	1.32265	.82056	1.23504
1.785	.18400	.67662	.68229	1.32504	.81924	1.23868
1.790	.18257	.67540	.67946	1.32745	.81792	1.24235
1.795	.18114	.67417	.67664	1.32986	.81660	1.24605
1.800	.17972	.67295	.67382	1.33228	.81529	1.24979
1.805	.17831	.67173	.67100	1.33470	.81398	1.25355
1.810	.17690	.67050	.66818	1.33713	.81268	1.25736
1.815	.17551	.66928	.66536	1.33957	.81138	1.26119
1.820	.17413	.66807	.66255	1.34201	.81008	1.26506
1.825	.17276	.66685	.65974	1.34446	.80879	1.26896
1.830	.17140	.66563	.65693	1.34692	.80750	1.27289
1.835	.17005	.66441	.65413	1.34939	.80621	1.27686
1.840	.16870	.66320	.65133	1.35186	.80492	1.28086
1.845	.16737	.66199	.64853	1.35433	.80365	1.28489
1.850	.16605	.66077	.64573	1.35682	.80237	1.28896
1.855	.16473	.65956	.64294	1.35931	.80110	1.29306
1.860	.16343	.65835	.64015	1.36181	.79983	1.29720
1.865	.16213	.65715	.63737	1.36431	.79856	1.30137
1.870	.16084	.65594	.63459	1.36682	.79730	1.30557
1.875	.15957	.65473	.63181	1.36934	.79605	1.30981
1.880	.15830	.65353	.62904	1.37187	.79479	1.31408
1.885	.15704	.65232	.62627	1.37440	.79354	1.31839
1.890	.15579	.65112	.62350	1.37694	.79230	1.32273
1.895	.15455	.64992	.62074	1.37948	.79105	1.32711
1.900	.15331	.64872	.61799	1.38203	.78982	1.33153
1.905	.15209	.64752	.61523	1.38459	.78858	1.33597
1.910	.15087	.64632	.61249	1.38716	.78735	1.34046
1.915	.14967	.64513	.60975	1.38973	.78612	1.34498
1.920	.14847	.64393	.60701	1.39231	.78490	1.34953
1.925	.14728	.64274	.60428	1.39489	.78368	1.35413
1.930	.14610	.64155	.60155	1.39749	.78247	1.35875
1.935	.14493	.64036	.59883	1.40009	.78126	1.36342
1.940	.14377	.63917	.59611	1.40269	.78005	1.36812
1.945	.14261	.63798	.59340	1.40530	.77885	1.37285
1.950	.14147	.63679	.59070	1.40792	.77765	1.37763
1.955	.14033	.63561	.58800	1.41055	.77645	1.38244
1.960	.13920	.63442	.58530	1.41318	.77526	1.38728
1.965	.13808	.63324	.58261	1.41582	.77407	1.39217
1.970	.13697	.63206	.57993	1.41847	.77289	1.39709
1.975	.13586	.63088	.57725	1.42112	.77171	1.40205
1.980	.13476	.62970	.57458	1.42378	.77053	1.40704
1.985	.13368	.62852	.57191	1.42644	.76936	1.41208
1.990	.13260	.62735	.56926	1.42912	.76819	1.41715
1.995	.13152	.62617	.56660	1.43180	.76703	1.42226

TABLE X. Isentropic Exponent = 1.30

M	p/p_t	T/T_t	GAMMA	ISOTHERMAL T_t/T_t^*	RAYLEIGH T_t/T_t^*	SHOCK p_{t1}/p_{t2}
2.000	.13046	.62500	.56396			
2.010	.12836	.62266	.55868	1.43448	.76587	1.42741
2.020	.12628	.62032	.55343	1.43988	.76356	1.43782
2.030	.12424	.61800	.54821	1.44530	.76127	1.44838
2.040	.12223	.61567	.54302	1.45074	.75899	1.45911
				1.45622	.75673	1.46999
2.050	.12025	.61336	.53785			
2.060	.11830	.61105	.53271	1.46172	.75449	1.48103
2.070	.11638	.60874	.52760	1.46724	.75226	1.49223
2.080	.11449	.60644	.52252	1.47280	.75004	1.50359
2.090	.11262	.60415	.51746	1.47838	.74785	1.51511
				1.48399	.74566	1.52680
2.100	.11079	.60187	.51244			
2.110	.10898	.59959	.50745	1.48962	.74350	1.53866
2.120	.10720	.59731	.50248	1.49528	.74135	1.55069
2.130	.10545	.59505	.49755	1.50097	.73921	1.56288
2.140	.10373	.59279	.49265	1.50669	.73709	1.57524
				1.51243	.73499	1.58778
2.150	.10203	.59054	.48778			
2.160	.10036	.58829	.48295	1.51820	.73290	1.60049
2.170	.09872	.58605	.47814	1.52399	.73083	1.61338
2.180	.09710	.58382	.47337	1.52982	.72877	1.62644
2.190	.09550	.58159	.46863	1.53567	.72673	1.63969
				1.54154	.72470	1.65311
2.200	.09393	.57937	.46392			
2.210	.09239	.57716	.45924	1.54745	.72269	1.66672
2.220	.09087	.57496	.45460	1.55338	.72069	1.68051
2.230	.08937	.57276	.44999	1.55934	.71871	1.69449
2.240	.08790	.57057	.44542	1.56532	.71674	1.70866
				1.57133	.71479	1.72301
2.250	.08645	.56838	.44088			
2.260	.08503	.56621	.43637	1.57737	.71285	1.73756
2.270	.08362	.56404	.43190	1.58344	.71093	1.75230
2.280	.08224	.56187	.42746	1.58953	.70902	1.76723
2.290	.08088	.55972	.42305	1.59565	.70713	1.78236
				1.60179	.70525	1.79770
2.300	.07955	.55757	.41868			
2.310	.07823	.55543	.41434	1.60797	.70339	1.81323
2.320	.07694	.55329	.41004	1.61416	.70154	1.82896
2.330	.07566	.55117	.40577	1.62039	.69970	1.84490
2.340	.07441	.54905	.40154	1.62664	.69788	1.86105
				1.63293	.69607	1.87741
2.350	.07318	.54693	.39734			
2.360	.07197	.54483	.39317	1.63923	.69428	1.89398
2.370	.07077	.54273	.38904	1.64557	.69250	1.91076
2.380	.06960	.54064	.38495	1.65193	.69073	1.92775
2.390	.06844	.53856	.38088	1.65832	.68898	1.94497
				1.66473	.68724	1.96240
2.400	.06731	.53648	.37686			
2.410	.06619	.53441	.37286	1.67117	.68551	1.98005
2.420	.06509	.53235	.36891	1.67764	.68380	1.99793
2.430	.06401	.53030	.36498	1.68414	.68210	2.01604
2.440	.06295	.52825	.36109	1.69066	.68042	2.03437
				1.69721	.67874	2.05293
2.450	.06190	.52621	.35724			
2.460	.06087	.52418	.35342	1.70378	.67709	2.07173
2.470	.05986	.52216	.34963	1.71039	.67544	2.09077
2.480	.05886	.52014	.34588	1.71702	.67381	2.11004
2.490	.05789	.51813	.34216	1.72367	.67218	2.12955
				1.73036	.67058	2.14930

TABLE X.   Isentropic Exponent = 1.30

M	p/p_t	T/T_t	GAMMA	ISOTHERMAL T_t/T_t^*	RAYLEIGH T_t/T_t^*	SHOCK p_{t1}/p_{t2}
2.500	.05692	.51613	.33847	1.73707	.66898	2.16930
2.510	.05598	.51413	.33482	1.74381	.66740	2.18954
2.520	.05504	.51215	.33120	1.75057	.66583	2.21004
2.530	.05413	.51017	.32762	1.75736	.66427	2.23079
2.540	.05323	.50820	.32407	1.76418	.66272	2.25179
2.550	.05234	.50623	.32055	1.77103	.66119	2.27305
2.560	.05147	.50428	.31706	1.77790	.65967	2.29458
2.570	.05061	.50233	.31361	1.78480	.65816	2.31636
2.580	.04977	.50039	.31019	1.79172	.65666	2.33841
2.590	.04894	.49845	.30680	1.79868	.65517	2.36073
2.600	.04813	.49652	.30345	1.80565	.65370	2.38332
2.610	.04733	.49461	.30013	1.81266	.65223	2.40618
2.620	.04654	.49269	.29684	1.81969	.65078	2.42932
2.630	.04577	.49079	.29358	1.82676	.64934	2.45274
2.640	.04500	.48889	.29036	1.83384	.64791	2.47644
2.650	.04426	.48700	.28716	1.84096	.64649	2.50043
2.660	.04352	.48512	.28400	1.84810	.64508	2.52470
2.670	.04280	.48325	.28087	1.85527	.64369	2.54927
2.680	.04208	.48138	.27777	1.86246	.64230	2.57413
2.690	.04138	.47952	.27470	1.86968	.64093	2.59928
2.700	.04070	.47767	.27166	1.87693	.63956	2.62474
2.710	.04002	.47582	.26865	1.88421	.63821	2.65050
2.720	.03935	.47399	.26568	1.89151	.63687	2.67656
2.730	.03870	.47216	.26273	1.89884	.63554	2.70294
2.740	.03806	.47034	.25981	1.90619	.63421	2.72962
2.750	.03742	.46852	.25692	1.91358	.63290	2.75662
2.760	.03680	.46671	.25407	1.92099	.63160	2.78394
2.770	.03619	.46491	.25124	1.92842	.63031	2.81158
2.780	.03559	.46312	.24844	1.93589	.62903	2.83955
2.790	.03500	.46134	.24567	1.94338	.62776	2.86784
2.800	.03442	.45956	.24293	1.95090	.62649	2.89647
2.810	.03385	.45779	.24021	1.95844	.62524	2.92543
2.820	.03329	.45603	.23753	1.96601	.62400	2.95473
2.830	.03274	.45427	.23487	1.97361	.62277	2.98437
2.840	.03219	.45252	.23224	1.98124	.62154	3.01436
2.850	.03166	.45078	.22964	1.98889	.62033	3.04469
2.860	.03114	.44905	.22707	1.99657	.61913	3.07538
2.870	.03062	.44732	.22452	2.00427	.61793	3.10642
2.880	.03011	.44560	.22200	2.01200	.61674	3.13783
2.890	.02962	.44389	.21951	2.01976	.61557	3.16959
2.900	.02913	.44218	.21705	2.02755	.61440	3.20173
2.910	.02864	.44049	.21461	2.03536	.61324	3.23423
2.920	.02817	.43880	.21219	2.04320	.61209	3.26710
2.930	.02771	.43711	.20981	2.05107	.61095	3.30036
2.940	.02725	.43544	.20745	2.05897	.60982	3.33400
2.950	.02680	.43377	.20511	2.06689	.60869	3.36802
2.960	.02636	.43211	.20280	2.07484	.60758	3.40243
2.970	.02592	.43045	.20051	2.08281	.60647	3.43723
2.980	.02550	.42881	.19825	2.09081	.60537	3.47243
2.990	.02508	.42717	.19602	2.09884	.60428	3.50804

TABLE X. Isentropic Exponent = 1.30

M	p/p_t	T/T_t	GAMMA	ISOTHERMAL T_t/T_t^*	RAYLEIGH T_t/T_t^*	SHOCK p_{tf}/p_{t2}
3.000	.02466	.42553	.19381	2.10690	.60320	3.54405
3.010	.02426	.42391	.19162	2.11498	.60213	3.58047
3.020	.02386	.42229	.18946	2.12309	.60106	3.61730
3.030	.02347	.42067	.18732	2.13122	.60000	3.65454
3.040	.02308	.41907	.18520	2.13939	.59895	3.69221
3.050	.02270	.41747	.18311	2.14758	.59791	3.73031
3.060	.02233	.41588	.18104	2.15579	.59688	3.76883
3.070	.02196	.41430	.17899	2.16404	.59585	3.80779
3.080	.02160	.41272	.17697	2.17231	.59483	3.84719
3.090	.02125	.41115	.17497	2.18061	.59382	3.88703
3.100	.02090	.40958	.17299	2.18893	.59282	3.92732
3.110	.02056	.40803	.17103	2.19728	.59182	3.96806
3.120	.02022	.40648	.16910	2.20566	.59083	4.00926
3.130	.01989	.40493	.16718	2.21407	.58985	4.05092
3.140	.01957	.40340	.16529	2.22250	.58888	4.09304
3.150	.01925	.40187	.16342	2.23096	.58791	4.13564
3.160	.01893	.40035	.16157	2.23944	.58695	4.17870
3.170	.01862	.39883	.15974	2.24796	.58600	4.22225
3.180	.01832	.39732	.15793	2.25649	.58505	4.26628
3.190	.01802	.39582	.15615	2.26506	.58412	4.31080
3.200	.01773	.39432	.15438	2.27365	.58319	4.35582
3.210	.01744	.39283	.15263	2.28228	.58226	4.40133
3.220	.01716	.39135	.15090	2.29092	.58134	4.44735
3.230	.01688	.38987	.14919	2.29960	.58043	4.49387
3.240	.01660	.38840	.14751	2.30830	.57953	4.54091
3.250	.01634	.38694	.14584	2.31703	.57863	4.58846
3.260	.01607	.38548	.14418	2.32578	.57774	4.63654
3.270	.01581	.38403	.14255	2.33456	.57685	4.68515
3.280	.01555	.38259	.14094	2.34337	.57598	4.73429
3.290	.01530	.38115	.13934	2.35221	.57510	4.78397
3.300	.01506	.37972	.13777	2.36107	.57424	4.83419
3.310	.01481	.37830	.13621	2.36996	.57338	4.88496
3.320	.01457	.37688	.13467	2.37887	.57253	4.93629
3.330	.01434	.37547	.13314	2.38782	.57168	4.98818
3.340	.01411	.37406	.13164	2.39679	.57084	5.04063
3.350	.01388	.37267	.13015	2.40578	.57000	5.09365
3.360	.01366	.37127	.12868	2.41481	.56918	5.14725
3.370	.01344	.36989	.12722	2.42386	.56835	5.20144
3.380	.01322	.36851	.12579	2.43294	.56754	5.25620
3.390	.01301	.36713	.12436	2.44204	.56673	5.31157
3.400	.01280	.36576	.12296	2.45117	.56592	5.36753
3.410	.01259	.36440	.12157	2.46033	.56512	5.42410
3.420	.01239	.36305	.12020	2.46952	.56433	5.48127
3.430	.01219	.36170	.11884	2.47873	.56354	5.53907
3.440	.01200	.36036	.11750	2.48797	.56276	5.59748
3.450	.01181	.35902	.11617	2.49723	.56198	5.65652
3.460	.01162	.35769	.11486	2.50653	.56121	5.71620
3.470	.01143	.35636	.11357	2.51584	.56044	5.77652
3.480	.01125	.35504	.11229	2.52519	.55968	5.83748
3.490	.01107	.35373	.11102	2.53456	.55893	5.89910

TABLE X.   Isentropic Exponent = 1.30

M	p/p_t	T/T_t	GAMMA	ISOTHERMAL T_t/T_t^*	RAYLEIGH T_t/T_t^*	SHOCK p_{t1}/p_{t2}
3.500	.01090	.35242	.10977	2.54397	.55818	5.96138
3.510	.01072	.35112	.10854	2.55339	.55743	6.02432
3.520	.01055	.34983	.10731	2.56285	.55669	6.08793
3.530	.01039	.34854	.10611	2.57233	.55596	6.15222
3.540	.01022	.34725	.10491	2.58184	.55523	6.21719
3.550	.01006	.34598	.10373	2.59137	.55451	6.28285
3.560	.00990	.34470	.10257	2.60093	.55379	6.34921
3.570	.00974	.34344	.10141	2.61052	.55307	6.41628
3.580	.00959	.34218	.10027	2.62014	.55237	6.48405
3.590	.00944	.34092	.09915	2.62978	.55166	6.55254
3.600	.00929	.33967	.09804	2.63945	.55096	6.62176
3.610	.00914	.33843	.09694	2.64914	.55027	6.69170
3.620	.00900	.33719	.09585	2.65887	.54958	6.76238
3.630	.00886	.33596	.09477	2.66862	.54889	6.83381
3.640	.00872	.33473	.09371	2.67839	.54821	6.90598
3.650	.00858	.33351	.09266	2.68820	.54754	6.97892
3.660	.00845	.33230	.09163	2.69803	.54687	7.05262
3.670	.00831	.33109	.09060	2.70789	.54620	7.12709
3.680	.00818	.32989	.08959	2.71777	.54554	7.20234
3.690	.00805	.32869	.08859	2.72768	.54488	7.27838
3.700	.00793	.32749	.08760	2.73762	.54423	7.35521
3.710	.00780	.32631	.08662	2.74759	.54358	7.43284
3.720	.00768	.32512	.08565	2.75758	.54294	7.51129
3.730	.00756	.32395	.08470	2.76760	.54230	7.59054
3.740	.00745	.32277	.08375	2.77764	.54166	7.67062
3.750	.00733	.32161	.08282	2.78771	.54103	7.75154
3.760	.00722	.32045	.08190	2.79781	.54041	7.83329
3.770	.00710	.31929	.08098	2.80794	.53978	7.91588
3.780	.00699	.31814	.08008	2.81809	.53917	7.99934
3.790	.00688	.31700	.07919	2.82827	.53855	8.08365
3.800	.00678	.31586	.07831	2.83848	.53794	8.16883
3.810	.00667	.31472	.07744	2.84872	.53734	8.25489
3.820	.00657	.31359	.07658	2.85898	.53673	8.34184
3.830	.00647	.31247	.07573	2.86927	.53614	8.42968
3.840	.00637	.31135	.07489	2.87958	.53554	8.51843
3.850	.00627	.31023	.07406	2.88992	.53495	8.60808
3.860	.00617	.30912	.07324	2.90029	.53437	8.69865
3.870	.00608	.30802	.07243	2.91069	.53378	8.79015
3.880	.00599	.30692	.07163	2.92111	.53321	8.88258
3.890	.00589	.30583	.07084	2.93156	.53263	8.97596
3.900	.00580	.30474	.07006	2.94203	.53206	9.07029
3.910	.00571	.30365	.06929	2.95254	.53149	9.16558
3.920	.00563	.30258	.06852	2.96307	.53093	9.26184
3.930	.00554	.30150	.06777	2.97362	.53037	9.35908
3.940	.00546	.30043	.06702	2.98421	.52982	9.45731
3.950	.00537	.29937	.06628	2.99482	.52926	9.55653
3.960	.00529	.29831	.06555	3.00546	.52871	9.65675
3.970	.00521	.29725	.06483	3.01612	.52817	9.75799
3.980	.00513	.29620	.06412	3.02681	.52763	9.86026
3.990	.00505	.29516	.06341	3.03753	.52709	9.96355

TABLE X. Isentropic Exponent = 1.30

M	p/p_t	T/T_t	GAMMA	ISOTHERMAL T_t/T_t^*	RAYLEIGH T_t/T_t^*	SHOCK p_{ti}/p_{t2}
4.000	.00498	.29412	.06272	3.04828	.52656	10.06790
4.010	.00490	.29308	.06203	3.05905	.52602	10.17328
4.020	.00483	.29205	.06135	3.06985	.52550	10.27973
4.030	.00475	.29102	.06068	3.08067	.52497	10.38724
4.040	.00468	.29000	.06002	3.09153	.52445	10.49583
4.050	.00461	.28899	.05936	3.10241	.52393	10.60551
4.060	.00454	.28797	.05871	3.11331	.52342	10.71629
4.070	.00447	.28697	.05807	3.12425	.52291	10.82818
4.080	.00441	.28596	.05744	3.13521	.52240	10.94118
4.090	.00434	.28496	.05681	3.14619	.52190	11.05531
4.100	.00427	.28397	.05619	3.15721	.52139	11.17058
4.110	.00421	.28298	.05558	3.16825	.52090	11.28699
4.120	.00415	.28200	.05498	3.17932	.52040	11.40456
4.130	.00408	.28101	.05438	3.19041	.51991	11.52330
4.140	.00402	.28004	.05379	3.20153	.51942	11.64321
4.150	.00396	.27907	.05320	3.21268	.51894	11.76431
4.160	.00390	.27810	.05263	3.22386	.51845	11.88661
4.170	.00385	.27714	.05206	3.23506	.51797	12.01012
4.180	.00379	.27618	.05149	3.24629	.51750	12.13484
4.190	.00373	.27522	.05093	3.25754	.51702	12.26080
4.200	.00368	.27427	.05038	3.26883	.51655	12.38799
4.210	.00362	.27333	.04984	3.28014	.51609	12.51644
4.220	.00357	.27239	.04930	3.29147	.51562	12.64614
4.230	.00352	.27145	.04877	3.30284	.51516	12.77712
4.240	.00346	.27052	.04824	3.31423	.51470	12.90938
4.250	.00341	.26959	.04772	3.32565	.51424	13.04294
4.260	.00336	.26866	.04721	3.33709	.51379	13.17780
4.270	.00331	.26774	.04670	3.34856	.51334	13.31398
4.280	.00326	.26683	.04620	3.36006	.51289	13.45149
4.290	.00322	.26591	.04571	3.37159	.51245	13.59033
4.300	.00317	.26501	.04522	3.38314	.51201	13.73053
4.310	.00312	.26410	.04473	3.39472	.51157	13.87209
4.320	.00308	.26320	.04425	3.40632	.51113	14.01502
4.330	.00303	.26231	.04378	3.41796	.51069	14.01502
4.340	.00299	.26141	.04331	3.42961	.51026	14.15934
						14.30505
4.350	.00294	.26053	.04285	3.44130	.50983	14.45218
4.360	.00290	.25964	.04239	3.45301	.50941	14.60072
4.370	.00286	.25876	.04194	3.46476	.50898	14.75070
4.380	.00282	.25789	.04149	3.47652	.50856	14.90213
4.390	.00277	.25702	.04105	3.48832	.50814	15.05501
4.400	.00273	.25615	.04061	3.50014	.50773	15.20936
4.410	.00269	.25528	.04018	3.51199	.50732	15.36519
4.420	.00266	.25442	.03976	3.52386	.50690	15.52252
4.430	.00262	.25357	.03933	3.53576	.50650	15.68135
4.440	.00258	.25271	.03892	3.54769	.50609	15.84171
4.450	.00254	.25187	.03850	3.55965	.50569	16.00360
4.460	.00250	.25102	.03810	3.57163	.50528	16.16703
4.470	.00247	.25018	.03769	3.58364	.50489	16.33202
4.480	.00243	.24934	.03730	3.59567	.50449	16.49858
4.490	.00240	.24851	.03690	3.60774	.50410	16.66673

TABLE X. Isentropic Exponent = 1.30

M	p/p_t	T/T_t	GAMMA	ISOTHERMAL T_t/T_t^*	RAYLEIGH T_t/T_t^*	SHOCK p_{ti}/p_{t2}
4.500	.00236	.24768	.03651	3.61983	.50370	16.83647
4.510	.00233	.24685	.03613	3.63194	.50331	17.00783
4.520	.00230	.24603	.03575	3.64409	.50293	17.18080
4.530	.00226	.24521	.03537	3.65626	.50254	17.35542
4.540	.00223	.24439	.03500	3.66846	.50216	17.53169
4.550	.00220	.24358	.03463	3.68068	.50178	17.70962
4.560	.00217	.24278	.03427	3.69293	.50140	17.88923
4.570	.00214	.24197	.03391	3.70521	.50103	18.07053
4.580	.00211	.24117	.03356	3.71752	.50065	18.25354
4.590	.00208	.24037	.03321	3.72985	.50028	18.43827
4.600	.00205	.23958	.03286	3.74221	.49991	18.62473
4.610	.00202	.23879	.03252	3.75459	.49954	18.81294
4.620	.00199	.23800	.03218	3.76700	.49918	19.00292
4.630	.00196	.23722	.03184	3.77944	.49882	19.19467
4.640	.00193	.23644	.03151	3.79191	.49845	19.38822
4.650	.00191	.23566	.03118	3.80440	.49810	19.58357
4.660	.00188	.23489	.03086	3.81692	.49774	19.78074
4.670	.00185	.23412	.03054	3.82947	.49738	19.97975
4.680	.00183	.23335	.03022	3.84205	.49703	20.18060
4.690	.00180	.23259	.02991	3.85465	.49668	20.38332
4.700	.00177	.23183	.02960	3.86727	.49633	20.58793
4.710	.00175	.23107	.02929	3.87993	.49599	20.79443
4.720	.00172	.23032	.02899	3.89261	.49564	21.00284
4.730	.00170	.22957	.02869	3.90532	.49530	21.21318
4.740	.00168	.22883	.02839	3.91806	.49496	21.42546
4.750	.00165	.22808	.02810	3.93082	.49462	21.63970
4.760	.00163	.22734	.02781	3.94361	.49428	21.85591
4.770	.00161	.22661	.02753	3.95642	.49395	22.07411
4.780	.00159	.22587	.02724	3.96927	.49362	22.29432
4.790	.00156	.22514	.02696	3.98214	.49329	22.51654
4.800	.00154	.22442	.02669	3.99503	.49296	22.74081
4.810	.00152	.22369	.02641	4.00796	.49263	22.96713
4.820	.00150	.22297	.02614	4.02091	.49230	23.19551
4.830	.00148	.22226	.02588	4.03389	.49198	23.42599
4.840	.00146	.22154	.02561	4.04689	.49166	23.65857
4.850	.00144	.22083	.02535	4.05992	.49134	23.89326
4.860	.00142	.22012	.02509	4.07298	.49102	24.13010
4.870	.00140	.21942	.02484	4.08606	.49070	24.36909
4.880	.00138	.21872	.02458	4.09918	.49039	24.61025
4.890	.00136	.21802	.02433	4.11232	.49008	24.85360
4.900	.00134	.21732	.02409	4.12548	.48976	25.09914
4.910	.00132	.21663	.02384	4.13867	.48945	25.34692
4.920	.00130	.21594	.02360	4.15189	.48915	25.59693
4.930	.00129	.21525	.02336	4.16514	.48884	25.84920
4.940	.00127	.21457	.02312	4.17841	.48854	26.10374
4.950	.00125	.21389	.02289	4.19171	.48823	26.36058
4.960	.00123	.21321	.02266	4.20504	.48793	26.61972
4.970	.00122	.21253	.02243	4.21840	.48763	26.88120
4.980	.00120	.21186	.02220	4.23178	.48734	27.14502
4.990	.00118	.21119	.02198	4.24518	.48704	27.41121

TABLE XI.   Isentropic Exponent = 1.40

M	p/p_t	T/T_t	GAMMA	ISOTHERMAL T_t/T^*	RAYLEIGH T_t/T_t^*	SHOCK p_{t1}/p_{t2}
.000	1.00000	1.00000	.00000	.87500	.00000	
.005	.99998	.00000	.00862	.87500	.00012	1.
.010	.99993	.99998	.01727	.87502	.00048	1.
.015	.99984	.99996	.02591	.87504	.00108	1.
.020	.99972	.99992	.03455	.87507	.00192	1.
						1.
.025	.99956	.99988	.04318	.87511	.00299	
.030	.99937	.99982	.05181	.87516	.00431	1.
.035	.99914	.99976	.06043	.87521	.00586	1.
.040	.99888	.99968	.06905	.87528	.00765	1.
.045	.99858	.99960	.07766	.87535	.00967	1.
						1.
.050	.99825	.99950	.08627	.87544	.01192	
.055	.99789	.99940	.09487	.87553	.01441	1.
.060	.99748	.99928	.10346	.87563	.01712	1.
.065	.99705	.99916	.11203	.87574	.02006	1.
.070	.99658	.99902	.12060	.87586	.02322	1.
						1.
.075	.99607	.99888	.12916	.87598	.02661	
.080	.99553	.99872	.13771	.87612	.03021	1.
.085	.99496	.99856	.14624	.87626	.03404	1.
.090	.99435	.99838	.15477	.87642	.03807	1.
.095	.99371	.99820	.16327	.87658	.04232	1.
						1.
.100	.99303	.99800	.17177	.87675	.04678	
.105	.99232	.99780	.18024	.87693	.05144	1.
.110	.99158	.99759	.18871	.87712	.05630	1.
.115	.99080	.99736	.19715	.87731	.06135	1.
.120	.98999	.99713	.20558	.87752	.06661	1.
						1.
.125	.98914	.99688	.21399	.87773	.07205	
.130	.98826	.99663	.22238	.87796	.07767	1.
.135	.98735	.99637	.23075	.87819	.08348	1.
.140	.98640	.99610	.23910	.87843	.08947	1.
.145	.98542	.99581	.24743	.87868	.09563	1.
						1.
.150	.98441	.99552	.25573	.87894	.10196	
.155	.98336	.99522	.26402	.87920	.10846	1.
.160	.98228	.99491	.27228	.87948	.11511	1.
.165	.98117	.99458	.28051	.87976	.12192	1.
.170	.98003	.99425	.28872	.88006	.12888	1.
						1.
.175	.97885	.99391	.29691	.88036	.13599	
.180	.97765	.99356	.30507	.88067	.14324	1.
.185	.97641	.99320	.31320	.88099	.15062	1.
.190	.97514	.99283	.32131	.88132	.15814	1.
.195	.97383	.99245	.32939	.88165	.16579	1.
						1.
.200	.97250	.99206	.33744	.88200	.17355	
.205	.97113	.99167	.34546	.88235	.18144	1.
.210	.96973	.99126	.35345	.88272	.18943	1.
.215	.96830	.99084	.36140	.88309	.19754	1.
.220	.96685	.99041	.36933	.88347	.20574	1.
						1.
.225	.96536	.98998	.37723	.88386	.21404	
.230	.96383	.98953	.38509	.88426	.22244	1.
.235	.96228	.98908	.39292	.88466	.23092	1.
.240	.96070	.98861	.40071	.88508	.23948	1.
.245	.95909	.98814	.40847	.88550	.24812	1.

216

TABLE XI. Isentropic Exponent = 1.40

M	p/p_t	T/T_t	GAMMA	ISOTHERMAL T_t/T_t^*	RAYLEIGH T_t/T_t^*	SHOCK p_{t1}/p_{t2}
.250	.95745	.98765	.41620	.88594	.25684	1.
.255	.95578	.98716	.42389	.88638	.26562	1.
.260	.95408	.98666	.43154	.88683	.27446	1.
.265	.95236	.98615	.43915	.88729	.28336	1.
.270	.95060	.98563	.44673	.88776	.29231	1.
.275	.94882	.98510	.45427	.88823	.30131	1.
.280	.94700	.98456	.46178	.88872	.31035	1.
.285	.94516	.98401	.46924	.88921	.31943	1.
.290	.94329	.98346	.47666	.88972	.32855	1.
.295	.94139	.98289	.48404	.89023	.33769	1.
.300	.93947	.98232	.49138	.89075	.34686	1.
.305	.93752	.98173	.49868	.89128	.35605	1.
.310	.93554	.98114	.50594	.89182	.36525	1.
.315	.93353	.98054	.51316	.89236	.37447	1.
.320	.93150	.97993	.52033	.89292	.38369	1.
.325	.92945	.97931	.52746	.89348	.39291	1.
.330	.92736	.97868	.53455	.89406	.40214	1.
.335	.92525	.97805	.54159	.89464	.41136	1.
.340	.92312	.97740	.54858	.89523	.42056	1.
.345	.92096	.97675	.55553	.89583	.42976	1.
.350	.91877	.97609	.56244	.89644	.43894	1.
.355	.91656	.97541	.56930	.89705	.44810	1.
.360	.91433	.97473	.57611	.89768	.45723	1.
.365	.91207	.97405	.58288	.89831	.46634	1.
.370	.90979	.97335	.58959	.89896	.47541	1.
.375	.90748	.97264	.59626	.89961	.48445	1.
.380	.90516	.97193	.60288	.90027	.49346	1.
.385	.90280	.97121	.60946	.90094	.50242	1.
.390	.90043	.97048	.61598	.90162	.51134	1.
.395	.89803	.96974	.62245	.90230	.52021	1.
.400	.89561	.96899	.62888	.90300	.52903	1.
.405	.89317	.96824	.63525	.90370	.53780	1.
.410	.89071	.96747	.64157	.90442	.54651	1.
.415	.88823	.96670	.64784	.90514	.55516	1.
.420	.88572	.96592	.65406	.90587	.56376	1.
.425	.88320	.96513	.66023	.90661	.57229	1.
.430	.88065	.96434	.66635	.90736	.58076	1.
.435	.87808	.96354	.67241	.90811	.58915	1.
.440	.87550	.96272	.67842	.90888	.59748	1.
.445	.87289	.96190	.68438	.90965	.60574	1.
.450	.87027	.96108	.69029	.91044	.61393	1.
.455	.86762	.96024	.69614	.91123	.62204	1.
.460	.86496	.95940	.70194	.91203	.63007	1.
.465	.86228	.95855	.70768	.91284	.63802	1.
.470	.85958	.95769	.71337	.91366	.64589	1.
.475	.85686	.95682	.71901	.91448	.65368	1.
.480	.85413	.95595	.72459	.91532	.66139	1.
.485	.85138	.95507	.73011	.91616	.66901	1.
.490	.84861	.95418	.73558	.91702	.67655	1.
.495	.84582	.95328	.74100	.91788	.68400	1.

TABLE XI. Isentropic Exponent = 1.40

M	p/p_t	T/T_t	GAMMA	ISOTHERMAL T_t/T_t^*	RAYLEIGH T_t/T_t^*	SHOCK p_{t1}/p_{t2}
.500	.84302	.95238	.74636	.91875	.69136	1.
.505	.84020	.95147	.75166	.91963	.69863	1.
.510	.83737	.95055	.75691	.92052	.70581	1.
.515	.83452	.94963	.76210	.92141	.71290	1.
.520	.83165	.94869	.76723	.92232	.71990	1.
.525	.82877	.94776	.77231	.92323	.72680	1.
.530	.82588	.94681	.77733	.92416	.73361	1.
.535	.82297	.94585	.78230	.92509	.74033	1.
.540	.82005	.94489	.78720	.92603	.74695	1.
.545	.81711	.94393	.79205	.92698	.75348	1.
.550	.81417	.94295	.79685	.92794	.75991	1.
.555	.81120	.94197	.80158	.92890	.76625	1.
.560	.80823	.94098	.80626	.92988	.77249	1.
.565	.80524	.93999	.81088	.93086	.77863	1.
.570	.80224	.93898	.81544	.93186	.78468	1.
.575	.79923	.93798	.81995	.93286	.79063	1.
.580	.79621	.93696	.82440	.93387	.79648	1.
.585	.79317	.93594	.82879	.93489	.80223	1.
.590	.79013	.93491	.83312	.93592	.80789	1.
.595	.78707	.93388	.83739	.93695	.81346	1.
.600	.78400	.93284	.84161	.93800	.81892	1.
.605	.78093	.93179	.84577	.93905	.82429	1.
.610	.77784	.93073	.84987	.94012	.82957	1.
.615	.77474	.92967	.85391	.94119	.83474	1.
.620	.77164	.92861	.85789	.94227	.83983	1.
.625	.76853	.92754	.86182	.94336	.84481	1.
.630	.76540	.92646	.86569	.94446	.84970	1.
.635	.76227	.92537	.86950	.94556	.85450	1.
.640	.75913	.92428	.87325	.94668	.85920	1.
.645	.75598	.92319	.87694	.94780	.86381	1.
.650	.75283	.92208	.88058	.94894	.86833	1.
.655	.74967	.92098	.88416	.95008	.87275	1.
.660	.74650	.91986	.88768	.95123	.87708	1.
.665	.74332	.91874	.89114	.95239	.88132	1.
.670	.74014	.91762	.89454	.95356	.88547	1.
.675	.73695	.91649	.89789	.95473	.88953	1.
.680	.73376	.91535	.90118	.95592	.89350	1.
.685	.73056	.91421	.90441	.95711	.89738	1.
.690	.72735	.91306	.90759	.95832	.90118	1.
.695	.72414	.91191	.91071	.95953	.90488	1.
.700	.72093	.91075	.91377	.96075	.90850	1.
.705	.71771	.90958	.91677	.96198	.91203	1.
.710	.71448	.90841	.91971	.96322	.91548	1.
.715	.71126	.90724	.92260	.96446	.91884	1.
.720	.70803	.90606	.92544	.96572	.92212	1.
.725	.70479	.90488	.92821	.96698	.92532	1.
.730	.70155	.90369	.93093	.96826	.92843	1.
.735	.69831	.90249	.93360	.96954	.93147	1.
.740	.69507	.90129	.93620	.97083	.93442	1.
.745	.69182	.90009	.93875	.97213	.93730	1.

TABLE XI. Isentropic Exponent = 1.40

M	p/p_t	T/T_t	GAMMA	ISOTHERMAL T_t/T_t^*	RAYLEIGH T_t/T_t^*	SHOCK p_{t1}/p_{t2}
.750	.68857	.89888	.94125	.97344	.94009	1.
.755	.68532	.89766	.94369	.97475	.94281	1.
.760	.68207	.89644	.94607	.97608	.94546	1.
.765	.67882	.89522	.94840	.97741	.94802	1.
.770	.67556	.89399	.95068	.97876	.95052	1.
.775	.67231	.89276	.95290	.98011	.95293	1.
.780	.66905	.89152	.95506	.98147	.95528	1.
.785	.66579	.89028	.95717	.98284	.95755	1.
.790	.66254	.88903	.95923	.98422	.95975	1.
.795	.65928	.88778	.96123	.98560	.96189	1.
.800	.65602	.88652	.96318	.98700	.96395	1.
.805	.65277	.88527	.96507	.98840	.96594	1.
.810	.64951	.88400	.96691	.98982	.96787	1.
.815	.64625	.88273	.96870	.99124	.96973	1.
.820	.64300	.88146	.97044	.99267	.97152	1.
.825	.63975	.88018	.97212	.99411	.97325	1.
.830	.63650	.87890	.97375	.99556	.97492	1.
.835	.63325	.87762	.97533	.99701	.97652	1.
.840	.63000	.87633	.97685	.99848	.97807	1.
.845	.62675	.87504	.97833	.99995	.97955	1.
.850	.62351	.87374	.97975	1.00144	.98097	1.
.855	.62027	.87244	.98112	1.00293	.98233	1.
.860	.61703	.87114	.98244	1.00443	.98363	1.
.865	.61380	.86983	.98371	1.00594	.98488	1.
.870	.61057	.86852	.98493	1.00746	.98607	1.
.875	.60734	.86721	.98610	1.00898	.98720	1.
.880	.60412	.86589	.98722	1.01052	.98828	1.
.885	.60090	.86457	.98830	1.01206	.98931	1.
.890	.59768	.86324	.98932	1.01362	.99028	1.
.895	.59447	.86192	.99029	1.01518	.99120	1.
.900	.59126	.86059	.99121	1.01675	.99207	1.
.905	.58806	.85925	.99209	1.01833	.99289	1.
.910	.58486	.85791	.99292	1.01992	.99366	1.
.915	.58166	.85657	.99370	1.02151	.99438	1.
.920	.57848	.85523	.99443	1.02312	.99506	1.
.925	.57529	.85388	.99512	1.02473	.99568	1.
.930	.57211	.85253	.99576	1.02636	.99627	1.
.935	.56894	.85118	.99635	1.02799	.99680	1.
.940	.56578	.84982	.99690	1.02963	.99729	1.
.945	.56261	.84846	.99740	1.03128	.99774	1.
.950	.55946	.84710	.99786	1.03294	.99814	1.
.955	.55631	.84573	.99827	1.03460	.99851	1.
.960	.55317	.84437	.99864	1.03628	.99883	1.
.965	.55003	.84300	.99896	1.03796	.99911	1.
.970	.54691	.84162	.99924	1.03966	.99935	1.
.975	.54378	.84025	.99947	1.04136	.99955	1.
.980	.54067	.83887	.99966	1.04307	.99971	1.
.985	.53756	.83749	.99981	1.04479	.99984	1.
.990	.53446	.83611	.99992	1.04652	.99993	1.
.995	.53137	.83472	.99998	1.04825	.99998	1.

TABLE XI. Isentropic Exponent = 1.40

M	p/p_t	T/T_t	GAMMA	ISOTHERMAL T_t/T_t^*	RAYLEIGH T_t/T_t^*	SHOCK p_{t1}/p_{t2}
1.000	.52828	.83333	1.00000	1.05000	1.00000	1.00000
1.005	.52520	.83194	.99998	1.05175	.99998	1.00000
1.010	.52213	.83055	.99992	1.05352	.99993	1.00000
1.015	.51907	.82916	.99981	1.05529	.99985	1.00000
1.020	.51602	.82776	.99967	1.05707	.99973	1.00001
1.025	.51297	.82636	.99949	1.05886	.99958	1.00002
1.030	.50994	.82496	.99926	1.06066	.99940	1.00003
1.035	.50691	.82356	.99900	1.06246	.99919	1.00005
1.040	.50389	.82215	.99870	1.06428	.99895	1.00008
1.045	.50087	.82075	.99836	1.06610	.99868	1.00011
1.050	.49787	.81934	.99798	1.06794	.99838	1.00015
1.055	.49488	.81793	.99756	1.06978	.99805	1.00019
1.060	.49189	.81651	.99710	1.07163	.99769	1.00025
1.065	.48892	.81510	.99661	1.07349	.99731	1.00031
1.070	.48595	.81368	.99608	1.07536	.99690	1.00039
1.075	.48299	.81227	.99551	1.07723	.99647	1.00047
1.080	.48005	.81085	.99491	1.07912	.99601	1.00057
1.085	.47711	.80942	.99427	1.08101	.99552	1.00068
1.090	.47418	.80800	.99359	1.08292	.99501	1.00080
1.095	.47126	.80658	.99288	1.08483	.99448	1.00093
1.100	.46835	.80515	.99214	1.08675	.99392	1.00107
1.105	.46546	.80373	.99136	1.08868	.99335	1.00123
1.110	.46257	.80230	.99054	1.09062	.99275	1.00140
1.115	.45969	.80087	.98970	1.09256	.99212	1.00159
1.120	.45682	.79944	.98881	1.09452	.99148	1.00179
1.125	.45396	.79801	.98790	1.09648	.99082	1.00201
1.130	.45111	.79657	.98695	1.09846	.99013	1.00224
1.135	.44828	.79514	.98597	1.10044	.98943	1.00248
1.140	.44545	.79370	.98496	1.10243	.98871	1.00275
1.145	.44263	.79226	.98392	1.10443	.98797	1.00303
1.150	.43983	.79083	.98285	1.10644	.98721	1.00332
1.155	.43703	.78939	.98174	1.10845	.98643	1.00363
1.160	.43425	.78795	.98060	1.11048	.98564	1.00396
1.165	.43148	.78651	.97944	1.11251	.98483	1.00431
1.170	.42872	.78506	.97824	1.11456	.98400	1.00468
1.175	.42596	.78362	.97702	1.11661	.98316	1.00506
1.180	.42323	.78218	.97576	1.11867	.98230	1.00546
1.185	.42050	.78073	.97448	1.12074	.98143	1.00588
1.190	.41778	.77929	.97317	1.12282	.98054	1.00632
1.195	.41507	.77784	.97183	1.12490	.97963	1.00678
1.200	.41238	.77640	.97046	1.12700	.97872	1.00725
1.205	.40969	.77495	.96906	1.12910	.97779	1.00775
1.210	.40702	.77350	.96764	1.13122	.97684	1.00827
1.215	.40436	.77205	.96619	1.13334	.97589	1.00880
1.220	.40171	.77061	.96472	1.13547	.97492	1.00936
1.225	.39907	.76916	.96322	1.13761	.97394	1.00993
1.230	.39645	.76771	.96169	1.13976	.97294	1.01053
1.235	.39383	.76626	.96014	1.14191	.97194	1.01114
1.240	.39123	.76481	.95856	1.14408	.97092	1.01176
1.245	.38864	.76336	.95696	1.14625	.96990	1.01244

TABLE XI. Isentropic Exponent = 1.40

M	p/p_t	T/T_t	GAMMA	ISOTHERMAL T_t/T_t^*	RAYLEIGH T_t/T_t^*	SHOCK p_{t1}/p_{t2}
1.250	.38606	.76190	.95534	1.14844	.96886	1.01311
1.255	.38349	.76045	.95369	1.15063	.96781	1.01381
1.260	.38093	.75900	.95201	1.15283	.96675	1.01453
1.265	.37839	.75755	.95032	1.15504	.96569	1.01527
1.270	.37586	.75610	.94860	1.15726	.96461	1.01603
1.275	.37334	.75465	.94685	1.15948	.96353	1.01682
1.280	.37083	.75319	.94509	1.16172	.96243	1.01762
1.285	.36833	.75174	.94330	1.16396	.96133	1.01845
1.290	.36585	.75029	.94150	1.16622	.96022	1.01930
1.295	.36337	.74884	.93967	1.16848	.95910	1.02017
1.300	.36091	.74738	.93782	1.17075	.95798	1.02106
1.305	.35847	.74593	.93595	1.17303	.95685	1.02197
1.310	.35603	.74448	.93406	1.17532	.95571	1.02291
1.315	.35360	.74303	.93215	1.17761	.95456	1.02387
1.320	.35119	.74158	.93022	1.17992	.95341	1.02485
1.325	.34879	.74012	.92827	1.18223	.95225	1.02585
1.330	.34640	.73867	.92630	1.18456	.95108	1.02688
1.335	.34403	.73722	.92431	1.18689	.94991	1.02793
1.340	.34166	.73577	.92231	1.18923	.94873	1.02900
1.345	.33931	.73432	.92028	1.19158	.94755	1.03009
1.350	.33697	.73287	.91824	1.19394	.94637	1.03121
1.355	.33464	.73142	.91618	1.19630	.94517	1.03235
1.360	.33233	.72997	.91411	1.19868	.94398	1.03351
1.365	.33002	.72852	.91201	1.20106	.94277	1.03469
1.370	.32773	.72707	.90991	1.20346	.94157	1.03590
1.375	.32545	.72562	.90778	1.20586	.94036	1.03713
1.380	.32319	.72418	.90564	1.20827	.93914	1.03838
1.385	.32093	.72273	.90348	1.21069	.93793	1.03966
1.390	.31869	.72128	.90131	1.21312	.93671	1.04096
1.395	.31646	.71984	.89912	1.21555	.93548	1.04228
1.400	.31424	.71839	.89692	1.21800	.93425	1.04363
1.405	.31203	.71695	.89470	1.22045	.93302	1.04500
1.410	.30984	.71550	.89247	1.22292	.93179	1.04639
1.415	.30766	.71406	.89023	1.22539	.93055	1.04781
1.420	.30549	.71262	.88797	1.22787	.92931	1.04925
1.425	.30333	.71117	.88570	1.23036	.92807	1.05071
1.430	.30119	.70973	.88342	1.23286	.92683	1.05220
1.435	.29905	.70829	.88112	1.23536	.92559	1.05371
1.440	.29693	.70685	.87881	1.23788	.92434	1.05524
1.445	.29482	.70542	.87649	1.24040	.92309	1.05680
1.450	.29272	.70398	.87415	1.24294	.92184	1.05838
1.455	.29064	.70254	.87181	1.24548	.92059	1.05999
1.460	.28856	.70111	.86945	1.24803	.91933	1.06162
1.465	.28650	.69967	.86708	1.25059	.91808	1.06327
1.470	.28445	.69824	.86471	1.25316	.91682	1.06495
1.475	.28241	.69680	.86232	1.25573	.91557	1.06665
1.480	.28039	.69537	.85992	1.25832	.91431	1.06837
1.485	.27837	.69394	.85751	1.26091	.91305	1.07012
1.490	.27637	.69251	.85509	1.26352	.91179	1.07190
1.495	.27438	.69108	.85266	1.26613	.91053	1.07369

TABLE XI. Isentropic Exponent = 1.40

M	p/p_t	T/T_t	GAMMA	ISOTHERMAL T_t/T_t^*	RAYLEIGH T_t/T_t^*	SHOCK p_{t1}/p_{t2}
1.500	.27240	.68966	.85022	1.26875	.90928	1.07552
1.505	.27044	.68823	.84777	1.27138	.90802	1.07736
1.510	.26848	.68680	.84532	1.27402	.90676	1.07923
1.515	.26654	.68538	.84285	1.27666	.90550	1.08112
1.520	.26461	.68396	.84038	1.27932	.90424	1.08304
1.525	.26269	.68254	.83789	1.28198	.90298	1.08499
1.530	.26078	.68112	.83541	1.28466	.90172	1.08695
1.535	.25888	.67970	.83291	1.28734	.90046	1.08894
1.540	.25700	.67828	.83040	1.29003	.89920	1.09096
1.545	.25512	.67686	.82789	1.29273	.89795	1.09300
1.550	.25326	.67545	.82537	1.29544	.89669	1.09506
1.555	.25141	.67403	.82285	1.29815	.89544	1.09715
1.560	.24957	.67262	.82032	1.30088	.89418	1.09927
1.565	.24775	.67121	.81778	1.30361	.89293	1.10141
1.570	.24593	.66980	.81523	1.30636	.89168	1.10357
1.575	.24413	.66839	.81268	1.30911	.89042	1.10576
1.580	.24233	.66699	.81013	1.31187	.88917	1.10797
1.585	.24055	.66558	.80756	1.31464	.88792	1.11021
1.590	.23878	.66418	.80500	1.31742	.88668	1.11247
1.595	.23702	.66278	.80243	1.32020	.88543	1.11476
1.600	.23527	.66138	.79985	1.32300	.88419	1.11707
1.605	.23353	.65998	.79727	1.32580	.88294	1.11941
1.610	.23181	.65858	.79468	1.32862	.88170	1.12177
1.615	.23009	.65718	.79209	1.33144	.88046	1.12415
1.620	.22839	.65579	.78950	1.33427	.87922	1.12657
1.625	.22670	.65440	.78690	1.33711	.87799	1.12900
1.630	.22501	.65301	.78430	1.33996	.87675	1.13147
1.635	.22334	.65162	.78170	1.34281	.87552	1.13395
1.640	.22168	.65023	.77909	1.34568	.87429	1.13647
1.645	.22003	.64884	.77648	1.34855	.87306	1.13900
1.650	.21839	.64746	.77386	1.35144	.87184	1.14157
1.655	.21677	.64608	.77125	1.35433	.87061	1.14416
1.660	.21515	.64470	.76863	1.35723	.86939	1.14677
1.665	.21354	.64332	.76601	1.36014	.86817	1.14941
1.670	.21195	.64194	.76338	1.36306	.86696	1.15208
1.675	.21036	.64056	.76076	1.36598	.86574	1.15477
1.680	.20879	.63919	.75813	1.36892	.86453	1.15748
1.685	.20722	.63782	.75550	1.37186	.86332	1.16022
1.690	.20567	.63645	.75287	1.37482	.86212	1.16299
1.695	.20413	.63508	.75024	1.37778	.86091	1.16579
1.700	.20259	.63371	.74760	1.38075	.85971	1.16860
1.705	.20107	.63235	.74497	1.38373	.85851	1.17145
1.710	.19956	.63099	.74234	1.38672	.85731	1.17432
1.715	.19806	.62963	.73970	1.38971	.85612	1.17722
1.720	.19656	.62827	.73706	1.39272	.85493	1.18014
1.725	.19508	.62691	.73443	1.39573	.85374	1.18309
1.730	.19361	.62556	.73179	1.39876	.85256	1.18607
1.735	.19215	.62420	.72915	1.40179	.85137	1.18907
1.740	.19070	.62285	.72652	1.40483	.85019	1.19209
1.745	.18926	.62150	.72388	1.40788	.84902	1.19515

TABLE XI. Isentropic Exponent = 1.40

M	p/p_t	T/T_t	GAMMA	ISOTHERMAL T_t/T_t^*	RAYLEIGH T_t/T_t^*	SHOCK p_{t1}/p_{t2}
1.750	.18782	.62016	.72124	1.41094	.84784	1.19823
1.755	.18640	.61881	.71861	1.41400	.84667	1.20133
1.760	.18499	.61747	.71597	1.41708	.84551	1.20447
1.765	.18359	.61613	.71334	1.42016	.84434	1.20763
1.770	.18219	.61479	.71071	1.42326	.84318	1.21081
1.775	.18081	.61345	.70808	1.42636	.84202	1.21403
1.780	.17944	.61212	.70544	1.42947	.84087	1.21727
1.785	.17807	.61078	.70282	1.43259	.83972	1.22053
1.790	.17672	.60945	.70019	1.43572	.83857	1.22382
1.795	.17538	.60812	.69756	1.43885	.83742	1.22714
1.800	.17404	.60680	.69494	1.44200	.83628	1.23049
1.805	.17271	.60547	.69231	1.44515	.83514	1.23386
1.810	.17140	.60415	.68969	1.44832	.83400	1.23726
1.815	.17009	.60283	.68707	1.45149	.83287	1.24069
1.820	.16879	.60151	.68446	1.45467	.83174	1.24415
1.825	.16750	.60020	.68184	1.45786	.83062	1.24763
1.830	.16622	.59888	.67923	1.46106	.82949	1.25114
1.835	.15495	.59757	.67662	1.46426	.82837	1.25468
1.840	.16369	.59626	.67401	1.46748	.82726	1.25824
1.845	.16244	.59495	.67141	1.47070	.82615	1.26183
1.850	.16120	.59365	.66881	1.47394	.82504	1.26545
1.855	.15996	.59235	.66621	1.47718	.82393	1.26910
1.860	.15873	.59104	.66362	1.48043	.82283	1.27277
1.865	.15752	.58975	.66103	1.48369	.82173	1.27647
1.870	.15631	.58845	.65844	1.48696	.82064	1.28020
1.875	.15511	.58716	.65585	1.49023	.81954	1.28396
1.880	.15392	.58586	.65327	1.49352	.81845	1.28775
1.885	.15274	.58457	.65069	1.49681	.81737	1.29156
1.890	.15156	.58329	.64812	1.50012	.81629	1.29541
1.895	.15040	.58200	.64555	1.50343	.81521	1.29928
1.900	.14924	.58072	.64298	1.50675	.81414	1.30317
1.905	.14809	.57944	.64042	1.51008	.81306	1.30710
1.910	.14695	.57816	.63786	1.51342	.81200	1.31106
1.915	.14582	.57689	.63531	1.51676	.81093	1.31504
1.920	.14470	.57561	.63276	1.52012	.80987	1.31905
1.925	.14358	.57434	.63021	1.52348	.80882	1.32309
1.930	.14247	.57307	.62767	1.52686	.80776	1.32716
1.935	.14137	.57181	.62513	1.53024	.80671	1.33126
1.940	.14028	.57054	.62260	1.53363	.80567	1.33539
1.945	.13920	.56928	.62007	1.53703	.80462	1.33955
1.950	.13813	.56802	.61755	1.54044	.80358	1.34373
1.955	.13706	.56676	.61503	1.54385	.80255	1.34795
1.960	.13600	.56551	.61252	1.54728	.80152	1.35219
1.965	.13495	.56426	.61001	1.55071	.80049	1.35647
1.970	.13390	.56301	.60750	1.55416	.79946	1.36077
1.975	.13287	.56176	.60501	1.55761	.79844	1.36510
1.980	.13184	.56051	.60251	1.56107	.79742	1.36946
1.985	.13082	.55927	.60002	1.56454	.79641	1.37385
1.990	.12981	.55803	.59754	1.56802	.79540	1.37827
1.995	.12880	.55679	.59506	1.57150	.79439	1.38272

TABLE XI. Isentropic Exponent = 1.40

M	p/p_t	T/T_t	GAMMA	ISOTHERMAL T_t/T_t^*	RAYLEIGH T_t/T_t^*	SHOCK p_{t1}/p_{t2}
2.000	.12780	.55556	.59259	1.57500	.79339	1.38721
2.010	.12583	.55309	.58767	1.58202	.79139	1.39626
2.020	.12389	.55064	.58276	1.58907	.78941	1.40543
2.030	.12197	.54819	.57788	1.59616	.78744	1.41472
2.040	.12009	.54576	.57302	1.60328	.78549	1.42414
2.050	.11823	.54333	.56819	1.61044	.78355	1.43367
2.060	.11640	.54091	.56338	1.61763	.78162	1.44334
2.070	.11460	.53851	.55859	1.62486	.77971	1.45312
2.080	.11282	.53611	.55383	1.63212	.77782	1.46303
2.090	.11107	.53373	.54909	1.63942	.77593	1.47307
2.100	.10935	.53135	.54438	1.64675	.77406	1.48323
2.110	.10766	.52898	.53970	1.65412	.77221	1.49352
2.120	.10599	.52663	.53504	1.66152	.77037	1.50394
2.130	.10434	.52428	.53041	1.66896	.76854	1.51449
2.140	.10273	.52194	.52581	1.67643	.76673	1.52517
2.150	.10113	.51962	.52123	1.68394	.76493	1.53598
2.160	.09956	.51730	.51668	1.69148	.76314	1.54692
2.170	.09802	.51499	.51216	1.69906	.76137	1.55799
2.180	.09650	.51269	.50766	1.70667	.75961	1.56920
2.190	.09500	.51041	.50320	1.71432	.75787	1.58054
2.200	.09352	.50813	.49876	1.72200	.75613	1.59201
2.210	.09207	.50586	.49435	1.72972	.75442	1.60362
2.220	.09064	.50361	.48997	1.73747	.75271	1.61537
2.230	.08923	.50136	.48562	1.74526	.75102	1.62726
2.240	.08785	.49912	.48129	1.75308	.74934	1.63928
2.250	.08648	.49689	.47700	1.76094	.74768	1.65145
2.260	.08514	.49468	.47274	1.76883	.74602	1.66375
2.270	.08382	.49247	.46850	1.77676	.74438	1.67620
2.280	.08252	.49027	.46429	1.78472	.74276	1.68879
2.290	.08123	.48809	.46012	1.79272	.74114	1.70152
2.300	.07997	.48591	.45597	1.80075	.73954	1.71440
2.310	.07873	.48374	.45185	1.80882	.73795	1.72742
2.320	.07751	.48158	.44776	1.81692	.73638	1.74059
2.330	.07631	.47944	.44370	1.82506	.73482	1.75391
2.340	.07512	.47730	.43968	1.83323	.73326	1.76738
2.350	.07396	.47517	.43568	1.84144	.73173	1.78099
2.360	.07281	.47305	.43171	1.84968	.73020	1.79476
2.370	.07168	.47095	.42777	1.85796	.72868	1.80868
2.380	.07057	.46885	.42386	1.86627	.72718	1.82275
2.390	.06948	.46676	.41998	1.87462	.72569	1.83698
2.400	.06840	.46468	.41613	1.88300	.72421	1.85136
2.410	.06734	.46262	.41231	1.89142	.72275	1.86589
2.420	.06630	.46056	.40852	1.89987	.72129	1.88059
2.430	.06527	.45851	.40476	1.90836	.71985	1.89544
2.440	.06426	.45647	.40103	1.91688	.71842	1.91045
2.450	.06327	.45444	.39733	1.92544	.71699	1.92563
2.460	.06229	.45242	.39365	1.93403	.71559	1.94096
2.470	.06133	.45041	.39001	1.94266	.71419	1.95646
2.480	.06038	.44841	.38640	1.95132	.71280	1.97212
2.490	.05945	.44642	.38281	1.96002	.71142	1.98795

TABLE XI. Isentropic Exponent = 1.40

M	p/p_t	T/T_t	GAMMA	ISOTHERMAL T_t/T_t^*	RAYLEIGH T_t/T_t^*	SHOCK p_{t1}/p_{t2}
2.500	.05853	.44444	.37926	1.96875	.71006	2.00395
2.510	.05762	.44247	.37573	1.97752	.70871	2.02011
2.520	.05674	.44051	.37224	1.98632	.70736	2.03644
2.530	.05586	.43856	.36877	1.99516	.70603	2.05294
2.540	.05500	.43662	.36533	2.00403	.70471	2.06962
2.550	.05415	.43469	.36192	2.01294	.70340	2.08646
2.560	.05332	.43277	.35854	2.02188	.70210	2.10348
2.570	.05250	.43085	.35519	2.03086	.70081	2.12068
2.580	.05169	.42895	.35187	2.03987	.69953	2.13805
2.590	.05090	.42705	.34857	2.04892	.69826	2.15560
2.600	.05012	.42517	.34531	2.05800	.69700	2.17333
2.610	.04935	.42329	.34207	2.06712	.69575	2.19124
2.620	.04859	.42143	.33886	2.07627	.69451	2.20933
2.630	.04784	.41957	.33568	2.08546	.69328	2.22761
2.640	.04711	.41773	.33252	2.09468	.69206	2.24607
2.650	.04639	.41589	.32939	2.10394	.69084	2.26471
2.660	.04568	.41406	.32629	2.11323	.68964	2.28354
2.670	.04498	.41224	.32322	2.12256	.68845	2.30256
2.680	.04429	.41043	.32018	2.13192	.68727	2.32177
2.690	.04362	.40863	.31716	2.14132	.68610	2.34118
2.700	.04295	.40683	.31417	2.15075	.68494	2.36077
2.710	.04229	.40505	.31120	2.16022	.68378	2.38056
2.720	.04165	.40328	.30827	2.16972	.68264	2.40054
2.730	.04102	.40151	.30536	2.17926	.68150	2.42072
2.740	.04039	.39976	.30247	2.18883	.68037	2.44110
2.750	.03978	.39801	.29961	2.19844	.67926	2.46168
2.760	.03917	.39627	.29678	2.20808	.67815	2.48246
2.770	.03858	.39454	.29397	2.21776	.67705	2.50345
2.780	.03799	.39282	.29119	2.22747	.67595	2.52464
2.790	.03742	.39111	.28843	2.23722	.67487	2.54603
2.800	.03685	.38941	.28570	2.24700	.67380	2.56763
2.810	.03629	.38771	.28300	2.25682	.67273	2.58944
2.820	.03574	.38603	.28032	2.26667	.67167	2.61146
2.830	.03520	.38435	.27766	2.27656	.67062	2.63369
2.840	.03467	.38268	.27503	2.28648	.66958	2.65614
2.850	.03415	.38103	.27243	2.29644	.66855	2.67880
2.860	.03363	.37937	.26984	2.30643	.66752	2.70167
2.870	.03312	.37773	.26729	2.31646	.66651	2.72477
2.880	.03263	.37610	.26475	2.32652	.66550	2.74808
2.890	.03213	.37447	.26224	2.33662	.66450	2.77162
2.900	.03165	.37286	.25976	2.34675	.66350	2.79538
2.910	.03118	.37125	.25729	2.35692	.66252	2.81936
2.920	.03071	.36965	.25485	2.36712	.66154	2.84357
2.930	.03025	.36806	.25244	2.37736	.66057	2.86801
2.940	.02980	.36647	.25004	2.38763	.65960	2.89267
2.950	.02935	.36490	.24767	2.39794	.65865	2.91757
2.960	.02891	.36333	.24532	2.40828	.65770	2.94270
2.970	.02848	.36177	.24300	2.41866	.65676	2.96807
2.980	.02805	.36022	.24069	2.42907	.65583	2.99367
2.990	.02764	.35868	.23841	2.43952	.65490	3.01951

TABLE XI. Isentropic Exponent = 1.40

M	p/p_t	T/T_t	GAMMA	ISOTHERMAL T_t/T^*	RAYLEIGH T_t/T^*	SHOCK p_{t1}/p_{t2}
3.000	.02722	.35714	.23615	2.45000	.65398	3.04559
3.010	.02682	.35562	.23391	2.46052	.65307	3.07191
3.020	.02642	.35410	.23170	2.47107	.65216	3.09847
3.030	.02603	.35259	.22950	2.48166	.65126	3.12528
3.040	.02564	.35108	.22733	2.49228	.65037	3.15233
3.050	.02526	.34959	.22517	2.50294	.64949	3.17963
3.060	.02489	.34810	.22304	2.51363	.64861	3.20718
3.070	.02452	.34662	.22093	2.52436	.64774	3.23499
3.080	.02416	.34515	.21884	2.53512	.64687	3.26305
3.090	.02380	.34369	.21677	2.54592	.64601	3.29136
3.100	.02345	.34223	.21472	2.55675	.64516	3.31993
3.110	.02310	.34078	.21269	2.56762	.64432	3.34876
3.120	.02276	.33934	.21067	2.57852	.64348	3.37785
3.130	.02243	.33791	.20868	2.58946	.64265	3.40720
3.140	.02210	.33648	.20671	2.60043	.64182	3.43682
3.150	.02177	.33506	.20476	2.61144	.64100	3.46670
3.160	.02146	.33365	.20282	2.62248	.64018	3.49686
3.170	.02114	.33225	.20091	2.63356	.63938	3.52728
3.180	.02083	.33085	.19901	2.64467	.63857	3.55797
3.190	.02053	.32947	.19714	2.65582	.63778	3.58894
3.200	.02023	.32808	.19528	2.66700	.63699	3.62019
3.210	.01993	.32671	.19344	2.67822	.63621	3.65171
3.220	.01964	.32534	.19161	2.68947	.63543	3.68352
3.230	.01936	.32398	.18981	2.70076	.63465	3.71560
3.240	.01908	.32263	.18802	2.71208	.63389	3.74797
3.250	.01880	.32129	.18625	2.72344	.63313	3.78063
3.260	.01853	.31995	.18450	2.73483	.63237	3.81357
3.270	.01826	.31862	.18276	2.74626	.63162	3.84680
3.280	.01799	.31729	.18105	2.75772	.63088	3.88033
3.290	.01773	.31597	.17935	2.76922	.63014	3.91415
3.300	.01748	.31466	.17766	2.78075	.62940	3.94826
3.310	.01722	.31336	.17600	2.79232	.62868	3.98268
3.320	.01698	.31206	.17434	2.80392	.62795	4.01739
3.330	.01673	.31077	.17271	2.81556	.62724	4.05240
3.340	.01649	.30949	.17109	2.82723	.62652	4.08773
3.350	.01625	.30821	.16949	2.83894	.62582	4.12335
3.360	.01602	.30694	.16790	2.85068	.62512	4.15929
3.370	.01579	.30568	.16633	2.86246	.62442	4.19553
3.380	.01557	.30443	.16478	2.87427	.62373	4.23209
3.390	.01534	.30318	.16324	2.88612	.62304	4.26897
3.400	.01512	.30193	.16172	2.89800	.62236	4.30616
3.410	.01491	.30070	.16021	2.90992	.62168	4.34367
3.420	.01470	.29947	.15871	2.92187	.62101	4.38150
3.430	.01449	.29824	.15723	2.93386	.62034	4.41965
3.440	.01428	.29703	.15577	2.94588	.61968	4.45813
3.450	.01408	.29581	.15432	2.95794	.61902	4.49694
3.460	.01388	.29461	.15288	2.97003	.61837	4.53608
3.470	.01368	.29341	.15146	2.98216	.61772	4.57555
3.480	.01349	.29222	.15006	2.99432	.61708	4.61536
3.490	.01330	.29103	.14866	3.00652	.61644	4.65551

TABLE XI.   Isentropic Exponent = 1.40

M	p/p_t	T/T_t	GAMMA	ISOTHERMAL T_t/T_t^*	RAYLEIGH T_t/T_t^*	SHOCK p_{t1}/p_{t2}
3.500	.01311	.28986	.14728	3.01875	.61580	4.69599
3.510	.01293	.28868	.14592	3.03102	.61517	4.73681
3.520	.01274	.28752	.14457	3.04332	.61455	4.77798
3.530	.01256	.28635	.14323	3.05566	.61393	4.81950
3.540	.01239	.28520	.14190	3.06803	.61331	4.86136
3.550	.01221	.28405	.14059	3.08044	.61270	4.90358
3.560	.01204	.28291	.13929	3.09288	.61209	4.94615
3.570	.01188	.28177	.13801	3.10536	.61149	4.98907
3.580	.01171	.28064	.13673	3.11787	.61089	5.03235
3.590	.01155	.27952	.13547	3.13042	.61029	5.07599
3.600	.01138	.27840	.13423	3.14300	.60970	5.12000
3.610	.01123	.27728	.13299	3.15562	.60911	5.16437
3.620	.01107	.27618	.13177	3.16827	.60853	5.20911
3.630	.01092	.27507	.13056	3.18096	.60795	5.25422
3.640	.01076	.27398	.12936	3.19368	.60738	5.29970
3.650	.01062	.27289	.12817	3.20644	.60681	5.34556
3.660	.01047	.27180	.12700	3.21923	.60624	5.39179
3.670	.01032	.27073	.12583	3.23206	.60568	5.43840
3.680	.01018	.26965	.12468	3.24492	.60512	5.48540
3.690	.01004	.26858	.12354	3.25782	.60456	5.53278
3.700	.00990	.26752	.12241	3.27075	.60401	5.58055
3.710	.00977	.26647	.12130	3.28372	.60346	5.62871
3.720	.00963	.26542	.12019	3.29672	.60292	5.67726
3.730	.00950	.26437	.11909	3.30976	.60238	5.72621
3.740	.00937	.26333	.11801	3.32283	.60184	5.77555
3.750	.00924	.26230	.11694	3.33594	.60131	5.82530
3.760	.00912	.26127	.11587	3.34908	.60078	5.87545
3.770	.00899	.26024	.11482	3.36226	.60025	5.92600
3.780	.00887	.25922	.11378	3.37547	.59973	5.97696
3.790	.00875	.25821	.11275	3.38872	.59921	6.02833
3.800	.00863	.25720	.11172	3.40200	.59870	6.08012
3.810	.00851	.25620	.11071	3.41532	.59819	6.13232
3.820	.00840	.25520	.10971	3.42867	.59768	6.18494
3.830	.00828	.25421	.10872	3.44206	.59717	6.23799
3.840	.00817	.25322	.10774	3.45548	.59667	6.29145
3.850	.00806	.25224	.10677	3.46894	.59617	6.34535
3.860	.00795	.25126	.10581	3.48243	.59568	6.39967
3.870	.00784	.25029	.10485	3.49596	.59519	6.45443
3.880	.00774	.24932	.10391	3.50952	.59470	6.50962
3.890	.00763	.24836	.10298	3.52312	.59421	6.56525
3.900	.00753	.24740	.10205	3.53675	.59373	6.62132
3.910	.00743	.24645	.10114	3.55042	.59325	6.67783
3.920	.00733	.24550	.10023	3.56412	.59278	6.73479
3.930	.00723	.24456	.09933	3.57786	.59231	6.79220
3.940	.00714	.24362	.09844	3.59163	.59184	6.85006
3.950	.00704	.24269	.09756	3.60544	.59137	6.90838
3.960	.00695	.24176	.09669	3.61928	.59091	6.96715
3.970	.00686	.24084	.09583	3.63316	.59045	7.02638
3.980	.00676	.23992	.09498	3.64707	.58999	7.08608
3.990	.00667	.23900	.09413	3.66102	.58954	7.14624

TABLE XI. Isentropic Exponent = 1.40

M	p/p_t	T/T_t	GAMMA	ISOTHERMAL T_t/T_t^*	RAYLEIGH T_t/T_t^*	SHOCK p_{tl}/p_{t2}
4.000	.00659	.23810	.09329	3.67500	.58909	7.20688
4.010	.00650	.23719	.09247	3.68902	.58864	7.26799
4.020	.00641	.23629	.09164	3.70307	.58819	7.32957
4.030	.00633	.23539	.09083	3.71716	.58775	7.39163
4.040	.00624	.23450	.09003	3.73128	.58731	7.45417
4.050	.00616	.23362	.08923	3.74544	.58687	7.51719
4.060	.00608	.23274	.08844	3.75963	.58644	7.58070
4.070	.00600	.23186	.08766	3.77386	.58601	7.64470
4.080	.00592	.23099	.08689	3.78812	.58558	7.70920
4.090	.00585	.23012	.08612	3.80242	.58516	7.77419
4.100	.00577	.22925	.08536	3.81675	.58473	7.83968
4.110	.00569	.22839	.08461	3.83112	.58431	7.90567
4.120	.00562	.22754	.08387	3.84552	.58390	7.97217
4.130	.00555	.22669	.08313	3.85996	.58348	8.03917
4.140	.00547	.22584	.08240	3.87443	.58307	8.10669
4.150	.00540	.22500	.08168	3.88894	.58266	8.17472
4.160	.00533	.22416	.08097	3.90348	.58225	8.24327
4.170	.00526	.22332	.08026	3.91806	.58185	8.31234
4.180	.00520	.22250	.07956	3.93267	.58145	8.38194
4.190	.00513	.22167	.07886	3.94732	.58105	8.45206
4.200	.00506	.22085	.07818	3.96200	.58065	8.52271
4.210	.00500	.22003	.07750	3.97672	.58026	8.59390
4.220	.00493	.21922	.07682	3.99147	.57987	8.66562
4.230	.00487	.21841	.07615	4.00626	.57948	8.73788
4.240	.00481	.21760	.07549	4.02108	.57909	8.81069
4.250	.00474	.21680	.07484	4.03594	.57870	8.88404
4.260	.00468	.21601	.07419	4.05083	.57832	8.95795
4.270	.00462	.21521	.07355	4.06576	.57794	9.03240
4.280	.00457	.21442	.07291	4.08072	.57757	9.10741
4.290	.00451	.21364	.07228	4.09572	.57719	9.18299
4.300	.00445	.21286	.07166	4.11075	.57682	9.25912
4.310	.00439	.21208	.07104	4.12582	.57645	9.33583
4.320	.00434	.21131	.07043	4.14092	.57608	9.41310
4.330	.00428	.21054	.06983	4.15606	.57571	9.49094
4.340	.00423	.20977	.06923	4.17123	.57535	9.56937
4.350	.00417	.20901	.06863	4.18644	.57499	9.64837
4.360	.00412	.20825	.06804	4.20168	.57463	9.72796
4.370	.00407	.20750	.06746	4.21696	.57427	9.80813
4.380	.00402	.20674	.06688	4.23227	.57392	9.88889
4.390	.00397	.20600	.06631	4.24762	.57357	9.97025
4.400	.00392	.20525	.06575	4.26300	.57322	10.05220
4.410	.00387	.20451	.06519	4.27842	.57287	10.13476
4.420	.00382	.20378	.06463	4.29387	.57252	10.21792
4.430	.00377	.20305	.06408	4.30936	.57218	10.30168
4.440	.00372	.20232	.06354	4.32488	.57183	10.38606
4.450	.00368	.20159	.06300	4.34044	.57149	10.47106
4.460	.00363	.20087	.06246	4.35603	.57116	10.55667
4.470	.00359	.20015	.06194	4.37166	.57082	10.64290
4.480	.00354	.19944	.06141	4.38732	.57049	10.72976
4.490	.00350	.19873	.06089	4.40302	.57015	10.81725

TABLE XI. Isentropic Exponent = 1.40

M	p/p_t	T/T_t	GAMMA	ISOTHERMAL T_t/T_t^*	RAYLEIGH T_t/T_t^*	SHOCK p_{t1}/p_{t2}
4.500	.00346	.19802	.06038	4.41875	.56982	10.90537
4.510	.00341	.19732	.05987	4.43452	.56950	10.99413
4.520	.00337	.19662	.05937	4.45032	.56917	11.08353
4.530	.00333	.19592	.05887	4.46616	.56885	11.17357
4.540	.00329	.19522	.05837	4.48203	.56852	11.26426
4.550	.00325	.19453	.05788	4.49794	.56820	11.35560
4.560	.00321	.19385	.05740	4.51388	.56789	11.44760
4.570	.00317	.19316	.05692	4.52986	.56757	11.54026
4.580	.00313	.19248	.05644	4.54587	.56726	11.63357
4.590	.00309	.19181	.05597	4.56192	.56694	11.72756
4.600	.00305	.19113	.05550	4.57800	.56663	11.82221
4.610	.00302	.19046	.05504	4.59412	.56632	11.91754
4.620	.00298	.18979	.05458	4.61027	.56602	12.01355
4.630	.00294	.18913	.05413	4.62646	.56571	12.11023
4.640	.00291	.18847	.05368	4.64268	.56541	12.20761
4.650	.00287	.18781	.05323	4.65894	.56510	12.30567
4.660	.00284	.18716	.05279	4.67523	.56480	12.40442
4.670	.00280	.18651	.05235	4.69156	.56451	12.50387
4.680	.00277	.18586	.05192	4.70792	.56421	12.60403
4.690	.00273	.18521	.05149	4.72432	.56391	12.70488
4.700	.00270	.18457	.05107	4.74075	.56362	12.80645
4.710	.00267	.18393	.05064	4.75722	.56333	12.90873
4.720	.00264	.18330	.05023	4.77372	.56304	13.01172
4.730	.00260	.18266	.04981	4.79026	.56275	13.11544
4.740	.00257	.18203	.04940	4.80683	.56246	13.21988
4.750	.00254	.18141	.04900	4.82344	.56218	13.32505
4.760	.00251	.18078	.04860	4.84008	.56190	13.43095
4.770	.00248	.18016	.04820	4.85676	.56161	13.53759
4.780	.00245	.17954	.04781	4.87347	.56133	13.64497
4.790	.00242	.17893	.04742	4.89022	.56106	13.75309
4.800	.00239	.17832	.04703	4.90700	.56078	13.86197
4.810	.00237	.17771	.04665	4.92382	.56050	13.97159
4.820	.00234	.17710	.04627	4.94067	.56023	14.08198
4.830	.00231	.17650	.04589	4.95756	.55996	14.19312
4.840	.00228	.17590	.04552	4.97448	.55969	14.30503
4.850	.00226	.17530	.04515	4.99144	.55942	14.41771
4.860	.00223	.17471	.04478	5.00843	.55915	14.53116
4.870	.00220	.17411	.04442	5.02546	.55888	14.64540
4.880	.00218	.17352	.04406	5.04252	.55862	14.76041
4.890	.00215	.17294	.04370	5.05962	.55836	14.87621
4.900	.00213	.17235	.04335	5.07675	.55809	14.99280
4.910	.00210	.17177	.04300	5.09392	.55783	15.11019
4.920	.00208	.17120	.04266	5.11112	.55758	15.22837
4.930	.00205	.17062	.04231	5.12836	.55732	15.34736
4.940	.00203	.17005	.04197	5.14563	.55706	15.46716
4.950	.00200	.16948	.04164	5.16294	.55681	15.58777
4.960	.00198	.16891	.04130	5.18028	.55655	15.70920
4.970	.00196	.16835	.04097	5.19766	.55630	15.83144
4.980	.00193	.16778	.04065	5.21507	.55605	15.95451
4.990	.00191	.16722	.04032	5.23252	.55580	16.07842

TABLE XII.   Isentropic Exponent = 1.67

M	p/p_t	T/T_t	GAMMA	ISOTHERMAL T_t/T_t^*	RAYLEIGH T_t/T_t^*	SHOCK p_{t1}/p_{t2}
.000	1.00000	1.00000	.00000	.83292	.00000	
.005	.99998	.99999	.00889	.83292	.00013	1.
.010	.99992	.99997	.01778	.83295	.00053	1.
.015	.99981	.99992	.02667	.83298	.00120	1.
.020	.99967	.99987	.03556	.83303	.00213	1.
						1.
.025	.99948	.99979	.04444	.83309	.00333	
.030	.99925	.99970	.05332	.83317	.00479	1.
.035	.99898	.99959	.06219	.83326	.00652	1.
.040	.99867	.99946	.07106	.83336	.00850	1.
.045	.99831	.99932	.07992	.83348	.01075	1.
						1.
.050	.99792	.99916	.08877	.83362	.01325	
.055	.99748	.99899	.09761	.83376	.01601	1.
.060	.99700	.99880	.10645	.83392	.01902	1.
.065	.99648	.99859	.11527	.83410	.02228	1.
.070	.99592	.99836	.12408	.83428	.02579	1.
						1.
.075	.99532	.99812	.13288	.83449	.02954	
.080	.99468	.99786	.14166	.83470	.03353	1.
.085	.99399	.99759	.15044	.83493	.03776	1.
.090	.99327	.99729	.15919	.83518	.04222	1.
.095	.99250	.99699	.16793	.83544	.04691	1.
						1.
.100	.99170	.99666	.17666	.83571	.05183	
.105	.99085	.99632	.18536	.83599	.05697	1.
.110	.98997	.99596	.19405	.83629	.06233	1.
.115	.98904	.99559	.20272	.83661	.06790	1.
.120	.98808	.99520	.21137	.83694	.07368	1.
						1.
.125	.98707	.99479	.22000	.83728	.07966	
.130	.98603	.99437	.22860	.83763	.08584	1.
.135	.98494	.99393	.23719	.83800	.09222	1.
.140	.98382	.99348	.24575	.83839	.09878	1.
.145	.98266	.99301	.25428	.83878	.10552	1.
						1.
.150	.98146	.99252	.26280	.83920	.11245	
.155	.98022	.99202	.27128	.83962	.11954	1.
.160	.97894	.99150	.27974	.84006	.12680	1.
.165	.97762	.99096	.28817	.84051	.13422	1.
.170	.97627	.99041	.29658	.84098	.14180	1.
						1.
.175	.97488	.98984	.30495	.84146	.14953	
.180	.97345	.98926	.31330	.84196	.15740	1.
.185	.97198	.98866	.32161	.84247	.16541	1.
.190	.97048	.98805	.32990	.84299	.17355	1.
.195	.96894	.98742	.33815	.84353	.18182	1.
						1.
.200	.96737	.98678	.34637	.84408	.19020	
.205	.96575	.98612	.35455	.84464	.19870	1.
.210	.96411	.98544	.36271	.84522	.20731	1.
.215	.96242	.98475	.37082	.84582	.21602	1.
.220	.96070	.98404	.37891	.84642	.22483	1.
						1.
.225	.95895	.98332	.38695	.84704	.23373	
.230	.95716	.98259	.39496	.84768	.24271	1.
.235	.95534	.98184	.40293	.84833	.25178	1.
.240	.95348	.98107	.41086	.84899	.26091	1.
.245	.95159	.98029	.41876	.84967	.27011	1.

TABLE XII.   Isentropic Exponent = 1.67

M	p/p_t	T/T_t	GAMMA	ISOTHERMAL T_t/T_t^*	RAYLEIGH T_t/T_t^*	SHOCK p_{t1}/p_{t2}
.250	.94966	.97949	.42661	.85036	.27937	1.
.255	.94770	.97868	.43443	.85106	.28869	1.
.260	.94571	.97786	.44220	.85178	.29806	1.
.265	.94369	.97702	.44994	.85251	.30748	1.
.270	.94163	.97616	.45763	.85326	.31693	1.
.275	.93954	.97529	.46527	.85402	.32641	1.
.280	.93742	.97441	.47288	.85479	.33593	1.
.285	.93527	.97351	.48044	.85558	.34547	1.
.290	.93309	.97260	.48796	.85638	.35502	1.
.295	.93088	.97167	.49543	.85720	.36459	1.
.300	.92864	.97073	.50285	.85803	.37416	1.
.305	.92636	.96978	.51024	.85887	.38374	1.
.310	.92406	.96881	.51757	.85973	.39332	1.
.315	.92173	.96783	.52486	.86060	.40289	1.
.320	.91937	.96683	.53209	.86149	.41245	1.
.325	.91698	.96582	.53929	.86239	.42199	1.
.330	.91456	.96480	.54643	.86330	.43152	1.
.335	.91211	.96377	.55352	.86423	.44101	1.
.340	.90964	.96272	.56057	.86517	.45049	1.
.345	.90714	.96166	.56756	.86613	.45992	1.
.350	.90462	.96058	.57450	.86710	.46933	1.
.355	.90206	.95949	.58139	.86808	.47869	1.
.360	.89949	.95839	.58824	.86908	.48801	1.
.365	.89688	.95728	.59502	.87009	.49728	1.
.370	.89425	.95615	.60176	.87112	.50650	1.
.375	.89160	.95501	.60845	.87216	.51567	1.
.380	.88892	.95386	.61508	.87321	.52478	1.
.385	.88622	.95269	.62166	.87428	.53383	1.
.390	.88349	.95152	.62818	.87536	.54282	1.
.395	.88074	.95033	.63465	.87645	.55174	1.
.400	.87797	.94913	.64107	.87756	.56059	1.
.405	.87518	.94791	.64743	.87869	.56937	1.
.410	.87236	.94669	.65374	.87982	.57808	1.
.415	.86952	.94545	.65999	.88097	.58671	1.
.420	.86666	.94420	.66618	.88214	.59527	1.
.425	.86378	.94294	.67232	.88332	.60374	1.
.430	.86088	.94167	.67840	.88451	.61213	1.
.435	.85796	.94039	.68443	.88572	.62044	1.
.440	.85502	.93909	.69040	.88694	.62866	1.
.445	.85206	.93779	.69631	.88817	.63679	1.
.450	.84908	.93647	.70217	.88942	.64483	1.
.455	.84609	.93514	.70797	.89068	.65278	1.
.460	.84307	.93381	.71371	.89196	.66064	1.
.465	.84004	.93246	.71939	.89325	.66841	1.
.470	.83699	.93110	.72501	.89455	.67608	1.
.475	.83392	.92973	.73058	.89587	.68365	1.
.480	.83084	.92835	.73609	.89721	.69113	1.
.485	.82774	.92696	.74153	.89855	.69851	1.
.490	.82462	.92555	.74692	.89991	.70579	1.
.495	.82149	.92414	.75226	.90129	.71297	1.

231

TABLE XII. Isentropic Exponent = 1.67

M	p/p_t	T/T_t	GAMMA	ISOTHERMAL T_t/T_t^*	RAYLEIGH T_t/T_t^*	SHOCK p_{t1}/p_{t2}
.500	.81835	.92272	.75753	.90267	.72005	
.505	.81519	.92129	.76274	.90408	.72703	1.
.510	.81201	.91985	.76789	.90549	.73391	1.
.515	.80883	.91840	.77299	.90692	.74069	1.
.520	.80563	.91694	.77802	.90837	.74736	1.
.525	.80241	.91547	.78300	.90982	.75394	1.
.530	.79919	.91399	.78791	.91130	.76041	1.
.535	.79595	.91250	.79277	.91278	.76677	1.
.540	.79270	.91101	.79756	.91428	.77304	1.
.545	.78943	.90950	.80230	.91580	.77920	1.
.550	.78616	.90799	.80698	.91732	.78526	1.
.555	.78288	.90646	.81159	.91887	.79122	1.
.560	.77958	.90493	.81615	.92042	.79707	1.
.565	.77628	.90339	.82065	.92199	.80282	1.
.570	.77297	.90184	.82508	.92357	.80847	1.
.575	.76965	.90029	.82946	.92517	.81402	1.
.580	.76631	.89872	.83377	.92678	.81946	1.
.585	.76297	.89715	.83803	.92841	.82481	1.
.590	.75963	.89557	.84223	.93005	.83005	1.
.595	.75627	.89398	.84637	.93170	.83519	1.
.600	.75291	.89238	.85044	.93337	.84024	1.
.605	.74954	.89077	.85446	.93505	.84519	1.
.610	.74616	.88916	.85842	.93674	.85003	1.
.615	.74278	.88754	.86232	.93845	.85478	1.
.620	.73939	.88592	.86616	.94018	.85943	1.
.625	.73600	.88428	.86994	.94191	.86399	1.
.630	.73260	.88264	.87366	.94366	.86845	1.
.635	.72920	.88099	.87732	.94543	.87282	1.
.640	.72579	.87934	.88092	.94721	.87709	1.
.645	.72237	.87768	.88446	.94900	.88127	1.
.650	.71896	.87601	.88795	.95081	.88535	1.
.655	.71554	.87434	.89138	.95263	.88935	1.
.660	.71211	.87266	.89474	.95446	.89325	1.
.665	.70869	.87097	.89805	.95631	.89707	1.
.670	.70526	.86928	.90130	.95817	.90079	1.
.675	.70183	.86758	.90450	.96005	.90443	1.
.680	.69840	.86587	.90763	.96194	.90798	1.
.685	.69496	.86416	.91071	.96384	.91144	1.
.690	.69153	.86245	.91373	.96576	.91482	1.
.695	.68809	.86072	.91669	.96769	.91812	1.
.700	.68465	.85900	.91960	.96964	.92133	1.
.705	.68122	.85726	.92245	.97160	.92446	1.
.710	.67778	.85552	.92524	.97358	.92751	1.
.715	.67434	.85378	.92798	.97556	.93048	1.
.720	.67090	.85203	.93066	.97757	.93337	1.
.725	.66747	.85028	.93328	.97958	.93618	1.
.730	.66403	.84852	.93585	.98161	.93892	1.
.735	.66060	.84676	.93836	.98366	.94158	1.
.740	.65717	.84499	.94082	.98571	.94416	1.
.745	.65374	.84322	.94322	.98778	.94667	1.

TABLE XII. Isentropic Exponent = 1.67

M	p/p_t	T/T_t	GAMMA	ISOTHERMAL T_t/T_t^*	RAYLEIGH T_t/T_t^*	SHOCK p_{t1}/p_{t2}
.750	.65031	.84144	.94557	.98987	.94911	1.
.755	.64688	.83966	.94786	.99197	.95148	1.
.760	.64346	.83787	.95010	.99408	.95377	1.
.765	.64004	.83609	.95229	.99621	.95600	1.
.770	.63662	.83429	.95442	.99835	.95816	1.
.775	.63321	.83249	.95650	1.00051	.96025	1.
.780	.62980	.83069	.95852	1.00268	.96227	1.
.785	.62639	.82889	.96050	1.00486	.96423	1.
.790	.62299	.82708	.96242	1.00706	.96612	1.
.795	.61960	.82527	.96428	1.00927	.96795	1.
.800	.61621	.82345	.96610	1.01150	.96972	1.
.805	.61282	.82163	.96786	1.01373	.97143	1.
.810	.60944	.81981	.96958	1.01599	.97307	1.
.815	.60606	.81799	.97124	1.01825	.97466	1.
.820	.60269	.81616	.97285	1.02054	.97619	1.
.825	.59932	.81433	.97441	1.02283	.97766	1.
.830	.59596	.81249	.97592	1.02514	.97907	1.
.835	.59261	.81066	.97738	1.02746	.98043	1.
.840	.58927	.80882	.97880	1.02980	.98173	1.
.845	.58593	.80697	.98016	1.03215	.98299	1.
.850	.58259	.80513	.98147	1.03451	.98418	1.
.855	.57927	.80328	.98274	1.03689	.98533	1.
.860	.57595	.80143	.98396	1.03929	.98643	1.
.865	.57264	.79958	.98513	1.04169	.98747	1.
.870	.56934	.79773	.98625	1.04411	.98847	1.
.875	.56604	.79587	.98732	1.04655	.98942	1.
.880	.56275	.79401	.98835	1.04900	.99032	1.
.885	.55947	.79215	.98934	1.05146	.99118	1.
.890	.55620	.79029	.99027	1.05394	.99199	1.
.895	.55294	.78843	.99116	1.05643	.99276	1.
.900	.54969	.78657	.99201	1.05893	.99348	1.
.905	.54644	.78470	.99281	1.06145	.99416	1.
.910	.54321	.78283	.99357	1.06398	.99480	1.
.915	.53998	.78096	.99428	1.06653	.99539	1.
.920	.53676	.77909	.99495	1.06909	.99595	1.
.925	.53355	.77722	.99558	1.07166	.99647	1.
.930	.53036	.77535	.99616	1.07425	.99695	1.
.935	.52717	.77348	.99670	1.07685	.99739	1.
.940	.52399	.77160	.99720	1.07947	.99779	1.
.945	.52082	.76973	.99765	1.08210	.99816	1.
.950	.51766	.76785	.99806	1.08474	.99849	1.
.955	.51452	.76597	.99844	1.08740	.99878	1.
.960	.51138	.76410	.99877	1.09007	.99905	1.
.965	.50825	.76222	.99906	1.09275	.99928	1.
.970	.50514	.76034	.99931	1.09545	.99947	1.
.975	.50203	.75846	.99952	1.09817	.99964	1.
.980	.49894	.75658	.99970	1.10090	.99977	1.
.985	.49585	.75470	.99983	1.10364	.99987	1.
.990	.49278	.75282	.99992	1.10639	.99994	1.
.995	.48972	.75094	.99998	1.10916	.99999	1.

TABLE XII. Isentropic Exponent = 1.67

M	p/p_t	T/T_t	GAMMA	ISOTHERMAL T_t/T_t^*	RAYLEIGH T_t/T_t^*	SHOCK p_{t1}/p_{t2}
1.000	.48667	.74906	1.00000	1.11195	1.00000	1.00000
1.005	.48363	.74718	.99998	1.11474	.99999	1.00000
1.010	.48060	.74530	.99993	1.11755	.99994	1.00000
1.015	.47759	.74343	.99983	1.12038	.99988	1.00000
1.020	.47459	.74155	.99970	1.12322	.99978	1.00001
1.025	.47159	.73967	.99954	1.12607	.99966	1.00002
1.030	.46861	.73779	.99934	1.12894	.99952	1.00003
1.035	.46565	.73591	.99910	1.13182	.99935	1.00005
1.040	.46269	.73403	.99883	1.13471	.99915	1.00007
1.045	.45975	.73216	.99853	1.13762	.99894	1.00010
1.050	.45682	.73028	.99819	1.14055	.99870	1.00014
1.055	.45390	.72840	.99782	1.14348	.99844	1.00018
1.060	.45099	.72653	.99741	1.14643	.99815	1.00024
1.065	.44810	.72466	.99697	1.14940	.99785	1.00030
1.070	.44522	.72278	.99650	1.15238	.99752	1.00037
1.075	.44235	.72091	.99600	1.15537	.99718	1.00045
1.080	.43949	.71904	.99546	1.15838	.99681	1.00054
1.085	.43665	.71717	.99489	1.16140	.99643	1.00064
1.090	.43382	.71530	.99430	1.16443	.99603	1.00075
1.095	.43100	.71343	.99367	1.16748	.99561	1.00087
1.100	.42820	.71157	.99301	1.17054	.99517	1.00101
1.105	.42540	.70970	.99232	1.17362	.99471	1.00116
1.110	.42262	.70784	.99160	1.17671	.99424	1.00132
1.115	.41986	.70598	.99085	1.17981	.99375	1.00149
1.120	.41710	.70411	.99007	1.18293	.99324	1.00168
1.125	.41436	.70225	.98927	1.18606	.99272	1.00188
1.130	.41164	.70040	.98843	1.18921	.99219	1.00209
1.135	.40892	.69854	.98757	1.19237	.99163	1.00232
1.140	.40622	.69669	.98668	1.19554	.99107	1.00256
1.145	.40354	.69483	.98576	1.19873	.99049	1.00282
1.150	.40086	.69298	.98482	1.20193	.98990	1.00309
1.155	.39820	.69113	.98385	1.20515	.98929	1.00338
1.160	.39555	.68929	.98285	1.20838	.98867	1.00369
1.165	.39292	.68744	.98183	1.21162	.98804	1.00400
1.170	.39030	.68560	.98079	1.21488	.98740	1.00434
1.175	.38769	.68376	.97971	1.21815	.98674	1.00469
1.180	.38510	.68192	.97862	1.22144	.98608	1.00506
1.185	.38252	.68008	.97749	1.22473	.98540	1.00544
1.190	.37995	.67825	.97635	1.22805	.98471	1.00584
1.195	.37739	.67641	.97518	1.23138	.98401	1.00626
1.200	.37485	.67458	.97399	1.23472	.98330	1.00669
1.205	.37233	.67275	.97277	1.23807	.98258	1.00714
1.210	.36981	.67093	.97153	1.24144	.98185	1.00761
1.215	.36731	.66910	.97027	1.24482	.98111	1.00809
1.220	.36483	.66728	.96899	1.24822	.98037	1.00860
1.225	.36235	.66546	.96768	1.25163	.97961	1.00912
1.230	.35989	.66365	.96635	1.25506	.97885	1.00965
1.235	.35745	.66184	.96501	1.25850	.97808	1.01021
1.240	.35501	.66002	.96364	1.26195	.97730	1.01078
1.245	.35260	.65822	.96225	1.26542	.97651	1.01137

TABLE XII.   Isentropic Exponent = 1.67

M	p/p_t	T/T_t	GAMMA	ISOTHERMAL T_t/T_t^*	RAYLEIGH T_t/T_t^*	SHOCK p_{t1}/p_{t2}
1.250	.35019	.65641	.96084	1.26890	.97571	1.01198
1.255	.34780	.65461	.95941	1.27239	.97491	1.01260
1.260	.34542	.65281	.95796	1.27590	.97410	1.01325
1.265	.34305	.65101	.95650	1.27942	.97329	1.01391
1.270	.34070	.64922	.95501	1.28296	.97246	1.01459
1.275	.33836	.64742	.95350	1.28651	.97164	1.01529
1.280	.33604	.64563	.95198	1.29008	.97080	1.01601
1.285	.33372	.64385	.95044	1.29365	.96996	1.01674
1.290	.33142	.64207	.94888	1.29725	.96912	1.01750
1.295	.32914	.64029	.94730	1.30085	.96827	1.01827
1.300	.32687	.63851	.94571	1.30447	.96741	1.01906
1.305	.32461	.63673	.94410	1.30811	.96655	1.01987
1.310	.32236	.63496	.94247	1.31176	.96569	1.02070
1.315	.32013	.63320	.94083	1.31542	.96482	1.02154
1.320	.31791	.63143	.93917	1.31909	.96395	1.02241
1.325	.31570	.62967	.93749	1.32278	.96307	1.02329
1.330	.31351	.62791	.93580	1.32649	.96219	1.02419
1.335	.31133	.62616	.93409	1.33021	.96130	1.02512
1.340	.30917	.62440	.93237	1.33394	.96041	1.02605
1.345	.30701	.62266	.93064	1.33769	.95952	1.02701
1.350	.30487	.62091	.92889	1.34144	.95862	1.02799
1.355	.30274	.61917	.92712	1.34522	.95773	1.02898
1.360	.30063	.61743	.92535	1.34901	.95683	1.03000
1.365	.29853	.61570	.92355	1.35281	.95592	1.03103
1.370	.29644	.61396	.92175	1.35662	.95501	1.03208
1.375	.29436	.61224	.91993	1.36045	.95411	1.03315
1.380	.29230	.61051	.91810	1.36430	.95319	1.03424
1.385	.29025	.60879	.91626	1.36815	.95228	1.03535
1.390	.28822	.60707	.91440	1.37203	.95137	1.03648
1.395	.28619	.60536	.91254	1.37591	.95045	1.03762
1.400	.28418	.60365	.91066	1.37981	.94953	1.03878
1.405	.28218	.60194	.90877	1.38372	.94861	1.03997
1.410	.28019	.60024	.90686	1.38765	.94769	1.04117
1.415	.27822	.59854	.90495	1.39159	.94676	1.04239
1.420	.27626	.59684	.90303	1.39555	.94584	1.04363
1.425	.27431	.59515	.90109	1.39952	.94491	1.04488
1.430	.27237	.59346	.89915	1.40350	.94399	1.04616
1.435	.27045	.59177	.89720	1.40750	.94306	1.04745
1.440	.26854	.59009	.89523	1.41151	.94213	1.04877
1.445	.26664	.58841	.89326	1.41553	.94120	1.05010
1.450	.26475	.58674	.89127	1.41957	.94027	1.05145
1.455	.26288	.58507	.88928	1.42363	.93934	1.05282
1.460	.26101	.58340	.88728	1.42769	.93841	1.05421
1.465	.25916	.58174	.88527	1.43177	.93748	1.05561
1.470	.25733	.58008	.88325	1.43587	.93655	1.05704
1.475	.25550	.57842	.88122	1.43998	.93562	1.05848
1.480	.25369	.57677	.87919	1.44410	.93469	1.05994
1.485	.25188	.57513	.87714	1.44824	.93376	1.06142
1.490	.25009	.57348	.87509	1.45239	.93283	1.06292
1.495	.24831	.57184	.87303	1.45655	.93190	1.06444

235

TABLE XII. Isentropic Exponent = 1.67

M	p/p_t	T/T_t	GAMMA	ISOTHERMAL T_t/T_t^*	RAYLEIGH T_t/T_t^*	SHOCK p_{t1}/p_{t2}
1.500	.24655	.57021	.87097	1.46073	.93097	1.06597
1.505	.24479	.56857	.86890	1.46492	.93004	1.06753
1.510	.24305	.56695	.86682	1.46913	.92911	1.06910
1.515	.24132	.56532	.86473	1.47335	.92818	1.07069
1.520	.23960	.56370	.86264	1.47758	.92725	1.07230
1.525	.23789	.56209	.86054	1.48183	.92632	1.07393
1.530	.23619	.56047	.85843	1.48609	.92540	1.07557
1.535	.23451	.55887	.85632	1.49037	.92447	1.07724
1.540	.23283	.55726	.85420	1.49466	.92355	1.07724
1.545	.23117	.55566	.85208	1.49896	.92262	1.08062
1.550	.22952	.55407	.84995	1.50328	.92170	1.08234
1.555	.22788	.55247	.84782	1.50761	.92078	1.08407
1.560	.22625	.55089	.84568	1.51196	.91986	1.08583
1.565	.22463	.54930	.84354	1.51632	.91894	1.08760
1.570	.22302	.54772	.84139	1.52069	.91802	1.08940
1.575	.22143	.54615	.83924	1.52508	.91711	1.09121
1.580	.21984	.54458	.83708	1.52948	.91619	1.09303
1.585	.21827	.54301	.83492	1.53390	.91528	1.09488
1.590	.21671	.54144	.83276	1.53833	.91437	1.09674
1.595	.21515	.53988	.83059	1.54277	.91346	1.09863
1.600	.21361	.53833	.82842	1.54723	.91255	1.10053
1.605	.21208	.53678	.82624	1.55170	.91164	1.10245
1.610	.21056	.53523	.82406	1.55618	.91074	1.10438
1.615	.20905	.53369	.82188	1.56068	.90983	1.10634
1.620	.20755	.53215	.81969	1.56520	.90893	1.10831
1.625	.20606	.53061	.81751	1.56972	.90803	1.11030
1.630	.20458	.52908	.81531	1.57427	.90713	1.11231
1.635	.20312	.52756	.81312	1.57882	.90623	1.11434
1.640	.20166	.52603	.81092	1.58339	.90534	1.11638
1.645	.20021	.52452	.80873	1.58797	.90445	1.11844
1.650	.19877	.52300	.80653	1.59257	.90356	1.12052
1.655	.19735	.52149	.80432	1.59718	.90267	1.12262
1.660	.19593	.51999	.80212	1.60181	.90178	1.12474
1.665	.19452	.51849	.79991	1.60644	.90090	1.12687
1.670	.19313	.51699	.79770	1.61110	.90002	1.12902
1.675	.19174	.51549	.79549	1.61576	.89914	1.13119
1.680	.19036	.51401	.79328	1.62044	.89826	1.13338
1.685	.18899	.51252	.79107	1.62514	.89738	1.13559
1.690	.18763	.51104	.78886	1.62985	.89651	1.13781
1.695	.18629	.50956	.78664	1.63457	.89564	1.14005
1.700	.18495	.50809	.78443	1.63931	.89477	1.14231
1.705	.18362	.50662	.78221	1.64406	.89390	1.14458
1.710	.18230	.50516	.78000	1.64882	.89304	1.14688
1.715	.18099	.50370	.77778	1.65360	.89218	1.14919
1.720	.17969	.50224	.77556	1.65839	.89132	1.15152
1.725	.17840	.50079	.77334	1.66320	.89046	1.15387
1.730	.17711	.49935	.77113	1.66802	.88961	1.15623
1.735	.17584	.49790	.76891	1.67285	.88876	1.15861
1.740	.17458	.49646	.76669	1.67770	.88791	1.16101
1.745	.17332	.49503	.76447	1.68256	.88706	1.16343

TABLE XII.   Isentropic Exponent = 1.67

M	p/p_t	T/T_t	GAMMA	ISOTHERMAL T_t/T_t^*	RAYLEIGH T_t/T_t^*	SHOCK p_{t1}/p_{t2}
1.750	.17208	.49360	.76226	1.68744	.88622	1.16587
1.755	.17084	.49217	.76004	1.69233	.88537	1.16832
1.760	.16961	.49075	.75782	1.69723	.88453	1.17079
1.765	.16839	.48933	.75561	1.70215	.88370	1.17328
1.770	.16718	.48792	.75339	1.70708	.88286	1.17578
1.775	.16598	.48651	.75118	1.71203	.88203	1.17830
1.780	.16479	.48510	.74896	1.71699	.88120	1.18085
1.785	.16361	.48370	.74675	1.72196	.88038	1.18340
1.790	.16243	.48231	.74454	1.72695	.87955	1.18598
1.795	.16126	.48091	.74233	1.73195	.87873	1.18857
1.800	.16011	.47952	.74012	1.73697	.87791	1.19118
1.805	.15896	.47814	.73791	1.74200	.87710	1.19381
1.810	.15782	.47676	.73571	1.74704	.87628	1.19646
1.815	.15668	.47538	.73350	1.75210	.87547	1.19912
1.820	.15556	.47401	.73130	1.75717	.87467	1.20180
1.825	.15444	.47264	.72910	1.76225	.87386	1.20450
1.830	.15333	.47128	.72690	1.76735	.87306	1.20721
1.835	.15223	.46992	.72470	1.77247	.87226	1.20995
1.840	.15114	.46857	.72250	1.77759	.87146	1.21270
1.845	.15006	.46721	.72031	1.78273	.87067	1.21546
1.850	.14898	.46587	.71812	1.78789	.86987	1.21825
1.855	.14791	.46452	.71593	1.79306	.86909	1.22105
1.860	.14685	.46318	.71374	1.79824	.86830	1.22387
1.865	.14580	.46185	.71156	1.80344	.86752	1.22671
1.870	.14475	.46052	.70938	1.80865	.86674	1.22956
1.875	.14372	.45919	.70720	1.81387	.86596	1.23244
1.880	.14269	.45787	.70502	1.81911	.86518	1.23533
1.885	.14167	.45655	.70284	1.82436	.86441	1.23823
1.890	.14065	.45524	.70067	1.82963	.86364	1.24116
1.895	.13965	.45393	.69850	1.83491	.86287	1.24410
1.900	.13865	.45262	.69634	1.84021	.86211	1.24706
1.905	.13765	.45132	.69417	1.84551	.86135	1.25004
1.910	.13667	.45002	.69201	1.85084	.86059	1.25303
1.915	.13569	.44873	.68986	1.85617	.85984	1.25604
1.920	.13472	.44744	.68770	1.86152	.85908	1.25907
1.925	.13376	.44615	.68555	1.86689	.85833	1.26212
1.930	.13280	.44487	.68340	1.87227	.85758	1.26518
1.935	.13186	.44359	.68126	1.87766	.85684	1.26826
1.940	.13091	.44232	.67912	1.88306	.85610	1.27136
1.945	.12998	.44105	.67698	1.88848	.85536	1.27447
1.950	.12905	.43979	.67485	1.89392	.85462	1.27761
1.955	.12813	.43852	.67271	1.89937	.85389	1.28076
1.960	.12722	.43727	.67059	1.90483	.85316	1.28392
1.965	.12631	.43601	.66846	1.91030	.85243	1.28711
1.970	.12541	.43476	.66634	1.91579	.85171	1.29031
1.975	.12452	.43352	.66423	1.92130	.85098	1.29353
1.980	.12363	.43228	.66211	1.92682	.85026	1.29677
1.985	.12275	.43104	.66001	1.93235	.84955	1.30002
1.990	.12188	.42981	.65790	1.93789	.84883	1.30329
1.995	.12101	.42858	.65580	1.94345	.84812	1.30658

TABLE XII.   Isentropic Exponent = 1.67

M	p/p_t	T/T_t	GAMMA	ISOTHERMAL T_t/T_t^*	RAYLEIGH T_t/T_t^*	SHOCK p_{t1}/p_{t2}
2.000	.12015	.42735	.65370	1.94903	.84741	1.30989
2.010	.11845	.42491	.64952	1.96022	.84600	1.31655
2.020	.11677	.42249	.64535	1.97146	.84460	1.32329
2.030	.11512	.42008	.64120	1.98276	.84322	1.33009
2.040	.11349	.41769	.63707	1.99412	.84184	1.33696
2.050	.11189	.41531	.63296	2.00553	.84048	1.34391
2.060	.11031	.41295	.62886	2.01700	.83912	1.35092
2.070	.10875	.41060	.62478	2.02852	.83778	1.35800
2.080	.10722	.40827	.62072	2.04010	.83644	1.36515
2.090	.10571	.40596	.61667	2.05174	.83512	1.37237
2.100	.10422	.40366	.61265	2.06343	.83381	1.37966
2.110	.10276	.40137	.60864	2.07518	.83251	1.38703
2.120	.10132	.39910	.60465	2.08698	.83122	1.39446
2.130	.09990	.39685	.60069	2.09884	.82994	1.40196
2.140	.09850	.39461	.59674	2.11075	.82867	1.40953
2.150	.09712	.39238	.59281	2.12272	.82741	1.41717
2.160	.09576	.39017	.58890	2.13475	.82616	1.42488
2.170	.09442	.38798	.58501	2.14683	.82492	1.43266
2.180	.09311	.38579	.58114	2.15897	.82369	1.44051
2.190	.09181	.38363	.57729	2.17116	.82247	1.44843
2.200	.09053	.38148	.57346	2.18341	.82126	1.45642
2.210	.08927	.37934	.56966	2.19572	.82006	1.46448
2.220	.08803	.37721	.56587	2.20808	.81887	1.47261
2.230	.08681	.37510	.56210	2.22049	.81769	1.48082
2.240	.08560	.37301	.55835	2.23297	.81652	1.48909
2.250	.08442	.37093	.55463	2.24549	.81536	1.49744
2.260	.08325	.36886	.55093	2.25808	.81421	1.50585
2.270	.08210	.36681	.54724	2.27072	.81307	1.51434
2.280	.08097	.36477	.54358	2.28341	.81194	1.52289
2.290	.07985	.36274	.53994	2.29617	.81082	1.53152
2.300	.07875	.36073	.53632	2.30897	.80970	1.54022
2.310	.07767	.35873	.53272	2.32184	.80860	1.54899
2.320	.07660	.35675	.52915	2.33475	.80751	1.55784
2.330	.07555	.35478	.52559	2.34773	.80642	1.56675
2.340	.07452	.35282	.52206	2.36076	.80535	1.57574
2.350	.07350	.35087	.51855	2.37385	.80428	1.58480
2.360	.07249	.34894	.51506	2.38699	.80322	1.59392
2.370	.07150	.34702	.51159	2.40019	.80217	1.60313
2.380	.07053	.34512	.50814	2.41344	.80113	1.61240
2.390	.06957	.34322	.50471	2.42675	.80010	1.62175
2.400	.06862	.34134	.50131	2.44012	.79908	1.63117
2.410	.06769	.33948	.49792	2.45354	.79806	1.64066
2.420	.06677	.33762	.49456	2.46701	.79706	1.65022
2.430	.06587	.33578	.49122	2.48055	.79606	1.65986
2.440	.06498	.33395	.48790	2.49413	.79507	1.66957
2.450	.06410	.33213	.48461	2.50778	.79409	1.67935
2.460	.06323	.33033	.48133	2.52148	.79311	1.68921
2.470	.06238	.32854	.47808	2.53524	.79215	1.69914
2.480	.06154	.32676	.47484	2.54905	.79119	1.70914
2.490	.06072	.32499	.47163	2.56291	.79024	1.71922

TABLE XII. Isentropic Exponent = 1.67

M	p/p_t	T/T_t	GAMMA	ISOTHERMAL T_t/T_t^*	RAYLEIGH T_t/T_t^*	SHOCK p_{t1}/p_{t2}
2.500	.05990	.32323	.46844	2.57684	.78930	1.72937
2.510	.05910	.32149	.46527	2.59082	.78837	1.73959
2.520	.05831	.31976	.46213	2.60485	.78745	1.74989
2.530	.05753	.31804	.45900	2.61894	.78653	1.76026
2.540	.05676	.31633	.45589	2.63309	.78562	1.77071
2.550	.05601	.31463	.45281	2.64729	.78472	1.78123
2.560	.05526	.31294	.44974	2.66155	.78382	1.79183
2.570	.05453	.31127	.44670	2.67587	.78293	1.80250
2.580	.05381	.30961	.44368	2.69024	.78205	1.81325
2.590	.05309	.30796	.44068	2.70466	.78118	1.82407
2.600	.05239	.30632	.43770	2.71914	.78032	1.83497
2.610	.05170	.30469	.43474	2.73368	.77946	1.84594
2.620	.05102	.30307	.43180	2.74827	.77861	1.85699
2.630	.05035	.30146	.42888	2.76292	.77776	1.86811
2.640	.04968	.29987	.42598	2.77763	.77693	1.87931
2.650	.04903	.29828	.42310	2.79239	.77610	1.89059
2.660	.04839	.29671	.42024	2.80720	.77527	1.90194
2.670	.04776	.29514	.41741	2.82208	.77446	1.91337
2.680	.04713	.29359	.41459	2.83700	.77365	1.92487
2.690	.04652	.29205	.41179	2.85199	.77284	1.93645
2.700	.04591	.29052	.40901	2.86703	.77205	1.94811
2.710	.04532	.28899	.40625	2.88212	.77126	1.95985
2.720	.04473	.28748	.40352	2.89727	.77047	1.97167
2.730	.04415	.28598	.40080	2.91248	.76970	1.98356
2.740	.04358	.28449	.39810	2.92774	.76893	1.99553
2.750	.04301	.28301	.39542	2.94306	.76816	2.00757
2.760	.04246	.28154	.39276	2.95844	.76740	2.01970
2.770	.04191	.28008	.39011	2.97387	.76665	2.03190
2.780	.04137	.27863	.38749	2.98935	.76590	2.04418
2.790	.04084	.27719	.38489	3.00489	.76517	2.05654
2.800	.04032	.27576	.38230	3.02049	.76443	2.06898
2.810	.03980	.27433	.37974	3.03614	.76370	2.08150
2.820	.03929	.27292	.37719	3.05185	.76298	2.09410
2.830	.03879	.27152	.37466	3.06762	.76227	2.10678
2.840	.03830	.27013	.37215	3.08344	.76156	2.11953
2.850	.03781	.26874	.36966	3.09932	.76085	2.13237
2.860	.03733	.26737	.36719	3.11525	.76015	2.14529
2.870	.03686	.26600	.36473	3.13124	.75946	2.15828
2.880	.03639	.26465	.36229	3.14728	.75877	2.17136
2.890	.03593	.26330	.35987	3.16338	.75809	2.18452
2.900	.03548	.26196	.35747	3.17954	.75741	2.19775
2.910	.03503	.26063	.35509	3.19575	.75674	2.21107
2.920	.03459	.25931	.35272	3.21202	.75607	2.22447
2.930	.03415	.25800	.35037	3.22834	.75541	2.23795
2.940	.03373	.25670	.34804	3.24472	.75476	2.25152
2.950	.03330	.25541	.34573	3.26115	.75411	2.26516
2.960	.03289	.25412	.34343	3.27764	.75346	2.27889
2.970	.03248	.25284	.34115	3.29419	.75282	2.29270
2.980	.03207	.25158	.33889	3.31079	.75219	2.30659
2.990	.03167	.25032	.33664	3.32745	.75156	2.32056

TABLE XII. Isentropic Exponent = 1.67

M	p/p_t	T/T_t	GAMMA	ISOTHERMAL T_t/T_t^*	RAYLEIGH T_t/T_t^*	SHOCK p_{t1}/p_{t2}
3.000	.03128	.24907	.33441	3.34416	.75093	2.33462
3.010	.03089	.24782	.33220	3.36093	.75032	2.34876
3.020	.03051	.24659	.33000	3.37776	.74970	2.36298
3.030	.03014	.24536	.32782	3.39464	.74909	2.37728
3.040	.02976	.24414	.32566	3.41158	.74848	2.39167
3.050	.02940	.24293	.32351	3.42857	.74788	2.40614
3.060	.02904	.24173	.32138	3.44562	.74729	2.42070
3.070	.02868	.24054	.31926	3.46272	.74670	2.43534
3.080	.02833	.23935	.31717	3.47988	.74611	2.45007
3.090	.02798	.23817	.31508	3.49710	.74553	2.46487
3.100	.02764	.23700	.31301	3.51437	.74495	2.47977
3.110	.02730	.23584	.31096	3.53170	.74438	2.49475
3.120	.02697	.23469	.30892	3.54908	.74381	2.50981
3.130	.02664	.23354	.30690	3.56652	.74325	2.52496
3.140	.02632	.23240	.30489	3.58402	.74269	2.54020
3.150	.02600	.23127	.30290	3.60157	.74213	2.55552
3.160	.02569	.23014	.30093	3.61917	.74158	2.57092
3.170	.02538	.22902	.29896	3.63684	.74103	2.58642
3.180	.02507	.22791	.29702	3.65455	.74049	2.60200
3.190	.02477	.22681	.29508	3.67233	.73995	2.61766
3.200	.02447	.22571	.29317	3.69016	.73942	2.63342
3.210	.02418	.22462	.29126	3.70804	.73889	2.64926
3.220	.02389	.22354	.28937	3.72599	.73836	2.66518
3.230	.02361	.22247	.28750	3.74398	.73784	2.68120
3.240	.02333	.22140	.28564	3.76204	.73732	2.69730
3.250	.02305	.22034	.28379	3.78014	.73680	2.71349
3.260	.02277	.21929	.28196	3.79831	.73629	2.72977
3.270	.02250	.21824	.28014	3.81653	.73579	2.74614
3.280	.02224	.21720	.27833	3.83481	.73528	2.76259
3.290	.02198	.21617	.27654	3.85314	.73478	2.77914
3.300	.02172	.21514	.27476	3.87153	.73429	2.79577
3.310	.02146	.21412	.27300	3.88997	.73380	2.81249
3.320	.02121	.21311	.27125	3.90847	.73331	2.82931
3.330	.02096	.21210	.26951	3.92702	.73282	2.84621
3.340	.02071	.21110	.26778	3.94564	.73234	2.86320
3.350	.02047	.21010	.26607	3.96430	.73187	2.88028
3.360	.02023	.20912	.26437	3.98303	.73139	2.89745
3.370	.02000	.20814	.26269	4.00180	.73092	2.91472
3.380	.01976	.20716	.26101	4.02064	.73046	2.93207
3.390	.01953	.20619	.25935	4.03953	.72999	2.94951
3.400	.01931	.20523	.25770	4.05847	.72953	2.96705
3.410	.01908	.20427	.25606	4.07748	.72908	2.98468
3.420	.01886	.20332	.25444	4.09653	.72862	3.00240
3.430	.01865	.20238	.25283	4.11565	.72817	3.02021
3.440	.01843	.20144	.25123	4.13482	.72773	3.03811
3.450	.01822	.20051	.24964	4.15404	.72729	3.05611
3.460	.01801	.19958	.24807	4.17332	.72685	3.07419
3.470	.01780	.19866	.24650	4.19266	.72641	3.09238
3.480	.01760	.19775	.24495	4.21205	.72598	3.11065
3.490	.01740	.19684	.24341	4.23150	.72554	3.12902

240

TABLE XII. Isentropic Exponent = 1.67

M	p/p_t	T/T_t	GAMMA	ISOTHERMAL T_t/T_t^*	RAYLEIGH T_t/T_t^*	SHOCK p_{t1}/p_{t2}
3.500	.01720	.19593	.24188	4.25100	.72512	3.14748
3.510	.01701	.19504	.24036	4.27056	.72469	3.16603
3.520	.01681	.19415	.23886	4.29018	.72427	3.18468
3.530	.01662	.19326	.23736	4.30985	.72385	3.20342
3.540	.01643	.19238	.23588	4.32958	.72344	3.22226
3.550	.01625	.19150	.23440	4.34936	.72303	3.24119
3.560	.01606	.19063	.23294	4.36920	.72262	3.26022
3.570	.01588	.18977	.23149	4.38909	.72221	3.27934
3.580	.01571	.18891	.23005	4.40904	.72181	3.29855
3.590	.01553	.18806	.22862	4.42905	.72141	3.31787
3.600	.01536	.18721	.22720	4.44911	.72101	3.33727
3.610	.01518	.18637	.22580	4.46923	.72062	3.35678
3.620	.01501	.18553	.22440	4.48940	.72022	3.37638
3.630	.01485	.18470	.22301	4.50963	.71984	3.39607
3.640	.01468	.18387	.22164	4.52992	.71945	3.41587
3.650	.01452	.18305	.22027	4.55026	.71907	3.43576
3.660	.01436	.18223	.21891	4.57066	.71868	3.45574
3.670	.01420	.18142	.21757	4.59111	.71831	3.47583
3.680	.01404	.18061	.21623	4.61162	.71793	3.49601
3.690	.01389	.17981	.21490	4.63218	.71756	3.51629
3.700	.01373	.17901	.21359	4.65280	.71719	3.53667
3.710	.01358	.17822	.21228	4.67348	.71682	3.55715
3.720	.01343	.17744	.21099	4.69421	.71645	3.57772
3.730	.01329	.17665	.20970	4.71500	.71609	3.59839
3.740	.01314	.17588	.20842	4.73584	.71573	3.61917
3.750	.01300	.17510	.20715	4.75674	.71537	3.64004
3.760	.01286	.17433	.20589	4.77769	.71502	3.66101
3.770	.01272	.17357	.20464	4.79870	.71467	3.68208
3.780	.01258	.17281	.20340	4.81977	.71432	3.70325
3.790	.01244	.17206	.20217	4.84089	.71397	3.72452
3.800	.01231	.17131	.20095	4.86207	.71362	3.74590
3.810	.01217	.17056	.19974	4.88331	.71328	3.76737
3.820	.01204	.16982	.19853	4.90460	.71294	3.78894
3.830	.01191	.16909	.19734	4.92594	.71260	3.81062
3.840	.01179	.16836	.19615	4.94734	.71226	3.83239
3.850	.01166	.16763	.19497	4.96880	.71193	3.85427
3.860	.01153	.16691	.19380	4.99031	.71160	3.87625
3.870	.01141	.16619	.19264	5.01188	.71127	3.89833
3.880	.01129	.16547	.19149	5.03351	.71094	3.92051
3.890	.01117	.16476	.19035	5.05519	.71062	3.94280
3.900	.01105	.16406	.18921	5.07692	.71029	3.96519
3.910	.01093	.16336	.18809	5.09871	.70997	3.98768
3.920	.01082	.16266	.18697	5.12056	.70966	4.01027
3.930	.01070	.16197	.18586	5.14247	.70934	4.03297
3.940	.01059	.16128	.18475	5.16443	.70902	4.05577
3.950	.01048	.16060	.18366	5.18644	.70871	4.07868
3.960	.01037	.15991	.18257	5.20851	.70840	4.10169
3.970	.01026	.15924	.18150	5.23064	.70809	4.12480
3.980	.01015	.15857	.18042	5.25282	.70779	4.14802
3.990	.01004	.15790	.17936	5.27506	.70749	4.17135

TABLE XII. Isentropic Exponent = 1.67

M	p/p_t	T/T_t	GAMMA	ISOTHERMAL T_t/T_t^*	RAYLEIGH T_t/T_t^*	SHOCK p_{t1}/p_{t2}
4.000	.00994	.15723	.17831	5.29736	.70718	4.19478
4.010	.00984	.15657	.17726	5.31971	.70688	4.21831
4.020	.00973	.15592	.17622	5.34211	.70659	4.24195
4.030	.00963	.15526	.17519	5.36457	.70629	4.26570
4.040	.00953	.15461	.17416	5.38709	.70600	4.28955
4.050	.00943	.15397	.17314	5.40966	.70570	4.31351
4.060	.00934	.15333	.17213	5.43229	.70541	4.33758
4.070	.00924	.15269	.17113	5.45498	.70513	4.36175
4.080	.00914	.15206	.17014	5.47772	.70484	4.38603
4.090	.00905	.15143	.16915	5.50052	.70455	4.41041
4.100	.00896	.15080	.16817	5.52337	.70427	4.43491
4.110	.00886	.15018	.16719	5.54628	.70399	4.45951
4.120	.00877	.14956	.16622	5.56924	.70371	4.48422
4.130	.00868	.14894	.16526	5.59226	.70343	4.50904
4.140	.00860	.14833	.16431	5.61534	.70316	4.53396
4.150	.00851	.14772	.16336	5.63847	.70289	4.55900
4.160	.00842	.14712	.16242	5.66165	.70261	4.58414
4.170	.00834	.14651	.16149	5.68490	.70234	4.60939
4.180	.00825	.14592	.16056	5.70820	.70208	4.63476
4.190	.00817	.14532	.15964	5.73155	.70181	4.66023
4.200	.00808	.14473	.15873	5.75496	.70154	4.68581
4.210	.00800	.14414	.15782	5.77843	.70128	4.71150
4.220	.00792	.14356	.15692	5.80195	.70102	4.73730
4.230	.00784	.14298	.15603	5.82553	.70076	4.76322
4.240	.00776	.14240	.15514	5.84916	.70050	4.78924
4.250	.00769	.14183	.15426	5.87285	.70024	4.81538
4.260	.00761	.14125	.15338	5.89660	.69999	4.84162
4.270	.00753	.14069	.15252	5.92040	.69974	4.86798
4.280	.00746	.14012	.15165	5.94425	.69948	4.89445
4.290	.00738	.13956	.15080	5.96817	.69923	4.92104
4.300	.00731	.13900	.14994	5.99213	.69898	4.94773
4.310	.00724	.13845	.14910	6.01616	.69874	4.97454
4.320	.00717	.13789	.14826	6.04024	.69849	5.00146
4.330	.00710	.13735	.14743	6.06437	.69825	5.02849
4.340	.00703	.13680	.14660	6.08857	.69800	5.05564
4.350	.00696	.13626	.14578	6.11281	.69776	5.08290
4.360	.00689	.13572	.14496	6.13712	.69752	5.11027
4.370	.00682	.13518	.14415	6.16148	.69729	5.13776
4.380	.00675	.13465	.14335	6.18589	.69705	5.16536
4.390	.00669	.13412	.14255	6.21036	.69681	5.19308
4.400	.00662	.13359	.14176	6.23489	.69658	5.22091
4.410	.00656	.13307	.14097	6.25947	.69635	5.24885
4.420	.00649	.13254	.14019	6.28411	.69612	5.27692
4.430	.00643	.13202	.13941	6.30880	.69589	5.30509
4.440	.00637	.13151	.13864	6.33355	.69566	5.33339
4.450	.00631	.13100	.13788	6.35836	.69543	5.36179
4.460	.00624	.13049	.13712	6.38322	.69521	5.39032
4.470	.00618	.12998	.13636	6.40814	.69498	5.41896
4.480	.00612	.12947	.13561	6.43311	.69476	5.44772
4.490	.00607	.12897	.13487	6.45814	.69454	5.47659

TABLE XII. Isentropic Exponent = 1.67

M	p/p_t	T/T_t	GAMMA	ISOTHERMAL T_t/T_t^*	RAYLEIGH T_t/T_t^*	SHOCK p_{t1}/p_{t2}
4.500	.00601	.12847	.13413	6.48322	.69432	5.50559
4.510	.00595	.12798	.13339	6.50836	.69410	5.53470
4.520	.00589	.12748	.13266	6.53356	.69388	5.56392
4.530	.00584	.12699	.13194	6.55881	.69367	5.59327
4.540	.00578	.12650	.13122	6.58412	.69345	5.62273
4.550	.00573	.12602	.13050	6.60948	.69324	5.65232
4.560	.00567	.12554	.12979	6.63490	.69303	5.68202
4.570	.00562	.12506	.12909	6.66038	.69282	5.71184
4.580	.00556	.12458	.12839	6.68591	.69261	5.74177
4.590	.00551	.12410	.12769	6.71149	.69240	5.77183
4.600	.00546	.12363	.12700	6.73714	.69219	5.80201
4.610	.00541	.12316	.12632	6.76283	.69199	5.83231
4.620	.00536	.12269	.12564	6.78859	.69178	5.86273
4.630	.00531	.12223	.12496	6.81440	.69158	5.89326
4.640	.00526	.12177	.12429	6.84026	.69138	5.92392
4.650	.00521	.12131	.12362	6.86619	.69118	5.95470
4.660	.00516	.12085	.12296	6.89216	.69098	5.98560
4.670	.00511	.12040	.12230	6.91820	.69078	6.01663
4.680	.00506	.11994	.12164	6.94429	.69058	6.04777
4.690	.00501	.11949	.12100	6.97043	.69038	6.07903
4.700	.00497	.11905	.12035	6.99663	.69019	6.11042
4.710	.00492	.11860	.11971	7.02289	.68999	6.14193
4.720	.00488	.11816	.11907	7.04920	.68980	6.17357
4.730	.00483	.11772	.11844	7.07557	.68961	6.20532
4.740	.00479	.11728	.11781	7.10199	.68942	6.23720
4.750	.00474	.11684	.11719	7.12847	.68923	6.26920
4.760	.00470	.11641	.11657	7.15501	.68904	6.30133
4.770	.00466	.11598	.11595	7.18160	.68885	6.33358
4.780	.00461	.11555	.11534	7.20825	.68867	6.36595
4.790	.00457	.11512	.11474	7.23495	.68848	6.39845
4.800	.00453	.11470	.11413	7.26171	.68830	6.43107
4.810	.00449	.11428	.11353	7.28852	.68812	6.46382
4.820	.00445	.11386	.11294	7.31539	.68793	6.49669
4.830	.00441	.11344	.11235	7.34232	.68775	6.52969
4.840	.00437	.11303	.11176	7.36930	.68757	6.56281
4.850	.00433	.11261	.11118	7.39634	.68739	6.59606
4.860	.00429	.11220	.11060	7.42343	.68722	6.62943
4.870	.00425	.11179	.11002	7.45058	.68704	6.66293
4.880	.00421	.11139	.10945	7.47779	.68686	6.69656
4.890	.00417	.11098	.10888	7.50505	.68669	6.73032
4.900	.00413	.11058	.10832	7.53236	.68651	6.76420
4.910	.00410	.11018	.10776	7.55974	.68634	6.79821
4.920	.00406	.10978	.10720	7.58716	.68617	6.83234
4.930	.00402	.10938	.10665	7.61465	.68600	6.86661
4.940	.00399	.10899	.10610	7.64219	.68583	6.90100
4.950	.00395	.10860	.10555	7.66978	.68566	6.93552
4.960	.00392	.10821	.10501	7.69744	.68549	6.97017
4.970	.00388	.10782	.10447	7.72514	.68532	7.00494
4.980	.00385	.10743	.10393	7.75291	.68516	7.03985
4.990	.00381	.10705	.10340	7.78073	.68499	7.07489